Energy Conversion Engineering

ENERGY SCIENCE AND TECHNOLOGY
A Series of Graduate Textbooks, Monographs, Research Papers

Series Editors: C. Y. WEN and E. STANLEY LEE

Consulting Editor: Richard Bellman

No. 1   RICHARD C. BAILIE
Energy Conversion Engineering, 1978

Other Numbers in preparation

Energy Conversion Engineering

Richard C. Bailie

West Virginia University
and
Environmental Energy Engineering, Inc.
Morgantown, Virginia

1978

Addison-Wesley Publishing Company
Advanced Book Program
Reading, Massachusetts

London · Amsterdam · Don Mills, Ontario · Sydney · Tokyo

Library of Congress Cataloging in Publication Data

Bailie, R. C.
Energy conversion engineering.

(Energy science and technology ; no. 1)
Includes index.
1. Power (Mechanics)   2. Power resources.
I. Title. II. Series.
TJ163.9.B34   621   78-11969
ISBN: 0-201-00840-8

Reproduced by Addison-Wesley Publishing Company, Inc., Advanced Book Program, Reading, Massachusetts, from camera-ready copy prepared by the office of the author.

Manufactured in the United States of America

ABCDEFGHIJ-MA-798

This book is dedicated to

Dr. Harold G. Donnelly

with thanks and gratitude

for all he has given

to his students.

CONTENTS

SERIES EDITORS' FOREWORD

Accelerated research and development for alternate energy
resources to augment the dwindling supply of petroleum and
natural gas is currently under way throughout the world. The
energy problem has suddenly created new demands on the effective dissemination of large volumes of information as well as
on science and engineering education. A serious gap already
exists between research publications and the more comprehensive
and systematic treatment of subjects found in a monographic
text. One of the purposes of this Series is to fill this gap.
A second purpose is to provide a vehicle for use as a teaching
text. In fact, the magnitude of the energy problem and its
solution through the creative application of scientific and
engineering principles requires that essentially all scientists
and engineers possess some background in the basic concepts of
energy production, distribution, utilization, and related
economics and social implications. However, most of the
currently available books and monographs are neither adequate
nor suitable for use as textbooks. Much of the material
published in the last several years addresses itself either to

the general public or to a group of specialists involved in
research and development.

Needed are books that are logically developed to emphasize
underlying principles and the applications of these principles to
energy, complete with sample problems suitable for use by
instructors and students in classes. To these ends, this Series
attempts to deal with subjects such as:

> Energy Conversion Engineering
>
> Coal Conversion Technology
>
> Geothermal Modeling
>
> Trace Elements and Pollution in Energy Production
>
> Water Resources in Energy Production
>
> Solar Energy: Materials and Models.

It is the sincere wish of the Editors that this Series will
be able to play a key role in filling the existing gaps in
information and in education.

"Energy Conversion Engineering," the first volume of the
new ENERGY SCIENCE AND TECHNOLOGY SERIES, presents the fundamental
background material needed to understand the basic principles
involved in energy conversion systems. The book also serves as
an introduction to much of the material that will be appearing
in subsequent volumes of the Series. In addition, "Energy
Conversion Engineering" provides an excellent review for
scientists and engineers who are either entering into or
redirecting their activities to the area of energy science and
technology.

C. Y. WEN

E. STANLEY LEE

PREFACE

This book is intended as a teaching text and not a state-of-the-art review or a data source book on energy conversion systems. It is directed toward students at the senior year in engineering or the physical sciences. The information presented is directed toward providing some of the analytical tools that will allow many of the arguments surrounding energy policy to be reduced to numerical values. The book does not dwell on particular systems or on today's energy priorities as these are in a state of rapid change. The book emphasizes and provides some basic principles that are invariant with time. The processes of today will be used as examples on how to apply some of the fundamental principles to real systems.

The book provides a discussion of the Laws of Thermodynamics, Chemical Equilibrium, Chemical Reaction Kinetics and applies them to energy conversion systems. The Laws of Thermodynamics, the science of energy conversion, serves as the foundation of the text and is the link that ties all the information together.

The book assumes no prior knowledge of Thermodynamics other
than that provided in basic physics and chemistry courses.

The text relies on example problems and on graphical
presentations to clarify the information being presented.
The many problems that have been provided in the text serve
a valuable role. They allow for a student to practice his
analytical skills, provide re-enforcement of statements provided in the text and provide new information not discussed
in the text. They are an extension of the discussion material
and an integral part of the text. For this reason, the index
refers to problems as well as material discussed.

The text has an emphasis on coal as the basic fossil
fuel. This was done because the authors are more familiar
with and qualified to speak in this area as well as the general
(but not universal) agreement that for the next century coal
is likely to play the major role in meeting the energy needs
of the nation.

The book does not emphasize economies. When dealing
with a problem that personally affects the lives of every
person and can threaten national security, economics may not
be the primary consideration. The cost of energy will be
influenced as much by political decision, at home and abroad,
as by the actual cost of production. This text is limited to
a discussion of scientific principles on energy conversion.
To venture any information regarding future energy policy is
not appropriate. It is hoped that the information provided
will be used to help those who must ultimately make the decisions on any future energy policy to come to a more rational
conclusion.

It is the young adults that are in our Universities today
that will be the ones that must live with the results of the
decisions to be made. They must look to the future, for they

are to spend their whole adult productive lives in the future.
To these people, we provide the dedication of <u>Energy & Man</u>,
edited by M. Granger Mergas, IEEE Press.

> For the grandparents...who found 110 KWh per capita
>> per day sufficient

> For the parents...who thought 150 KWh per capita
>> per day sufficient

> For the brothers and sisters...who have been doing
>> well on 250 KWh per capita per day sufficient

> And for our children...who will probably feel they
>> have to skimp at 350 KWh per day per capita

We add to this

> For our grandchildren who <u>may have to</u> find
>> 110 KWh capita per day sufficient.

These grandchildren we speak of are yours, the reader's,
children.

Before embarking into the text material, the authors
wish to warn the readers that both British and International
units are used. It is assumed that the student of today will
have to be equally familiar with both sets of units. The mixed
use of units in the text may best be described as planned chaos.

RICHARD C. BAILIE

1.0 ENERGY USES AND RESOURCES

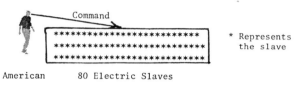

American   80 Electric Slaves

FIGURE 1.1   An American Averages 80 Electric Slaves
to do His Work

Man has learned to use energy to perform much of his work
activity. Figure 1.1 shows a person in the United States with
eighty energy slaves at his command. They respond to his
every wish. They provide water when he turns a faucet,
they give quick get-away mobile power by depressing a gas pedal,
they provide warmth for his home merely by setting a thermostat.
They do all manners of useful tasks and provide man with a high
standard of living. A short history of how man developed
and uses energy is provided in this chapter.

It was not until man learned to harness heat to do work
that he moved away from an agricultural-based society to the
industrial-based society. Prior to the industrial society
man used energy primarily as a source of heat and food for

himself and his animals (or slaves/serfs). Food was the
primary fuel need, and there was little demand to use the
energy stored within the earth.

Man had learned to use animals to do work, to plow
his field, to grind his grain, to irrigate his land, etc.
This work was done under the direction of man.

With the time released from manual tasks, man was
able to explore and discover the world around him. This
contributed to the discoveries that brought about the industrial
revolution. The discovery of the steam engine is generally
recognized as the beginning of what is called the industrial
revolution. What was it that the steam engine was able to do?
The steam engine was able to convert (at man's command) <u>chemical</u>
<u>stored</u> <u>energy</u> <u>into</u> <u>work</u>. No longer was it necessary to
depend on the biological system of sun→food→animal→work.

Materials such as coal, petroleum, and gas could now be
enslaved to do all manners of tasks and replace the slave
and the animal. Prior to this time in history, little use
had been made of this energy resource stored within the
earth. Some of these energy resources were used as heat
for blacksmiths or brewers, but this use was quite small.
The resource used remained that of a replenishable fuel, a
direct product of solar energy via photosynthesis. To
this point in time, there was no real depletion of the
stored irreplaceable energy resources. Man lived on the energy
received from the sun.

When was it that the United States shifted away from
this energy?

Sir James Watt perfected the modern steam engine
in 1765, a little over 200 years ago, about as long as the
United States has been a nation - a small time Δt, in the

history of man.   Table 1.1 below provides some of the important dates in our development of the industrial society,
beginning with the invention of the steam engine.

Table 1.1

Significant Events of the Industrial Age

First 100 Years	1765	Modern steam engine conceived by Sir James Watt.
	1786	Steam engine used to power paper mill, corn mill, supply energy to iron works and breweries.
	1788	Steam engine powered steam boat and locomotives.
	1801	Steam automobile.
	1820–1860	Principles of thermodynamics.
	1823	Production of fuel gas from coal for illumination.
	1855	Bessemer steel converter that allowed large quantities of steel to be made at low cost.
	1857	First oil well drilled in Pennsylvania.
Second 100 Years	1876	Otto designed the four-stroke internal combustion engine.
	1880	Institute of refrigeration was founded hailing the frozen food industry.
	1882	First incandescent lighting for New York.
	1903	Wright brothers first flight.
	1913	First synthetic fiber rayon produced in large quantities.
	1928	DuPont manufactured first synthetic fiber.
	1930	Model T Ford--birth of traffic problems.
	1941	First jet plane flight.
	1942	Fermi demonstrates nuclear power in Chicago.
	1945	First atomic bomb.
	1952	First hydrogen bomb.
	1954	First nuclear powered submarine Sputnik I, first artificial sattelite. First nuclear power plant in United States.
	1969	Man lands on moon.

Table 1.1 shows but a few events of the past 200 years. It
is provided to show how short a period of time has passed
since our conquest and enslavement of energy to provide for
our needs.

This has brought about many changes. The dependence
on energy resulted in the switch from renewable energy
supplies to non-renewable supplies. People moved from the
rural areas to the cities.

The U. S. populace have become so dependent upon energy
today that one might question if we have not become the
slave of energy. Consider the consequences of a loss of
energy to New York City. No heat for buildings, no transportation to deliver goods (including food), no lights,
no way to leave town, no refrigerators to preserve food,
and the list is endless. For a rural population, life
could continue, but in a city much life would soon be
snuffed out. Energy is an essential ingredient to modern
life.

The comments that have been made about changing
from non-renewable to renewable energy are graphically
illustrated in Figure 1.2, In 1850 wood, a renewable fuel,
accounted for 90 percent of the energy consumed. By 1900
it accounted for 20 percent (2Q out of 10Q).* After 1900
it became a minor contribution , and the nation depended almost
solely upon non-renewable resources. Petroleum does not
even appear until 1860 and within 100 years became the primary
fuel (almost 50 percent).

Figure 1.2 is plotted on a semi-log scale. The time
in years follows a linear scale and each division represents
one year. The amount of energy used is plotted on a logarithmic

---

*Q is 10^{15} Btu.

scale. The division between 0.1 to 0.2Q is the same as
the distance between 1.0 to 2.0Q or between 10 to 20Q.

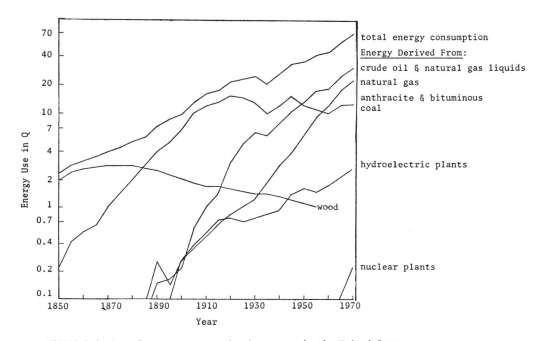

FIGURE 1.2   Annual energy consumption by source in the United States
             for the period 1850 to 1970; based on data of the United
             States Department of Commerce. Hydroelectric and nuclear
             energy are converted to thermal energy at the average
             efficiency of fossil-fuel plants operating during the
             specified year. The data are plotted for five-year intervals.

The equation for a straight line in Figure 1.2 is given by

$$Q \text{ (energy use at time T)} = Q_o e^{\lambda T} \qquad (1.1)$$

$$\text{where } e = 2.718$$
$$\lambda = \text{constant (year}^{-1})$$
$$T = \text{time A.D.}$$

For the curve labeled Total Energy,
$$Q_T = 1.85 \times 10^{-21} \, e^{0.0282T} \tag{1.2}$$

where Q_T is the yearly total energy consumption
in 10^{15} Btu/year

Equation 1.2 is an important equation. Converting this
equation to words it may be stated "The annual energy
use in the United States increases exponentially." The
term exponential is a result of raising the value of e(2.718)
to a power. There are several important characteristics of
this equation.

  a.   Doubling time--the time it takes an exponentially
       increasing function to double is given by

$$T_{1/2} = \frac{0.693}{\lambda}$$

       For total energy consumption, Equation 1.2 gives

$$\frac{0.693}{0.0282} = 25 \text{ years}$$

  b.   Total consumption--the total amount of energy used
       during one doubling period is equal to the total
       amount of energy used in all time prior to the
       doubling time.

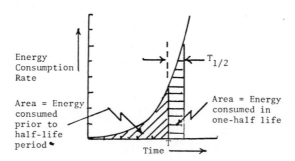

  c.   Yearly fractional increase--the yearly fractional
       increase is given by

$$\text{Fractional increase} = e^{\lambda} - 1$$

The percent increase is $(e^\lambda -1) \times 100$ and is the same as the interest in a compound interest formula. It is known as the rate of increase. The rate of increase of the total energy use is

$(e^{0.0282} - 1)\ (100) = 2.86\%$

(i.e., 2.86% per year)

The use of exponential curves to extrapolate for a significant time into the future is absurdity. This is shown in the following example.

EXAMPLE 1.1   Using the total energy & the natural gas consumption during 1940 to 1970, determine the exponential equation for each. Then determine when the amount of gas energy used will exceed the total energy used.

SOLUTION 1.1

From Figure 1.1

	Q Gas
1950	6
1970	23

The exponential equation (1.1) becomes

$Q = Q_o\ e^{\lambda T}$

Applying this equation to the gas

@ 1950 $6 = Q_o\ e^{1950\lambda}$ (i)

@ 1970 $23 = Q_o\ e^{1970\lambda}$ (ii)

Dividing (i) by (ii) and solving for λ

$6/23 = e^{\lambda(1950-1970)}$ $\longrightarrow$ $\boxed{\lambda = 0.067 (yr)^{-1}}$ Substituting into (i) and solving for Q.

$6 = Q_o\ e^{0.067(1950)} = 5.5 \times 10^{56}\ Q_o$

$Q_o = 1.1 \times 10^{-56}$

The equation (1.1) for the use of natural gas becomes

$Q_G = 1.1 \times 10^{-56}\ e^{+0.067T}$ [Quads/yr]

In a similar manner for the total energy used (See Eq. 1.2)

$Q_T = 1.85 \times 10^{-21} e^{0.0282T}$ [Quads/yr]

Letting $Q_T = Q_G$

$1.1 \times 10^{-56}\ e^{+0.067T} = 1.85 \times 10^{-21}\ e^{+0.0282T}$

and solving for T , gives

$\boxed{T = 2090}$

The purpose of this example is to point out that an
exponential increase can only be maintained for a limited
period of time. It can be seen in Figure 1.2 that coal increased
exponentially between 1850 and 1910 and leveled off. Prior to
this wood behaved in the same manner. It had leveled off by
1850, and the exponential increase is not on the graph.

EXAMPLE 1.2 Various estimates of the coal reserves are discussed
later. One estimate provides the energy in the total
U. S. coal reserves at 10,347Quads. At the present
rate of coal consumption, 15 Quads a year the coal would
last 10,746/15 = 716years. Assume that the coal use did
not level off in 1910 but continued at a rate consistent
with the 1850-1910 period. When would all of the coal
be depleted? The equation for Q_{coal} derived in a similar
manner to example 1.1 gives

$$Q_o = 1.4 \times 10^{-53} e^{0.065T}$$

SOLUTION 1.2 The total fuel consumption is a product of the rate of
consumption over time. When the rate is changing with
time the rate must be integrated over the time span.
For this problem

$$10,746 = \int_{1910}^{T} Q_{coal}\, dT = \int_{1910}^{T} Q_o e^{\lambda T}\, dT = (Q_o/\lambda) e^{\lambda T} \Big|_{1910}^{T}$$

$$= \frac{1.4 \times 10^{-53}}{0.065} \left[e^{0.065T} - e^{0.065(1910)} \right]$$

$$\boxed{T = 1973}$$

Coal would have been exhausted by 1973 if the rate had
continued to increase.

The use of coal was destined to level off and could not
increase exponentially. Gas has now leveled off. Oil must
soon level off. Unless a new energy source appears, the total
energy use must also level off. Figure 1.2 shows such a new
source appearing--nuclear power. There are serious questions,
however, as to whether this will become the new source of
energy that can allow the total energy curve to increase as
it has.

Often projections for future use are based upon exponential increase into the future. Such projections are not provided in this text. The period of leveling is felt to be at hand. How fast these energy growth curves will flatten out is not known. Extreme caution should be exercised in interpreting the various projections into the future. This text emphasizes judgment based upon Laws of Nature rather than on projections based upon a less scientific basis.

1.1 ENERGY RESOURCES

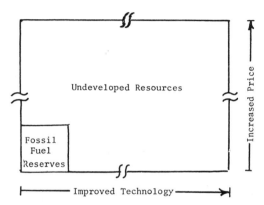

FIGURE 1.3 Fossil Fuel Resources Change with Price and Technology

One of the most controversial aspects of any energy decision is how to determine the energy resources available. There is no simple or single response. The amount of energy resources that are available depend upon the price that will be paid by the customer. When the price rises, more effort is made to locate new undiscovered sources. Sources that were known, but could not be exploited because of high cost of development will become available. Even sources being held by speculators waiting for better prices will become available. There is nothing unusual about this. For example, when rents for

apartments rise, more apartments are built. When rents are
controlled at low values, new apartments are not developed.
The amount of resources depends upon the price that is paid.
The higher the price, the greater the resources. This is shown
in Figure 1.3. For a given technology, as the price rises, the
area (which represents energy resources) increases.

A second factor that influences the amount of resources
reported is the status of technology. For example, before it
was technically possible to obtain off-shore oil, these oil de-
posits were not available. As the technology for off-shore
drilling was developed, then these sources became available.
The available energy reserves at a particular economic value
depend upon technology to extract it. The value of a resource
is not constant in time but continues to change as the price of
fuel increases or as new technology develops. A few examples
as to how this might work are discussed below.

Using today's technology, it is not uncommon in underground
mining to leave about one-third of the coal in place because it
is not economically attractive to remove it. Part of the coal
is left in to make pillars in the coal mine. If the cost of
coal would rise, then it would become economically attractive
to place concrete pillars in the mine and not use the coal for
any support. Then a larger fraction of the coal would be removed.
A major breakthrough in technology of drilling for oil or in
the technology of coal mining would result in a substantial
increase in the amount of fuel that is available. It is clear
that when we talk about proven reserves or available reserves
this is not a fixed number but will float with the value of the
fuel and the state of the technology.

The final factor in determining energy reserves is knowing
how much fuel is available. These reserves were deposited

underground billions of years ago and no record as to how
much or where they are deposited was made. There is no
way of knowing where all of these fuels are deposited.
New fields that were unknown several years ago are being
found. The North Sea finds and the Alaskan finds were
not known twenty years ago and were not included in the
value of the known or proven reserves. It is not until
after the field has been tapped that a reasonable estimate
as to how much fuel is available in any of these areas can
be determined.

Figure 1.4 shows the amount of reserves available
in a manner similar to that used in Figure 1.3. In this
case, both technology and prices are shown together. It
shows how the reserves (area) increases as the reserves
reported move from Measured →Indicated →Inferred →Undiscovered.

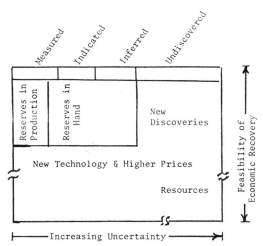

FIGURE 1.4   Fossil Fuel Resources Change with
             Uncertainty and Economics

Any estimate of future reserves are subject to large
uncertainty and change with time. Table 1.2 provides one
estimate of "Proven Reserves."

TABLE 1.2

Proven Reserves From F.E.A. Project Independence Report

Source	Fuel Units	Quad-rillion Btu's	Years Left at 1972 Consumption Levels
Coal			
high sulfur (more than 1%	273 billion tons	6908	
low sulfur (less than 1%)	160 billion tons	3838	
TOTAL	433 billion tons	10746	823
Oil			
lower 48 (crude)	30 billion barrels	176	
natural gas liquids	6 billion barrels	37	
Alaska	10 billion barrels	59	
TOTAL	46 billion barrels	272	8
Gas			
lower 48	218 TCF	225	
Alaska	32 TCF	32	
TOTAL		257	11
Shale	20-170 billion barrels	116-986	3-28
Tar Sands	29 billion barrels	168	28

Table 1.2 shows that the United States has abundant
coal reserves that would last a long time under current
levels of usage. The amount of coal that is actually
available is less than the figures shown. Only about 90%
of the stripped coal can be used, and about 50% of the underground coal. Large amounts of sulfur in the coal may rule
it out as a fuel source because of environmental reasons.
In spite of these problems, it remains far more abundant
than other fuel resources. Minable coal is not uniformly distributed but is concentrated in some Midwest, Great Plains
and Appalachian regions.

Oil and gas reserves are more limited. The proven
reserves are low, but new discoveries are almost sure to
exist in Alaska and off-shore.

In addition to the conventional oil fields, there is a large supply of oil contained in oil-saturated sand and oil impregnated rocks known as tar sands. Almost all of these are located in the West.

Nuclear energy provides another alternate energy resource. The amount of uranium available as a nuclear fuel depends upon the price.

Table 1.3

Uranium Reserves (U^{235})

	Tons of U_3O_8	Electric Capacity (Total)	Years of supply using Uranium to provide all electric needs at 1970 electric consumption rates
Low Cost	390,000	130Q	7.9
Medium Cost	160,000	54.4Q	3.3
High Cost	300,000	102Q	6.2
TOTAL	850,000	289Q	17.6

EXAMPLE 1.3   A 1,000 megawatt U^{235} enriched nuclear reactor at an eighty percent load factor (fraction of full power system operates) uses about 200 tons of U_3O_8/yr. The reactor operates at about a 35% efficiency (i.e., for every .35 units of energy leaving as electricity, 1.00 units of fuel was consumed). Determine the amount of energy available from this energy source and compare to the energy from coal.

SOLUTION 1.3*

1,000[MW electricity] x 1,000 [KW/MW] x 3413[Btu electricity/KW-Hr] x 24[Hr/day] x 365[day-yr] x

$0.80 = 24 \times 10^{13}$[Btu electricity/yr]

24×10^{12}[Btu electricity/yr] x 1/0.35[Btu fuel/
Btu electricity] $= 6.9 \times 10^{13}$[Btu fuel/yr]

6.9×10^{13}[Btu fuel/yr] x 1/200[yr/ton U_3O_8] $=$
3.4×10^{11}[Btu/ton U_3O_8]

From Table 1.3 the total Uranium reserves is given
as 850,000 tons U_3O_8

850,000 [tons U_3O_8] x 3.4×10^{11}[Btu/ton U_3O_8] x
$1/10^{15}$[Q/Btu] = 289[Q]

From Table 1.2 coal reserves are 212,000 Q

The potential of . . Uranium^{-235} is small compared
to coal as an energy resource.

_____

*See Appendix C and Table 1.6 for conversion factors

It is evident from Table 1.3 that even at present
utilization rates that the uranium resources would not
provide the energy needed in the future and would only
provide temporary relief. The table has considered the
use of only the U^{235} as the fuel. It is possible to use
U^{238} as a fuel by converting it to plutonium, a nuclear
fuel. If this is done, then the resource base given in
Table 1.3 may be multiplied by a factor of 50. The
production of nuclear fuel from non-nuclear fuel is called
breeding. A reactor that produces fuel is a breeder
reactor.

U^{238} is one material that can be converted to a nuclear
fuel. Another is thorium, Th^{234}. The amount of thorium
available in the same cost range as uranium is of the same
magnitude. If breeder reactors were developed the values
for total electrical generating capacity would be multiplied by a factor of 100. This could provide the needs for
a significant period (but finite time). It would likely
increase rapidly for some time, as have other fuels, and
then flatten out.

Even if the values given for the available resources
of fossil and fussion fuels are low, it does not alter
the need for energy options to be developed.

As already discussed, the use of renewable resources
was once common practice. This came from the burning of
wood, use of wind, use of water power, etc. Windmills and
water wheels were common sights. Figure 1.5 shows a pendulum

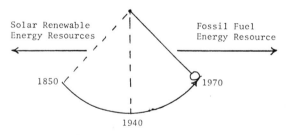

FIGURE 1.5 Pendulum Swinging from Renewable
           to Nonrenewable Resources

that swings from dependence on renewable (in 1856) to dependence
on non-renewable resources (in 1976). From the survey of the
resources of non-renewable resources it is a question as to how
much further the pendulum may swing. When will it reach the
end of its travel into the region of non-renewable resources
and begin to swing back toward renewable resources?

The amount of solar energy that is converted into plant
matter is a small fraction of that reaching the earth, but
even so represents 40 times the amount of energy consumed
in the world (in 1970). Table 1.4 gives some values for
renewable energy resources.

TABLE 1.4

Renewable Resources
(From S. S. Denver and L. Icerman. "Demands, Resources Impact,
Technology, and Policy", Vol. 1, Addison-Wesley Publishing Co., (1974).

	Q/y
solar input to earth	5.3×10^6
solar input used in photosynthesis	1,200
solar input per 10^6 acres on earth	21
total geothermal heat flux outward	800
heat stored in geothermal systems to 10 km depth	40,000
useful geothermal heat from hydrothermal areas	4
tidal dissipation on earth	90
usable tidal energy	1.9
power corresponding to total hydrologic run-off	2,700
useful hydroelectric power	860×10^{-2}

1.2   ENERGY USE PATTERNS

Figure 1.6 provides a flow sheet showing the input-output energy patterns. This diagram provides a great deal of useful information and is discussed briefly. About 15.6Q of fuel are shown to be used to provide for the production of electrical energy. Only about 5.5Q of this energy is converted to electricity and 10.1Q is lost and put to no effective use. The energy uses shown in addition to 15.6Q going to production of electrical energy about 13.1Q go to household and commercial use, 17.2Q go to transportation, and 19.6Q go into industrial use.

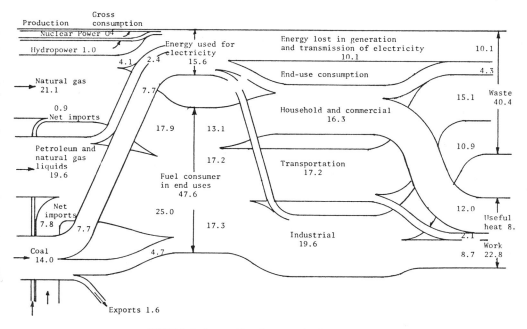

FIGURE 1.6   Energy Flow Patterns in the U.S.

(Courtesy of Earl Cook, Texas A & M, 1973)

The transportation sector uses a major amount of energy,
17.2Q and is the major contributor to waste. 15.1Q of
the energy is lost and only 2.1 goes into a useful product.
Looking at the area of energy waste, 15.1Q came from transportation, 10.9Q from industry, 10.1Q from electric generation
and 4.3Q from commercial and household use. The areas of
power generation and transportation are large contributors to
the waste. They have also been the largest growing sectors
for energy consumption and are looked at a little more
closely.

Figure 1.7 shows the relative energy use of the
electric power industry and transportation industry. The
problems associated with an exponential growth were discussed earlier. In the electric generating sector the
growth is not only growing exponentially but the
exponential rate is continuously increasing. In 1955 the
growth was considerably less than 7 percent; it grew to 8.6
percent/year by 1965 and 9.25 percent by 1975. The growth
rate must soon flatten out.

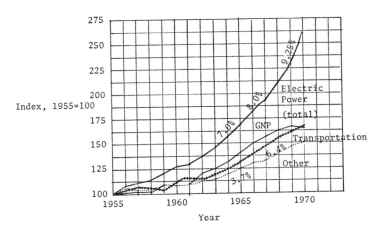

FIGURE 1.7 Relative Increase in Energy for
Major Energy Uses.

Table 1.5 Energy Utilization in Transportation

Intercity Passenger Transport, 1970

Mode	Passenger Miles x 10^9	Percentage of Total Passenger Miles	Energy Consumed (kcal x 10^{12})	Efficiency (kcal/ Passenger Mile)
Automobiles	1020	88.1	940	920
Airplanes	104	9.0	306	2940
Buses	25	2.2	7.6	308
Railroads	8	0.7	1.9	240

Freight Transport, 1970

Mode	Volume (Ton-Miles x 10^9)	Percentage of Total	Energy Consumed (kcal x 10^{12})	Efficiency (kcal/ Ton-Mile)
Railroads	768	40	131	170
Trucks	412	21.4	370	900
Pipelines	431	22.4	49	114
Water carriers	307	16	42	136
Airways	3.4	0.2	34	10,000

Source: Statistical Abstracts of the United States; Interstate
Commerce Commission, Federal Aviation Administration, and
Department of Transportation gross data.

Table 1.5 shows the energy required to transport freight and
passengers. The trend has been toward the less energy efficient
transportation system. Railroads that use about 24 percent as
much energy as trucks are being replaced by trucks. Bus and
rail passenger systems have been taken over by private autos and
air travel.

1.3 ENERGY CALCULATIONS

This section is provided to familiarize the reader with
energy units and the manipulation of these units. Up to this
point, the calculations have involved only English units. The
rest of the book will deal primarily with the metric units.

The two terms that must be understood in the industrial
society are work and power.

Work--work is a form of energy and is defined as force
x distance and has the units of ft-lb$_f$ in English
units, or joules (dyne-cm) in international units.

Power--power is the work done per unit time. The energy
unit representing 1 joule/sec is the watt. In
English, 1.356 ft-lb$_f$/sec = 1 watt.

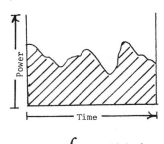

Work = $\int$ Power d(time)

Work is the area of a curve of power vs time

Table 1.6 provides the relationship between various
energy units.

EXAMPLE 1.4   Determine the heating value for a typical natural
               gas (1,000 Btu/ft) in terms of joules/m^3.

SOLUTION 1.4

$$1,000[\text{Btu/ft}^3] \times (3.28[\text{ft/m}])^3 \times 4.186/0.00397[\text{joules/}$$
$$\text{Btu}] = 37.2 \times 10^6[\text{joules/m}^3]$$

i.e., $1[\text{Btu/ft}^3] = 37.2 \times 10^3[\text{joules/m}^3]$

Table 1.6

Selected Dimensional Equivalents

Length	1 m = 3.280 ft = 39.37 in 1 cm ≡ 10^{-2} m = 0.394 in = 0.0328 ft 1 mm ≡ 10^{-3} m 1 micron (μ) ≡ 10^{-6} m 1 angstrom (Å) ≡ 10^{-10} m
Time	1 hr ≡ 3,600 sec = 60 min 1 millisec ≡ 10^{-3} sec 1 microsec (μsec) ≡ 10^{-6} sec 1 nanosec (nsec) = 10^{-9} sec
Mass	1 kg ≡ 1,000g = 2,2046 lbm = 6,8521 x 10^{-2} slugs 1 slug ≡ 1 lbf-sec^2/ft = 32.174 lbm
Force	1 newton ≡ 1 kg-m/sec^2 1 dyne ≡ 1 g-cm/sec^2 1 lbf = 4.448 x 10^5 dynes = 4.448 newtons
Energy	1 joule ≡ 1 kg-m^2/sec^2 1 Btu ≡ 778.16ft-lbf = 1.055 x 10^{10} ergs = 252 cal 1 cal ≡ 4.186 joules = 0.00397 Btu 1 kcal ≡ 4,186 joules = 1,000 cal 1 erg ≡ 1 g-cm^2/sec^2 1 ev ≡ 1.602 x 10^{-19} joules
Power	1 watt ≡ 1 kg-m^2/sec^3 = 1 joule/sec 1 hp ≡ 550 ft-lbf/sec = 33,000 ft-lbf/min 1 hp = 2,545 Btu/hr = 746 watts 1 kw ≡ 1,000 watts = 3,413 Btu/hr = 1,3405 hp
Pressure	1 atm ≡ 14,696 lbf/in^2 1 mm Hg = 0.01934 lbf/in^2 1 dyne/cm^2 = 145.04 x 10^{-7} lbf/in^2 1 bar = 14.504 lbf/in^2 ≡ 10^6 dynes/cm^2 1 micron (μ) ≡ 10^{-6} m Hg = 10^{-3} mm Hg
Volume	1 gal ≡ 0.13368 ft^3 1 liter ≡ 1000.028 cm^3

Table 1.7 provides some values that may be useful in working problems in the text. It is a handy reference of miscellaneous information.

EXAMPLE 1.5   Determine the rate of energy consumption by a 100 horsepower motor that is 20% efficient (in joules/hr). What is the rate in terms of kilowatts?

SOLUTION 1.5

$$100[hp] \times \frac{100}{20}\left[\frac{hp\text{-input}}{hp\text{-output}}\right] \times \frac{1.0}{1.3405}\left[\frac{kw}{hp}\right] = 373[kw]$$

$$373[hp] \times \frac{1,000}{1}\left[\frac{w}{kw}\right] \times \frac{1}{1}\left[\frac{joule}{w\text{-sec}}\right] \times \frac{3600}{1}\left[\frac{sec}{hr}\right] =$$

$$1.34 \times 10^9 [joule/sec]$$

TABLE 1.7

Values of Useful Information in Energy Studies
($Q = 10^{15}$ Btu)

Earth Characteristics	Q/yr
approximate heat capacity of the oceans	5.4×10^{6} °C
power absorbed in global evaporation	9.8×10^{5}
energy released in condensation of atmospheric water	3×10^{3} Q
approximate heat capacity of the atmosphere	5×10^{3} °C
heat release associated with the production of all of the atmospheric CO_2 from fossil fuels	3×10^{3}
heat release associated with the production of all of the atmospheric CO_2 that is exchanged between the atmosphere and the oceans	3×10^{3}

Energy-Use Rates	Q/yr
metabolic processes for 3.592×10^{9} people in 1970 at an average of 2.350×10^{3} kcal/day	$1.2 \times 10^{+1}$
U.S. 1970 energy consumption rate	$6.8 \times 10^{+1}$
world 1970 consumption rate	$1.9 \times 10^{+2}$
world use rate compounded at 3%/y to the year 2000	$4.6 \times 10^{+2}$
world use rate in the year 2000 if the per capita use rate equals the U.S. 1970 per capita use rate of 3.8×10^{7} Q/y	2.1×10^{3}

Adapted from: M. R. Gustavson, Dimension of World Energy, Mitre
                  Corporation, McLean, VA.

EXAMPLE 1.6   The fission of a uranium atom releases 931 ev of
                  energy. What is this in terms of kw-hr?

SOLUTION 1.6

$$931 \left[\frac{ev}{fission}\right] \times \frac{1.602 \times 10^{-19}}{1} \left[\frac{joules}{ev}\right] \times 1 \left[\frac{w\text{-}sec}{joule}\right]$$

$$\times \frac{1}{1,000} \left[\frac{kw}{w}\right] \times \frac{1}{3600} \left[\frac{hr}{sec}\right] = 4.14 \times 10^{-23} \left[\frac{kw\text{-}hr}{fission}\right]$$

Problems for Chapter 1.0

1. Starting in 1974, assume in order to keep increasing
    electrical capacity at a rate of 6.5 percent per year
    that 1,000 MW all nuclear electric power plants were built
    for all new capacity. Make a plot of number of nuclear
    plants vs. time and determine at what time:
    (a) That breeder reactors must become available.
    (b) That we run out of all nuclear fuels.

2. Solid waste represents a potential source of energy.
    On the average, each American discards about 5 lbs/person/day.
    It has a heating value of about 1/2 that of coal
    (6,500 Btu/lb). Determine the amount of energy avail-
    able from this source. What fraction of the energy needs
    could be met by this energy source?

3. Figure P-3 shows the average monthly electrical energy
    use for the southwestern region of the U.S. for a typical
    August day (Aug. 6, 1965). This is the peak month. P-3
    also gives the average monthly power generation.

Figure P-3

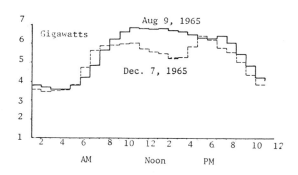

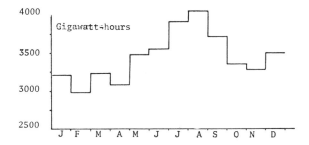

The Electrical Usage in Southwest U.S. Region
Drawing A: Hourly Average for Aug 6, and Dec 7, 1965.
Drawing B: Monthly Average 1965.

Determine:
- (a) The amount of work done for the one-year period.
- (b) The average load factor of the plants serving the area assuming 95% of capacity for noon on August 6.

4. Prove that the amount of energy consumed in one half-life is equal to the amount consumed in all time if the increase follows an exponential growth curve.

5. Suppose conservation measures can result in a slowing down in electrical consumption according to the following schedule. The doubling period doubles every doubling period. Assume the 1970 consumption rate to be 1.5 x 10^{12}KWhr and the doubling period is 10 years. Compare the demand in 2050 to the demand if no action is taken to decrease use. Also, determine the total amount of electricity used in each case.

6. Assume that only 1% of the solar energy reaching the earth is converted to useful energy. Assume that it was necessary for the U.S. to live on the share that it receives.
- (a) How much total energy is received?
- (b) How much of our current energy expenditure could be met by this energy source?

7. A name to remember as a pioneer in considering the use of fuels and minerals that are available in finite amounts is King Hubbert. A study of the mineral industry in Europe showed that consumption rose rapidly and reached a peak and declined rapidly as the minerals became in short supply. The following problem is based upon the logic provided by King Hubbert. Consider that the production rate of coal will follow a normal distribution curve.

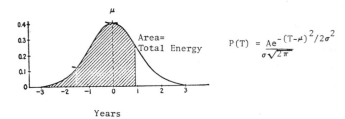

$$P(T) = \frac{Ae^{-(T-\mu)^2/2\sigma^2}}{\sigma\sqrt{2\pi}}$$

P-Production Rate

Area = Total Energy

Years

Normal Distribution Curve

The area under this curve is $\int$ Pdt the total coal reserves.
Assume that the maximum production will not exceed 8 times
the production rate. Assume that 500×10^7 metric tons
of coal is available and the rate of increase in 1967 is
6%/year. Determine when (year) the maximum production
will occur and when 80% of the coal is utilized. For
further discussion, see M. King Hubbert, Can. Mining &
Met. Review, 1973, 66, pp. 37-53.

8. The Club of Rome has published a book entitled, The Limits
    to Growth. A study of five basic factors that will
    ultimately limit growth--population, agricultural produc-
    tion, natural resources, industrial production, and pollu-
    tion--are discussed in terms of reserves and exponential
    growth. For the case of chromium, reserves are predicted
    to last 420 years if it is used at the present rate.
    (a) How long will it last at a growth rate of 2.6%/year?
    (b) What is the doubling time in years?
    (c) Assume that the amount of reserves are doubled
        because of new technology. At the 2.6% growth
        curve, how long will the reserves last?
    (d) Assume that 1/2 the chromium is recycled, how
        long will the reserves last?

    A popular (not highly technical) version of the study
    is published in Limits to Growth by Meadows, Meadows,
    Randers, and Behrens, Potomac Associates Books, 1972.

9. The Ph.D.s awarded in physics grew at an exponential
    rate between 1950 and 1960. The population of people
    and dogs also grew exponentially. Using this information
    it is possible to predict that by 2250 every person
    and dog in the country would be expected to have a
    Ph.D. in physics. Show graphically how such an absurd
    prediction could be made.

10. Assume that some planet contains 10^9 tons of a mineral
    resource (e.g., coal) and the inhabitants of that planet
    initially use the resource at a rate of 1,000 tons/year.
    (a) How long will the resource last at a constant use
        rate?
    (b) If the usage doubles every 20 years, how long
        will it last?
    (c) Fifty years before the resource is exhausted
        what percent of the supplies are left?

2.0 THERMODYNAMICS

2.1 INTRODUCTION

This chapter provides the analytical tools that are necessary to make quantitative evaluations for process systems
that will convert chemical, nuclear, solar, or geothermal
energy to work or alternate energy forms. Such a system is
shown in Figure 2.1. In this process, fossil fuel (Chemical
Stored Energy) is converted into electrical energy. This
conversion represents the most significant invention leading
to the industrial revolution. Chemical energy has little inherent value. Once converted to electrical energy it can
light our homes, power our industry and propel our transportation system. Electrical energy may be considered an "industrial slave." It faithfully carries out tasks previously performed by man or animal prior to the industrial revolution.

The Science of Thermodynamics deals with the conversion
of energy from one form to another. It provides a scientific
basis for making calculations and establishing limits

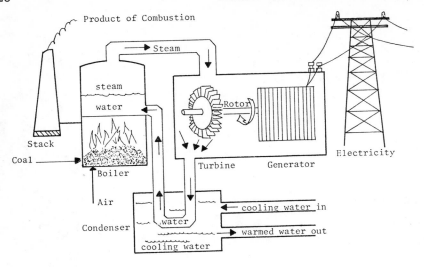

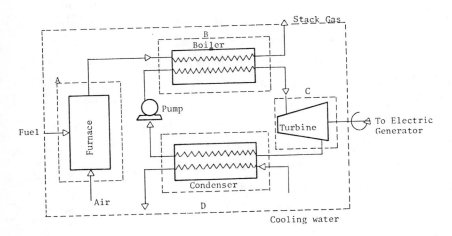

FIGURE 2.1 Schematic Diagram of Fossil Fuel Electric Power Plant

for energy conversion. It is not surprising that thermodynamics laws were developed as the industrial revolution
began less than 200 years ago. The first turbine-generator
to produce electricity was designed about 1840. Joule performed the basic experiments that led to the First Law of
Thermodynamics only one year earlier.

This chapter provides sufficient thermodynamics to allow for analysis and solution of simple energy conversion problems and an appreciation of the more complex analysis provided in the text. Full appreciation for and understanding of thermodynamics is beyond the scope of this book.

Thermodynamics deals with the transformation of energy from one form to another and from one system to another. Figure 2.1 shows several energy changes occurring. Figure 2.1 is subdivided into several separate units as sub-units (called systems). These are shown surrounded by a dashed line that separates a given unit from other units in this figure and the surroundings. A study of the inputs and outputs to these units provides information on the energy transformation that takes place. These are described below:

A. Furnace--Fuel is burned with air to produce a hot gas (flue-gas). The chemical energy available in the coal is converted into a high temperature gas (called sensible heat).

B. Boiler--Some of the heat from the flue gas is transferred to water to produce high pressure steam. The temperature of the gas is lowered. This is a heat transfer operation.

C. Turbine--The high pressure steam is reduced in pressure in a turbine and shaft work is produced.

D. Generator--The shaft work runs an electrical generator to produce electrical energy.

E. Condenser--The low pressure steam from the turbine is condensed to liquid water by removing heat and transferring it to a cooling water stream.

F. System--This refers to all of the operations
    working together. Coal, air, and cooling water
    are taken in by the system and electrical
    energy, flue gas and water discharged as the
    products of the system. This system converts
    coal to electricity, stack gas and water dis-
    charge.

This chapter shows how to reduce these various energy
transformations into numerical terms. It will show how to
determine which energy conversion schemes are possible and
which are not possible, which chemical reactions will occur
and which ones will not. <u>It will establish the limits to
what can be accomplished</u>.

Thermodynamics can only provide insight into what is
possible but does not allow the determination of how long
it takes to achieve a given conversion. Chemical kinetics
must be considered along with thermodynamics <u>to design</u> any
system that involves a chemical reaction.

2.2 FIRST LAW OF THERMODYNAMICS

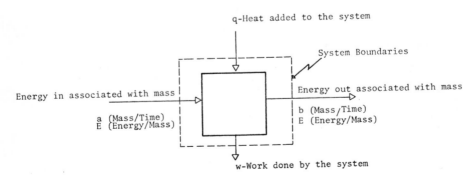

FIGURE 2.2 Typical Flow System

The First Law of Thermodynamics (also called an energy
balance) is a statement that energy is conserved or "the
energy in equals the energy out." Energy may change forms
but is conserved. Figure 2.2 shows a typical flow system.

The solid box represents an energy conversion system.
The dotted line defines a boundary between the system and
everything external to the system, i.e., the surroundings.
The arrows that cut this boundary show that the system exchanges energy with the surroundings. Only steady-state flow
systems will be discussed in this section. A steady-state
system is one where the system and the surroundings do not
change with time. A flow-system is one where mass flows
between the system and the surroundings.

Energy may enter or leave the system with the mass
that flows into or out of the system. Each unit of mass
contains energy that is associated with the temperature,
pressure, phase and chemical configuration of the mass

Total Energy = m $\overline{E}$
m is the mass of the substance
$\overline{E}$ is the internal energy of a flow stream/unit mass

The total internal energy of a stream is the product of
the mass flow rate and the internal energy per unit of mass.

Energy that is not associated with mass may be exchanged
between the system and surroundings. Work and heat are
energy terms that are not associated with mass. Because q
and w are not associated with mass they cannot be stored.
It is not correct to say a mass contains heat. Mass
contains energy. Some of this energy may be exchanged
between systems as heat or work.

The energy associated with mass flowing into or out of
a system will be defined by the symbol $\overline{H}$ (energy/mass)
and is called enthalpy. From Figure 2.2 the First Law of
Thermodynamics (energy balance) may be written

Energy Associated with Entering Mass + Heat =
Energy Associated with Exiting Mass + Work   (2.1)

When kinetic energy and potential energy may be neglected,
Equation 2.1 may be written

$$a\overline{H}_A + q = b\overline{H}_B + w \qquad\qquad (2.2)$$

The lower case a and b represents the mass flow rate of a chemical
specie identified by A and B respectively. q and w represent heat
and work respectively. There may be more than one stream entering
or leaving the system and 2.2 may be more generally written as

$$\sum_i a_i\overline{H}_{A_i} + \sum_i q_i = \sum_e b_e\overline{H}_{B_e} + \sum_e w_e \qquad (2.3)$$

The sign convention for q and w is as shown in Figure 2.2;
heat added to the system and work done by the system is
always taken as positive.

The value of $\overline{H}$ depends upon temperature, pressure and
physical state of a material. The value does not depend upon
how the condition of temperature, pressure and physical state
were achieved. The values of work and heat, q and w do
depend upon how a process takes place. To visualize this
difference, consider the answer to the following questions:

(a) What is the potential energy change that occurs
    when a 3,000 lb automobile drives from Chicago(C)

(Elevation 200 ft.) to Denver (D) (Elevation
5,200 ft.)?

(b) How many miles were put on the car in driving
between these same two cities?

The answer to question (a) is

$$P.E. = m(Z_D - Z_C) \frac{g}{g_c} = 2,000 [lb_m](5,200 - 200)[ft.]\times$$
$$1[lb_f/1\ lb_m] \doteq 10,000,000 [ft\text{-}lb_f]$$

To obtain this answer, the factors that were needed were

(a) Mass of auto, m

(b) Elevation of Chicago, Z_C and

(c) Elevation of Denver, Z_0

It was not necessary to know how the automobile traveled from
Chicago to Denver. Potential energy is a point property
and is independent of path.

The answer to (b) requires that the route (or path) used
to travel from Chicago to Denver be known. Different routes
(paths) provide different mileage. Distance is a path property.

Changes that are independent of the path followed are
point changes and are obtained by subtracting the values of
a function evaluated at two points $(Z_2 - Z_1)$. Changes that
are dependent upon the path are path changes, and the path
for the change must be known. Enthalpy, $\bar{H}$, is a point property
while q and w are path properties. The value of $\bar{H}$ is associated with mass (energy/mass) and is termed an intensive
property. q and w are not associated with mass and are
termed extensive properties.

Equation 2.3, the First Law of Thermodynamics provides for energy to be exchanged between the system and the surroundings:

  (a) Along with mass that enters or leaves. This requires a knowledge of the enthalpy of each stream.

  (b) Through the system boundaries. This is not associated with mass and requires knowledge of the path followed.

When there is no chemical reaction taking place within the system, the mass and chemical composition of stream "a" is identical to stream "b" and Equation 2.3 may be written

$$\sum a_i (\overline{H}_{A_e} - \overline{H}_{A_i}) + \sum q = \sum w \qquad (2.3a)$$

where $\overline{H}_{A_e}$ and $\overline{H}_{A_i}$ are the enthalpy of exit and inlet streams respectively.

The equation for the First Law of Thermodynamics is most often written with those terms associated with the mass flow and independent of path on the left hand side of the equation and those not associated with mass and dependent on the path on the right hand side of the equation.

$$\sum a_i \Delta \overline{H}_A = q - w \qquad (2.3b)$$

The value of $\Delta \overline{H}_A$ is obtained by subtracting the enthalpy of the exit stream, $\overline{H}_{A_e}$ from the enthalpy of the inlet stream $\overline{H}_{A_i}$.

The energy balance equations given in Equations 2.3, 2.3a, 2.3b are all statements of the First Law of Thermodynamics.

EXAMPLE 2.1   Water is to be heated by steam. The water enters the
heater at 21°C and is heated to 92°C. The enthalpy values
for each stream are given. The water flow rate is fixed
at 100 Kg/min. What is the flow rate of the steam?

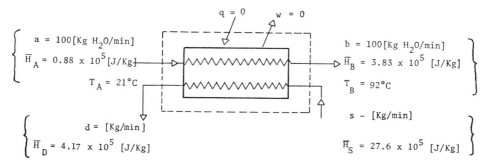

SOLUTION 2.1   Material balance over water stream: $a = b = 100$ [kg/min]
                Material balance over steam stream: $s = d$

Solution (A)   Using system given by --- lines

$a\overline{H}_A + s\overline{H}_S = b\overline{H}_B + d\overline{H}_D$

100 [Kg H_2O/min] $\times 0.88 \times 10^5$ [J/Kg] + s [Kg/min] $\times 27.6 \times 10^5$ [J/Kg]

$= 100$ [Kg H_2O/min] $\times 3.83 \times 10^5$ [J/Kg] + s [Kg/min] $\times 4.17 \times 10^5$ [J/Kg]

Solving for s

$s = 12.6$ [Kg/min]

Solution (B)   The heat exchanger is divided into two systems. Heat
is exchanged between systems.

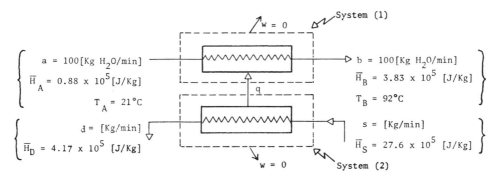

About System (1) (Shown by --- lines)

$a\overline{H}_A + q = b\overline{H}_B$ (2.3)

$100[Kg\ H_2O/min]\ x\ 0.88\ x\ 10^5[J/Kg] + q$

$= 100[Kg\ H_2O/min]\ x\ 3.83\ x\ 10^5[J/Kg]$

$q = 2.95\ x\ 10^7[J/min]$

About System (2) (Shown by --- lines)

$s\overline{H}_S = d\overline{H}_D + q$

$d[Kg/min]\ x\ 27.6\ x\ 10^5[J/Kg] =$

$d[Kg/min]\ x\ 4.17\ x\ 10^5[J/Kg] + 2.95\ x\ 10^7[J/min]$

$d = 12.6[Kg/min]$

This example problem illustrates that a problem may be solved in more than one way depending upon the choice of system boundaries. There are times where the choice of system boundaries can simplify the problem solution significantly. In the above example, breaking down the original problem into two sections did not simplify the calculation but showed more clearly what was happening. <u>The steam was transferring heat to the water.</u> In solution B, this transfer process is more apparent.

EXAMPLE 2.2   A hot water stream is to be obtained by injecting live steam into a cold water stream. The flow sheet and the enthalpy values for each stream are given in the diagram below.

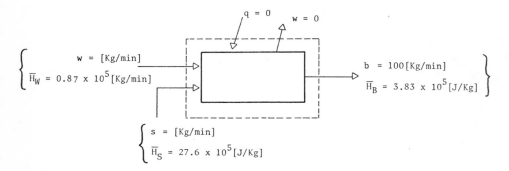

SOLUTION 2.2 First Law of Thermodynamics   (Eq. 2.3)

$$\sum b\overline{H}_B - \sum a\overline{H}_A = q - w = 0 \quad (q=0; \; w=0)$$

$$b\overline{H}_B - w\overline{H}_W - s\overline{H}_S = 0$$

$$100 \; [Kg/min] \times (3.83 \times 10^5)[Kg/min] \; -w[Kg/min] \times (0.87 \times 10^5)[J/Kg]$$

$$-s[Kg/min] \times (27.6 \times 10^5) \; [J/Kg] = 0$$

This equation has two unknowns: s & w
The additional relationship comes from a material balance

$$b = s + w$$

Solution to the energy and material balance yields

$$s = 11.3 \, [Kg/min]; \; w = 89.7 \, [Kg/min]$$

In this problem there were two unknowns, w and s. The energy balance gave one equation, and the other was supplied by a mass balance. (Mass Into System = Mass Out of System).

For those unfamiliar with mass balances for problems involving chemical reactions, Appendix B provides the necessary background material.

In the first two example problems the values of $\overline{H}$ were provided with the problem statement. This information is seldom available from the problem statement.

In order to use the energy balance it is necessary to be able to obtain values for $\overline{H}$, q and w. The value of $\overline{H}$ is easiest to evaluate because it is a "point" property. Once the conditions of Temperature, Pressure, Composition and State (vapor, liquid, solid) are identified there is a fixed value of $\overline{H}$. Because $\overline{H}$ is a point property it can be plotted against temperature and pressure. The next section shows how to obtain values for enthalpy. This is followed by a discussion on how to evaluate q and w for a few common paths.

2.3   EVALUATION OF ENTHALPY $\overline{H}$ (Energy/Unit Mass)

Enthalpy is a point property. Just as the datum elevation must be chosen before the potential energy can be stated,
datum conditions must be specified before the enthalpy, $\overline{H}$,
can be stated. Once the datum conditions have been specified,
the value of $\overline{H}$ is fixed and may be tabulated.

Enthalpy was described earlier as the amount of energy
associated with mass that flows into or out of a process.
Several characteristics of enthalpy, $\overline{H}$, can be deduced from
everyday observations.

1. Hot water carries with it more energy than cold
   water. Energy is added to cold water to make hot
   water. This means $\overline{H}$ for hot water is greater than
   $\overline{H}$ for cold water.

2. Steam @ 212°F carries with it more energy than
   water @ 212°F. Energy is added to water at 212°F
   to make steam at 212°F. This means $\overline{H}$ for steam
   is greater than $\overline{H}$ for water at 212°F.

3. A fossil fuel @ 25°C carries with it more energy
   than the combustion products @ 25°C. The fossil
   fuel can be reacted to produce energy to heat a
   home (combustion products cannot). This means
   fossil fuel has more $\overline{H}$ than the $\overline{H}$ of the products
   of combustion at 25°C. For methane at room tem-
   perature, the $\overline{H}$ results from its chemical configuration.

From these observations, it can be seen that the value
of $\overline{H}$ depends upon

a. The chemical nature of the mass
$$\equiv \overline{H}_C$$

   b. The temperature of the mass

$$\equiv \overline{H}_T$$

   c. The physical phase of the mass

$$\equiv \overline{H}_\lambda$$

In addition the value may depend upon the pressure

   d. The pressure of the mass

$$\equiv \overline{H}_P$$

The total value of $\overline{H}$ is the sum of these individual contributions.

$$\overline{H} = \overline{H}_C + \overline{H}_T + \overline{H}_\lambda + \overline{H}_P \tag{2.4}$$

The absolute value of these terms depends upon the choice
of datum conditions. For the purpose of this text (except
where noted) the ground state will be the products of combustion CO_2 (gas), H_2O (liquid) and SO_2 (gas) for C, H, and
S at 1.01 bar (i.e., 1 atm). This choice of datum was
chosen as it represents the lowest energy level a fuel can
be reduced to by combustion and is the ultimate end product
from using fossil fuel.

   Once the base conditions are chosen the values of $\overline{H}_C$,
$\overline{H}_T$, $\overline{H}_\lambda$, and $\overline{H}_P$ may be tabulated. Table 2.1 provides values
of $\overline{H}_C$ for many materials. The values of $\overline{H}_T$ are plotted in
Figure 2.3 (more accurate values may be obtained from
Appendix A). The value $\overline{H}_\lambda$ for water is given on Figure
2.3. The value of $\overline{H}_P$ is small and will be neglected in this
discussion.

   The procedure used in Example 2.3 is general and may
be used for problems that have chemical reactions or phase
changes occurring.

TABLE 2.1

$\overline{H}_C$ (Values) Note: Values are

-ΔH(Combustion) given in most texts

		$\overline{H}_c$ [Cal/g-mole]	x 10^{-3} [J/Kg-mole]
Methane	CH_4 (g)	212,800	890,888
Ethane	C_2H_6 (g)	372,820	1,560,813
Propane	C_3H_8 (g)	530,600	2,221,361
n-Butane	C_4H_{10} (g)	687,640	2,878,810
n-Pentane	C_5H_{12} (g)	845,160	3,538,270
n-Hexane	C_6H_{14} (g)	1,002,570	4,197,267
n-Heptane	C_6H_{10} (g)	---	---
n-Octane	C_8H_{18} (g)	---	---
Ethylene	C_2H_{11} (g)	337,230	1,141,816
Propylene	C_3H_6 (g)	491,990	2,059,720
1-Butene	C_4H_8 (g)	649,450	2,718,927
1-Pentene	C_5H_{10} (g)	800,850	3,377,884
1-Hexene	C_6H_{12} (g)	964,260	4,036,882
Acetylene	C_2H_2	310,620	1,300,413
Benzene	C_6H_6 (g)	789,080	3,303,489
Benzene	C_6H_6 (l)	780,980	3,269,579
Ethanol	C_2H_6O(g)	---	---
Ethanol	C_2H_6O (l)	---	---
Methanol	CH_4O (g)	---	---
Methanol	CH_4O (l)	---	---
Toluene	C_7H_8 (g)	943,580	3,950,305
Toluene	C_7H_8 (l)	934,500	3,912,292
Carbon Monoxide	CO (g)	67,636	283,124
Hydrogen Sulfide	H_2S (g)	---	---
Oxygen	O_2 (g)	0	0
Carbon	C (s)	94,052	393,749
Hydrogen	H_2 (g)	68,317	286,009
Nitrogen	N_2 (g)	0	0
Carbon Dioxide	CO_2 (g)	0	0
Sulfur Dioxide	SO_2 (g)	0	0

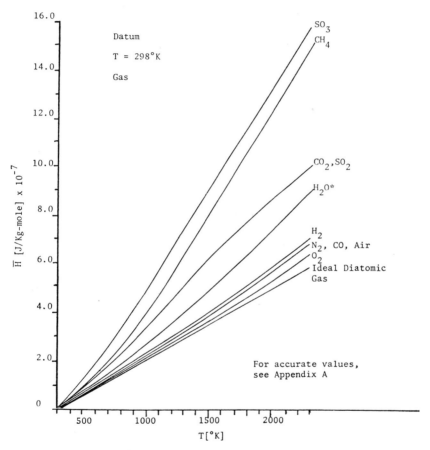

FIGURE 2.3 $\overline{H}_T$ vs. T for Common Gases (H values in going from
298°K to T)

Note: $\overline{H}_\lambda$ = 4.4 x 10^7[J/Kg-mole] for water

EXAMPLE 2.3   Methane is to be heated from 298°K to 900°K using a hot
CO_2 gas available at 1200°K. The CO_2 gas is cooled to
500°K. There is 100 Kg-mole of CO_2 available for this
purpose. How much methane may be heated?

SOLUTION 2.3

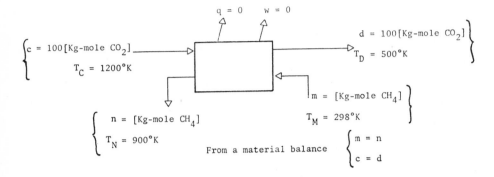

$$\left\{\begin{array}{l} c = 100[\text{Kg-mole } CO_2] \\ T_C = 1200°K \end{array}\right.$$

$$q = 0 \qquad w = 0$$

$$\left.\begin{array}{l} d = 100[\text{Kg-mole } CO_2] \\ T_D = 500°K \end{array}\right\}$$

$$\left.\begin{array}{l} \dot m = [\text{Kg-mole } CH_4] \\ T_M = 298°K \end{array}\right\}$$

$$\left\{\begin{array}{l} n = [\text{Kg-mole } CH_4] \\ T_N = 900°K \end{array}\right.$$

From a material balance $\left\{\begin{array}{l} m = n \\ c = d \end{array}\right.$

Evaluation of $\overline{H}$ Values

$\overline{H} = \overline{H}_C + \overline{H}_T + \overline{H}_\lambda + \overline{H}_P$

$\overline{H}_C = 0 + 4.5 \times 10^7 + 0 + 0 = 4.5 \times 10^7 [\text{J/Kg-mole}]$

$\overline{H}_D = 0 + 0.9 \times 10^7 + 0 + 0 = 0.9 \times 10^7 [\text{J/Kg-mole}]$

$\overline{H}_N = 89.088 \times 10^7 + 3.2 \times 10^7 + 0 + 0 = 92.29 \times 10^7 [\text{J/Kg-mole}]$

$\overline{H}_M = 89.088 \times 10^7 + 0 + 0 + 0 = 89.088 \times 10^7 [\text{J/Kg-mole}]$

Energy Balance   (Eq. 2.3)

$c\overline{H}_C + m\overline{H}_M = d\overline{H}_D + n\overline{H}_N$

$100[\text{Kg/mole } CO_2] \times 4.5 \times 10^7 [\text{J/Kg-mole}] + m[\text{Kg-mole } CH_4] \times$

$89.088 \times 10^7 [\text{J/Kg-mole}] = 100[\text{Kg-mole } CO_2] \times 0.9 \times 10^7 [\text{J/Kg-mole}] +$

$n[\text{Kg-mole } CH_4] \times 92.29 \times 10^7 [\text{J/Kg-mole}]$ [Also n = m]

$100[4.5 - 0.9] \times 10^7 = n[92.29 \times 10^7 - 89.088 \times 10^7]$

$n = 112.5[\text{Kg-mole } CH_4]$

It may be noted that $\overline{H}_N$ is subtracted from $\overline{H}_M$ and the energy due
to chemical configuration i.e., 89.088×10^7, has no effect on the
difference. If a material does not undergo a chemical reaction,
$\overline{H}_C$ need not be considered.

There are several other important thermodynamic functions that will be developed throughout this chapter that may all be evaluated in a manner similar to that used for $\overline{H}$. The contributions of the chemical configuration, temperature, phase and pressure are added together. The most useful calculation is that of a Δ term where Δ represents (for the enthalpy function) $\sum b_i \overline{H}_{B_i} - \sum a_i \overline{H}_{A_i}$ the total enthalpy of all exit streams - total enthalpy of all inlet streams. Equation 2.3

$$\sum a_i \overline{H}_{A_i} + \sum q_i = \sum b_i \overline{H}_{B_i} + \sum w_e$$

may be written in a Δ form

$$\Delta H = \sum q_i - \sum w_e \qquad (2.5)$$

where $\Delta H = \sum b_i \overline{H}_{B_i} - \sum a_i \overline{H}_{A_i}$

Table 2.2 provides a formalized way to evaluate the value of ΔH. It is convenient to use, provides a standard way to perform calculations, and helps assure that all of the terms are considered.

TABLE 2.2
Evaluation Form for ΔH

Stream Identifi- cation	Mass Flow	T	P	$\overline{H}_C$	$\overline{H}_T$	$\overline{H}_\lambda$	$\overline{H}_P$	$\overline{H}$	$H = m\overline{H}$
Units on Terms	[Kg-mole]	[°K]	[bar]	[J/Kg-mole]					[J]
Inlet ⎯ ⎯									
$\Sigma a_i \overline{H}_{A_i}$ =									
Exit ⎯ ⎯									
$\Sigma b_i H_{B_i}$ =									
$\Delta H = \Sigma b_i \overline{H}_{B_i} - \Sigma a_i \overline{H}_i$ = _____ - _____ =									

$\overline{H}_C$ provided in Table 2.1
$\overline{H}_T$ provided in Figure 2.3 or Appendix A
$\overline{H}_\lambda$ shown on Figure 2.3 for water
$\overline{H}_P$ neglected

EXAMPLE 2.4   Repeat Example 2.3 using the Δ form of Equation 2.3 and Table 2.2 to evaluate ΔH.

SOLUTION 2.4

Stream Identification	Mass Flow	T	P	$\overline{H}_C$	$\overline{H}_T$	$\overline{H}_\lambda$	$\overline{H}_P$	$\overline{H}$	$H = m\overline{H}$
Units on Terms	[Kg-mole]	[°K]				[J/Kg-mole]			[J]
Inlet									
c	100	1200	-	0	4.5×10^7	-	-	4.5×10^7	4.5×10^9
m	m	298	-	89.1×10^7	0	-	-	89.1×10^7	$89.1\times10^7 m$
$\sum a_i \overline{H}_{A_i}$ = 89.1 x 10^7m + 4.5 x 10^9									
Exit									
d	100	500	-	0	0.9×10^7	-	-	0.9×10^7	0.9×10^9
n	m	900	-	89.1×10^7	3.2×10^7	-	-	92.3×10^7	$92.3\times10^7 m$
$\sum b_i \overline{H}_{B_i}$ = 0.9x10^9 + 92.3x10^7m									

$\Delta H = 0.9 \times 10^9 + 92.3 \times 10^7 m - 89.1 \times 10^7 - 4.5 \times 10^9$

$= 3.2 \times 10^7 m - 3.6 \times 10^9$

First Law Thermodynamics (Eq. 2.5)

$\Delta H = \sum q - \sum w = 0 - 0 = 3.2 \times 10^7 m - 3.6 \times 10^9 = 0$

m = 112.5 [Kg-mole]

The previous problems did not include a chemical reacting system. The next example shows how a chemical reaction is handled.

EXAMPLE 2.5   Determine the amount of heat that will be released from
the combustion of one kilogram mole of methane in air.
The methane enters the combustion at 25°C and the air at
600°K. The gases exit at 1300°K. The flow sheet shown
below provides the flow rates of each stream.

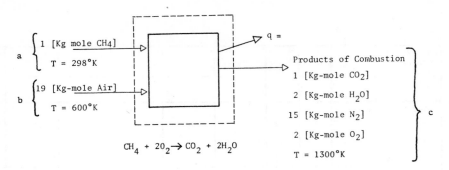

$$CH_4 + 2O_2 \rightarrow CO_2 + 2H_2O$$

SOLUTION 2.5

Stream Identification	Mass Flow	T	P	$\bar{H}_C$	$\bar{H}_T$	$\bar{H}_\lambda$	$\bar{H}_P$	$\bar{H}$	$H = m\bar{H}$
Units on Terms	[Kg-mole]	[°K]				[J/Kg-mole]			[J]
Inlet									
a(CH_4)	1	298	–	89.1×10^7	0	0	0	89.1×10^7	89.1×10^7
b(air)	19	600	–	0	9×10^6	0	0	9.6×10^6	18.2×10^7
$\sum a_i \bar{H}_{A_i}$ =									107.3×10^7
Exit									
c(CO_2)	1	1300	–	0	5.02×10^7		–	5.02×10^7	5.02×10^7
c(H_2O)	2	1300	–	0	3.89×10^7	4.4×10^7	–	8.29×10^7	16.56×10^7
c(N_2)	15	1300	–	0	3.15×10^7	–	–	3.15×10^7	47.25×10^7
c(O_2)	2	1300	–	0	3.32×10^7	–	–	3.32×10^7	6.64×10^7
$\sum b_i \bar{H}_{B_i}$ =									75.47×10^7
$\Delta H = \sum b_i \bar{H}_{B_i} - \sum a_i \bar{H}_{A_i}$ = $75.47 \times 10^7 [J] - 107.3 \times 10^7 [J] = -31.83 \times 10^7 [J]$									

First Law of Thermodynamics (Eq. 2.5)

$$\Delta H = q = -31.83 \times 10^7 [J]$$

The minus sign indicates heat was removed and not added to the system.

There were no complications in determining the value of
ΔH. The problem that is added when chemical reactions occur
is associated with determining the exit product and product
compositions. For combustion problems, the material
balances may be solved using Table 2.3. Any material balance
for a fossil fuel that can be represented by $C_\alpha H_\beta O_\gamma S_\delta$ that
is completely burned can be determined using Table 2.3.

TABLE 2.3

Chemical Reaction Material Balance
for Complete Combustion of Fuel with Composition

$$C_\alpha \, H_\beta \, O_\gamma \, S_\delta$$

1. Molecular Weight = 12 (α) + 1 (β) + 16 (γ) + 32 (δ)

2. Balanced Chemical Reaction

$$C_\alpha \, H_\beta \, O_\gamma \, S \; + \; (\alpha + \tfrac{\beta + \delta - \gamma}{4 \quad 2})O_2 = \alpha CO_2 + \beta/2 \, H_2O + \delta \, SO_2$$

3. Theoretical $(\alpha + \beta/4 + \delta - \gamma/2)O_2$

3a. Theoretical Air = Theoretical air/0.21

4. Actual O_2 = (% excess O_2 + 100) (Theoretical air)/100

4a. Actual Air = Actual Air/0.21

5. Exit Gas

     CO_2 α

     H_2O β

     SO_2 δ

     N_2 0.79 (Actual Air)

     O_2 Actual O_2 - Theoretical Air

              TOTAL

6. Exit Gas Composition

     CO_2 α/Total

     H_2O (β/2)/Total

     SO_2 δ/Total

     N_2 0.79 (Actual Air)/Total

     O_2 Actual O_2 - Theoretical Air/Total

EXAMPLE 2.6   Propane is burned with 200% excess air. The hot gases
               produced are transported to a dryer. The temperature at
               the dryer was measured and found to be 1,000°K. How much
               heat was lost in the transportation system (assume there
               was no loss in the furnace)? What is the gas composition?

SOLUTION 2.6

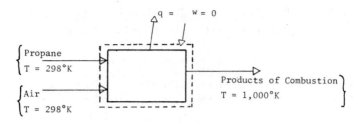

Using Table 2.3 C_3H_8 ($\alpha = 3$; $\beta = 8$)

1. Molecular weight: 12(3) + 1(8) = 44

2. Balance chemical reaction C_3H_8 + (3 + 8/4) O_2 = $3CO_2$ + 8/2 H_2O

    C_3H_8 + $5O_2$ → $3CO_2$ + $4H_2O$

3. O_2 (Theo) = (3 + 8/4) = 5 Kg-mole

3a. Air (Theo) = 5/0.21 = 23.8 Kg-mole

4. O_2 (Actual) = (200 + 100) (5)/100 = 15 Kg-mole

4a. Air (Actual) = 15/0.21 = 71.4 Kg-mole

5. & 6.

Moles	Composition
CO_2 = 3	3/73.4 = 0.04
H_2O = 8/2 = 4	4/73.4 = 0.05
SO_2 = 0	--- ---
N_2 = (71.4)(.79) = 56.4	56.4/73.4 = 0.77
O_2 = (15-5) = 10	10/73.4 = 0.14
TOTAL 73.4	TOTAL 1.00

Stream Identification	Mass Flow	T	P	$\bar{H}_C$	$\bar{H}_T$	$\bar{H}_\lambda$	$\bar{H}_p$	$\bar{H}$	$H = m\bar{H}$
Units on Terms	[Kg-mole]	[°K]		[J/Kg-mole]					[J]
Inlet									
1(C_3H_8)	1	298	-	2.2×10^9	-	-	-	2.2×10^9	2.2×10^9
2(air)	71.4	298	-	-	0	-	-	0	0
$\sum a_i \bar{H}_{A_i}$ =									2.2×10^9
Exit									
3(CO_2)	3	1000	-	0	3.3×10^7	-	-	3.3×10^7	9.9×10^7
3(N_2)	56.4	1000	-	0	2.2×10^7	-	-	2.2×10^7	1.24×10^9
3(O_2)	10.0	1000	-	0	2.3×10^7	-	-	2.3×10^7	2.3×10^8
3(H_2O)	4.0	1000	-	0	2.6×10^7	4.4×10^7	-	7.0×10^7	2.8×10^8
$\sum b_i \bar{H}_{B_i}$ =									1.85×10^9
$\Delta H = \sum b_i \bar{H}_{B_i} - \sum a_i \bar{H}_{A_i}$ = 1.85×10^9 [J] $- 2.2 \times 10^9$ [J] $= -3.5 \times 10^9$ [J]									

First Law of Thermodynamics (Eq. 2.5)

$\Delta H = q = -3.5 \times 10^9$ [J]

When the combustion reaction is involved the procedure given in Table 2.3 may be useful. For other reactions which are common in coal conversion, this table cannot be used. The next example involves the very important water-gas-shift reaction. The procedure followed in this example is more general and may be used for any chemical reaction including combustion. (See Appendix B.)

EXAMPLE 2.7   The water-gas-shift reaction is given by

$H_2O + CO \rightleftharpoons H_2 + CO_2$ (all gases)

Assume that a stream of gas containing 50% H_2 and 50% CO is to be reacted as shown below

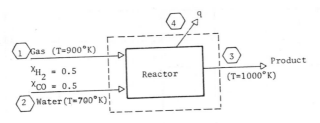

Fifty percent of the CO reacts. Determine the composition of the product gas and the amount of heat to be added or removed from the reactor.

SOLUTION 2.7   Let α be the amount of CO that reacts (from the problem statement 50% reacts). Basis: 1 Kg-mole gas

$$\alpha = (0.5 \text{ Kg-moles CO in}) (.5) = .25 \text{ moles CO reacts}$$

Component	Input	Generated	Consumed	Product	%	*
CO	0.5	---	.25	.25	12.5	
H_2	0.5	.25	0	.75	37.5	
H_2O	1.0	---	.25	.75	37.5	
CO_2	0	.25	0	.25	12.5	
				2.00 moles		

*See Appendix B for solution procedure used.

Stream Identification	Mass Flow	T	P	$\overline{H}_C$	$\overline{H}_T$	$\overline{H}_\lambda$	$\overline{H}_p$	$\overline{H}$	$H = m\overline{H}$
Units on Terms	[Kg-mole]	[°K]	-	[J/Kg-mole]					[J]
Inlet									
1 (CO)	0.5	900	-	2.8×10^8	1.85×10^7	-	-	2.99×10^8	1.50×10^8
1 (H_2)	0.5	900	-	2.9×10^8	1.77×10^7	-	-	3.08×10^8	1.54×10^8
2 (H_2O)	1.0	700	-	-	1.42×10^7	4.4×10^7	-	0.58×10^8	0.58×10^8
$\sum a_i \overline{H}_{A_i} =$									3.62×10^8
Exit									
3 (CO)	0.25	1000	-	2.8×10^8	2.18×10^7	-	-	3.02×10^8	0.76×10^8
3 (H_2)	0.75	1000	-	2.9×10^8	2.07×10^7	-	-	3.11×10^8	2.33×10^8
3 (H_2O)	0.75	1000	-	-	2.61×10^7	4.4×10^7	-	7.01×10^7	0.52×10^8
3 (CO_2)	0.25	1000	-	-	3.33×10^7	-	-	3.33×10^7	0.08×10^8
$\sum b_i \overline{H}_{B_i} =$									3.69×10^8
$\Delta H = \sum b_i \overline{H}_{B_i} - \sum a_i \overline{H}_{A_i} = 3.69 \times 10^8 [J] - 3.62 \times 10^7 [J] = 0.07 \times 10^8$									

First Law of Thermodynamics (Eq. 2.5)

$$\Delta H = q = 0.07 \times 10^8 [J]$$

The most common phase change that occurs is the vaporization of liquid water to gas (steam) or the condensation of steam to liquid water. This is important in power generation. Elaborate tables and figures of enthalpy values have been developed. The discussion below will focus on the water-steam system but will hold equally well for other phase changes. Figure 2.4 is an enthalpy-temperature diagram for H_2O. The following enthalpy information is shown on the diagram.

1. Saturated Liquid--this line represents the enthalpy of liquid water at its boiling point.

2. Saturated Vapor--this line represents the enthalpy of vapor at its condensing point.

3. Steam Region--the enthalpy of steam at various T and P can be read.

4. Liquid Region--the enthalpy of liquid is essentially independent of pressure and can be read with knowledge of temperature only.

5. Vapor Pressure--the vapor pressure at any temperature may be determined from the intersection of the constant temperature line with the saturated vapor or liquid line. This figure is presented in British Units because they are still in common usage in the power industry in the U.S.

When data is available in either graphical or tabular form the procedure for calculating $\Delta \overline{H}$ for a system that undergoes a phase change is no more difficult than calculating $\Delta \overline{H}$ for systems that do not undergo a change. It remains only to read two values and subtract.

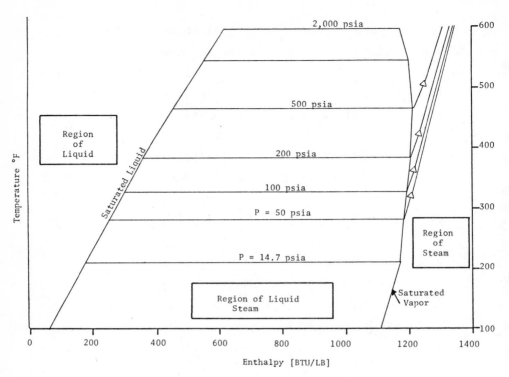

FIGURE 2.4   Enthalpy Temperature Diagram for H_2O

EXAMPLE 2.8   Determine the amount of heat needed to generate 100
               lbs/hour of steam at 100 psia and 600°F from water at
               100°F and 14.7psia.   What is the boiling point of water
               at 100 psia?

SOLUTION 2.8

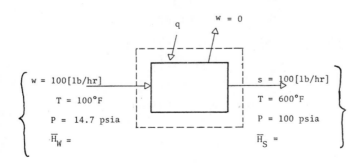

With only one stream in and out of the system, the use of
Table 2.2 to evaluate ΔH will not be used. Also as there
is no chemical reaction, it is not necessary to include
$\overline{H}_C$ in determining $\overline{H}$.

$\overline{H} = \overline{H}$ (read from Figure 2.4)

From Figure 2.4

@ 100°F and 14.7 psia $\overline{H}_T$ = 70[Btu/lb]

@ 600°F and 100 psia $\overline{H}_T$ = 1330[Btu/lb]

First Law of Thermodynamics (Eq. 2.5)

100[lb/hr](1330 - 70)[Btu/lb] = 1260[Btu/hr] = q

Boiling point @ 100 psia = 326°F

For a more accurate evaluation, tabulated values of $\overline{H}$ are
given in Appendix A.

When a single non-reacting flow stream is involved
the use of Table 2.2 as a format to evaluate ΔH is unneces-
sarily cumbersome and the procedure may be simplified (the
Table will yield the correct value). The value $\overline{H}_T$ shown
in Figure 2.3 and tabulated in Appendix A may be calculated from the equation

$$\overline{H}_T = \overline{C}_p(T - 298) \text{ where T is in °K.} \qquad (2.6)$$

Rewriting Equation 2.5 (First Law of Thermodynamics) for a
single component:

$$\Delta H = q - w \qquad (2.5)$$

$$\Delta H = m[\overline{H}_{exit} - \overline{H}_{inlet}]$$

Substituting Equation 2.6 into the expression for ΔH

$$\Delta H = m[\overline{C}_p(T - 298)_{exit} - \overline{C}_p(T - 298)_{inlet}]$$

If $\overline{C}_p$ is constant

$$\Delta H = m\overline{C}_p(T_{exit} - 298 - T_{inlet} + 298)$$
$$\Delta H = m\overline{C}_p(T_{exit} - T_{inlet}) \qquad (2.7)$$

Substituting equation 2.7 into 2.5

$$\Delta H = m\overline{C}_p(T_{exit} - T_{inlet}) = q - w \qquad (2.8)$$

For an ideal diatomic gas $\overline{C}_p$ is constant and equal to 29,100 [J/Kg-mole°K]. In this case

$$\Delta H(\text{Ideal diatomic gas}) =$$
$$29,100m[T_{exit} - T_{inlet}][J] \qquad (2.9)$$

2.4   EVALUATION OF q_R, w_R AND ΔS

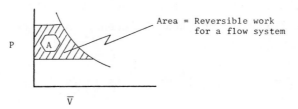

P

Area = Reversible work
for a flow system

A

$\overline{V}$

FIGURE 2.5   Reversible Work in a Flow System

In the previous section the discussion was limited to those problems where there was no work exchanged with the sur-roundings ($w = 0$). In this case equation 2.5 reduces to

$$\Delta H = q$$

The value of q was obtained by determining the ΔH. When there is a work term, then the evaluation of ΔH provides only the difference between q and w ($\Delta H = q - w$) and does not allow for either q or w to be evaluated. q <u>or</u> w must be evaluated independently.

2.4.1 Reversible Work

The most important product resulting from the consumption of fuel is the production of work, w. This section is directed to the numerical evaluation of the reversible work w_R. The definition of work, as found in the study of basic physics, is given by the relationship

$$\text{Work = Force x Distance or} \tag{2.10}$$

$$w = \int F dS$$

This relationship for work may be applied to the compression or expansion of gas obtained in an enclosed system. Consider the system shown below. A mass is contained in a rigid cylinder with a frictionless piston.

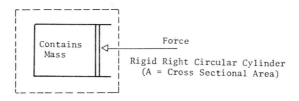

The equation for work given in equation 2.10 may be multiplied and divided by the cross sectional area of the piston.

$$w = \int F dS = \int \frac{F}{A} (A dS) = \int P dV \tag{2.11}$$

The product AdS (area times distance) is the volume, dV. The ratio F/A (force per unit area) is the pressure P. The

work may be evaluated by the area under the curve of P versus
V.

$$w = \int P dV$$

This is the total work. It is often desirable to write the
work equation in terms of a unit of mass. Dividing both
sides of the work equation by the mass of the system (M_s)

$$w/M_s = \int P(dV/M_s)$$

or

$$\overline{w} = \int P d\overline{V}$$

where $\overline{w}$ is work per unit mass and $\overline{V}$ is volume per unit mass.
    The force to be used in the work equations is the one
shown passing between the system and the surroundings. When
the pressure exerted by the mass within the cylinder is equal
to the pressure caused by the external force the system pressure, P_s, may be used in the work equation

$$\overline{w}_R = \int P_s d\overline{V}$$

When these pressures are balanced the change that occurs
is termed reversible and is designated by the subscript R.
    The derivation given above was obtained for a closed
system. For an open system the proper relationship is
given by

$$\overline{w}_R = -\int_{P_1}^{P_2} \overline{V}dP_s \qquad\qquad (2.12)$$

In order to obtain a value of $\overline{w}_R$ for a flow system it is necessary to perform the integration of Equation 2.12. If there is an analytical relationship that exists between $\overline{V}$ and P_s the integration may be performed. If the mass contained in the cylinder behaves as an ideal gas, i.e.,

$$P_s\overline{V} = RT \qquad\qquad (2.13)$$

where P_s is system pressure

$\overline{V}$ is volume/mole

T is the absolute temperature

R is a constant

TABLE 2.4

R-Values (Gas Law Constants)

Numerical Value	Units
1.99	Btu/lb-mole °R
1.99	Cal/g-mole °K
8314	Joule/Kg-mole °K
0.730	ft^3 atm/lb-mole °R
0.084	m^3 bar/Kg-mole °K

Equation 2.12 becomes

$$\overline{w}_R = -\int_{P_1}^{P_2} RT \frac{dP_s}{P_s} \qquad\qquad (2.14)$$

For an isothermal process (a process change where the temperature remains constant) Equation 2.11 may be integrated

$$\bar{w}_R = -\int_{P_2}^{P_1} RT \frac{dP_s}{P_s} = -RT \ln \frac{P_2}{P_1} = +RT \ln \frac{P_1}{P_2} \qquad (2.15)$$

As a matter of convention the following symbols will be used to represent compression or expansion in a flow process.

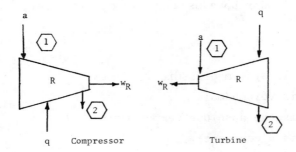

The R within the process unit indicates that the process is reversible (the system's internal pressure may be substituted for the external pressure). The subscript ②is the down-stream side and①the upstream side (output and input conditions respectively). The directions shown for q and w are always directed as shown. If the numerical value is negative this indicates the direction shown for q or w is opposite to that shown. The short side of the symbol is always the high pressure stream.

EXAMPLE 2.9   Evaluate the work required to compress 100 Kg-moles/hr of an ideal diatomic gas from 2.02 bars to 10.1 bars. The compression is reversible and isothermal (30°C).

SOLUTION 2.9

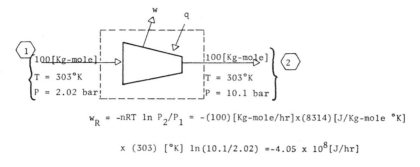

$$w_R = -nRT \ln P_2/P_1 = -(100)[Kg\text{-mole/hr}] \times (8314)[J/Kg\text{-mole }^\circ K]$$

$$\times (303)[^\circ K] \ln(10.1/2.02) = -4.05 \times 10^8 [J/hr]$$

The negative sign indicates work was done on the system.

There are several additional processes other than an isothermal process that are commonly encountered. Table 2.5 provides the relationship for the following reversible changes.

1. Isothermal (Constant Temperature)

2. Isobaric (Constant Pressure)

3. Isometric (Constant Volume)

4. Adiabatic Change (No Heat Exchange)

EXAMPLE 2.10   Evaluate the work required to adiabatically compress 100 Kg-moles of an ideal diatomic gas from 2.02 bar to 10.1 bar. Determine the horse power required and the cost. Compare to the isothermal process. (Cost of electricity is 3¢/KWhr). The inlet temperature is 294°K, $\gamma = 1.4$.

SOLUTION 2.10

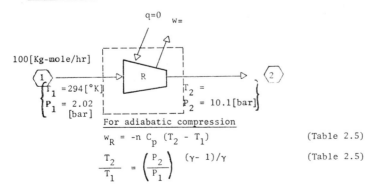

For adiabatic compression

$$w_R = -n C_p (T_2 - T_1) \qquad \text{(Table 2.5)}$$

$$\frac{T_2}{T_1} = \left(\frac{P_2}{P_1}\right)^{(\gamma-1)/\gamma} \qquad \text{(Table 2.5)}$$

$$T_2 = 294[°K] \left(\frac{10.1 \text{ bar}}{2.02 \text{ bar}}\right)^{0.4/1.4} = 466°K$$

$$w_R = -100[\text{Kg-mole/hr}] \times 29,100 \text{ [J/Kg-mole °K]} \times (466 - 294)[°K]$$

$$= -5.0 \times 10^8 [\text{J/hr}]$$

$$= -5.0 \times 10^8 [\text{J/hr}] \times 9.49 \times 10^{-4} \text{ [BTU/J]} \times 3.93 \times 10^{-4} [\text{Hp-hr/BTU}]$$

$$= 187 [\text{Hp}]$$

$$\$ = 187[\text{Hp}] \times 0.746 \text{ [KW/Hp]} \times 0.03 \text{ [\$/KW-hr]} = 4.17 \text{ [\$/hr]}$$

For isothermal compression

$$w_R = -nRT\ln(P_2/P_1) = 100[\text{Kg/mole/hr}] \times 8314 \text{ [J/Kg-mole °K]} \times$$

$$294 [°K] \times \ln (10.1 \text{ bar}/2.02 \text{ bar}) = -3.93 \times 10^8 [\text{J/hr}]$$

Comparison

$$\frac{5.0 \times 10^8 - 3.93 \times 10^8}{3.93 \times 10^8} \times 100 = 27.2\% \text{ more power for adiabatic compression}$$

TABLE 2.5

EVALUATION OF $\overline{w}_R$

	Isobaric	Isometric	Isothermal	Adiabatic
Constant Factor	P=C dP=0	V=C	T=C	q=0
Integration of $w_R =$ $- \int V dP_s$ for any material	$\overline{w}_R = 0$	$\overline{w}_R = -\overline{V}(P_2-P_1)$	$\overline{w}_R = \overline{V}dP$	$\overline{w}_R = \overline{V}dP$
Integration of $w_V =$ $- \int V dP$ for an ideal gas	$\overline{w}_R = 0$	$\overline{w}_R = -\overline{V}(P_2-P_1)$ $= -R(T_2-T_1)$	$\overline{w}_R = -RT \ln \dfrac{P_2}{P_1}$	$\overline{w}_R = -\dfrac{\gamma R(T_2-T_1)}{\gamma-1}$ $- \dfrac{\gamma(p_2V_2-P_1V_1)}{\gamma-1}$ $= C_p (T_2-T_1)$
Special Relationships (applicable to ideal gas)	$\dfrac{\overline{V}_1}{T_1} = \dfrac{\overline{V}_2}{T_2}$	$\dfrac{P_1}{P_2} = \dfrac{T_1}{T_2}$	$\dfrac{P_1}{P_2} = \dfrac{\overline{V}_2}{V_1}$	$\dfrac{T_2}{T_1} = \left[\dfrac{P_2}{P_1}\right]^{(\gamma-1)/\gamma}$ $= \left[\dfrac{V_1}{V_2}\right]^{\gamma-1}$

where: $\gamma = C_p/C_v$

C_p = 29,100 J/Kg-mole°K(7.0 BTU/LB-mole°R)for ideal diatomic gas

C_v = 20,600 J/Kg-mole°K(5.0 BTU/LB-mole°F)for ideal diatomic gas

$C_v = C_p - R$

The equations shown in Table 2.5 may be used to evaluate w_R (the ideal or reversible work). <u>When a gas is compressed the value of w_R is the minimum work requirement. When the gas is expanded, it represents the maximum work attainable.</u>

EXAMPLE 2.11   Consider the underground storage of compressed air. Air at 70°F is compressed adiabatically and reversibly to 10 atm. It is held underground at 10 atm until needed. While underground it cools to 140°F. When removed from the ground at 10 atm it is expanded to 1 atm adiabatically and reversibly. Determine the ratio of w_{out}/w_{in} (as an exercise this problem is presented and worked in British units).

SOLUTION 2.11

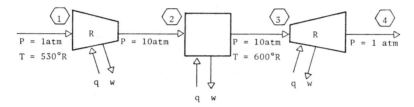

For compression stage

$T_2 = 530[°R] \times (10/1)^{(1.4-1)/1.4} = 1024[°R]$

$(w_R)_{in} = -C_p(T_2 - T_1) = -7.0[Btu/lb\text{-}mole°R] \times (1024 - 530)[°R]$

$\qquad = -3456[Btu/lb\text{-}mole]$

For expansion stage

$T_4 = 600[°R] \times (10/1)^{(1.4-1)/1.4} = 311[°R]$

$(w_R)_{out} = -C_p(T_4 - T_3) = 7.0[Btu/lb\text{-}mole°R] \times (311 - 600)[°R]$

$\qquad = 2023[Btu/lb\text{-}mole]$

Comparison

$w_{out}/w_{in} = 2023/3456 = 59\%$

Only 59% of the work done to compress the gas was recovered.

2.4.2 Reversible Heat Transfer

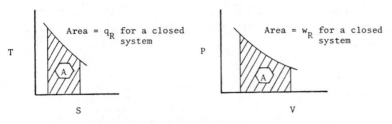

FIGURE 2.6 w_R and q_R for a Closed System

As previously discussed, the value of q is easily obtained by a $\Delta H = q$ when there is no work term involved. When work is involved q cannot be determined solely by evaluating a ΔH. In the previous section the basic equation for evaluating work for a closed system was developed.

$$\overline{w}_R = \int_{\overline{V}_1}^{\overline{V}_2} P_S d\overline{V}$$

The pressure was obtained from the value of the force acting upon the system. If the external force is greater than system force the system will compress and <u>work will be exchanged</u>.

In an analogous manner when the temperature of the surroundings are at a greater temperature than that of the system then the <u>heat will be exchanged</u>. Figure 2.7 shows this analogy between work and heat transfer.

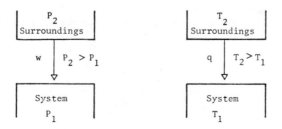

FIGURE 2.7   Transfer of q and w Between System and Surroundings

In either case a form of energy is exchanged between the
system and the sourroundings. By analogy with the equation
for work

$$\overline{w}_R = \int_{\overline{V}_1}^{\overline{V}_2} P d\overline{V}$$

an equation for $\overline{q}_R$ is written

$$\overline{q}_R = \int_{\overline{S}_1}^{\overline{S}_2} T d\overline{S} \tag{2.16}$$

Whereas the value $\overline{w}_R$ is the area on a $P\overline{V}$ diagram the value
$\overline{q}_R$ is the area on a $T\overline{S}$ diagram. This is shown in Figure 2.6.
The expression given by equation 2.16 may be used as a defi-
nition of a quantity called entropy ($\overline{S}$). It is a point
property (like $\overline{H}$). It has far reaching consequences that
will be discussed in the next section.

Table 2.6 provides working equations for the evaluation
of $\overline{q}_R$ for an ideal gas for several common processes.

TABLE 2.6

Evaluation of $\overline{q}_R$, $\Delta\overline{S}$ (gas)

	Isobaric	Isometric	Isothermal	Adiabatic
Constant Factor	P=C, dP=0	V=C	T=C	q=0
$\overline{q}_R$ for ideal gas	$\overline{q}_R = C_p(T_2-T_1)$	$\overline{q}_R = C_v(T_2-T_1)$	$\overline{q}_R = -RT \ln\left(\dfrac{P_2}{P_1}\right)$	0
See Table 2.5 for C_p & C_v				
$\Delta\overline{S}$ for ideal gas	$\Delta\overline{S} = C_p \ln\left(\dfrac{T_2}{T_1}\right)$	$\Delta\overline{S} = C_v \ln\left(\dfrac{T_2}{T_1}\right)$	$\Delta\overline{S} = -R \ln\left(\dfrac{P_2}{P_1}\right)$	0

EXAMPLE 2.12   Evaluate $\bar{q}_R$ in compressing an ideal diatomic gas from 1.01 [bar] and 298[°K] to 10.1 [bar] and 700[°K]. Also evaluate $\Delta\bar{S}$ and $\Delta\bar{H}$.

SOLUTION 2.12

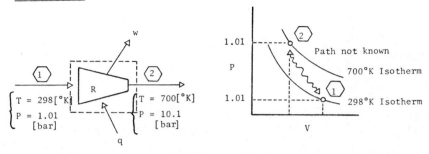

Because $\bar{S}$ is a point property $\Delta\bar{S}$ can be obtained by any path. The calculation path used will be to compress from 1.01 bar to 10.1 bar isothermally and then heat (or cool) isobarically to 700°K.

$\Delta\bar{S}_{1-2}$(compress isothermally) $= -R\ln[P_2/P_1]$ (Table 2.6)

$= -8314$ [J/Kg-mole °K] ln (10.1/1.01)

$= -1.91 \times 10^4$ [J/Kg-mole °K]

$\Delta\bar{S}_{2-3}$(heat isobarically) $= C_p\ln(T_2/T_1)$ (Table 2.6)

$= 29,100$ [J/Kg-mole°K] ln (700/298)

$= 2.49 \times 10^4$ [J/Kg-mole°K]

$\Delta\bar{S} = \Delta\bar{S}_{1-2} + \Delta\bar{S}_{2-3}$

$\Delta\bar{S} = [-1.91\times10^4 + 2.49\times10^4]$[J/Kg-mole°K] $= +0.58\times10^4$[J/Kg-mole°K]

$\Delta\bar{H} = C_p(T_3 - T_2) = 29,100$[J/Kg-mole°K](700 - 298)[°K]

$= 1.16\times10^7$[J/Kg-mole]

Values of $\bar{q}_R$ and $\bar{w}_R$ cannot be calculated. This is explained in the next paragraph.

The values of $\Delta\bar{H}$ and $\Delta\bar{S}$ were evaluated using point values, and they may be calculated by any convenient path. In Example 2.12 the path was an isothermal compression followed by isobaric heating. If the opposite order was used,

i.e., isobaric heating followed by an isothermal compression,
the same numerical values of $\Delta \overline{S}$ and $\Delta \overline{H}$ would be obtained.
The values for $\overline{q}_R$ and $\overline{w}_R$ would not be the same, as they depend upon path. This is shown in Example 2.13.

EXAMPLE 2.13 Evaluate $\overline{q}_R$ and $\overline{w}_R$ for the following two paths starting
and ending at the same condition as Example 2.11.

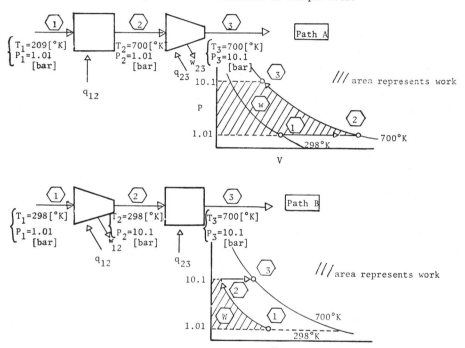

SOLUTION 2.13 For path A

$$q_R = C_p (T_2 - T_1) - RT_2 \ln (P_3/P_2) \qquad \text{(Table 2.6)}$$

$$= 29,100 \; [\text{J/Kg-mole} \; °K] \times (700-298) \; [°K]$$

$$-8,314 \; [\text{J/Kg-mole} \; °K] \times (700) \; [°K] \times \ln (10.1 \; \text{bar}/1.01 \text{bar})$$

$$= (1.17 \times 10^7 - 1.34 \times 10^7) \; [\text{J/Kg-mole}] = -0.17 \times 10^7 \; [\text{J/Kg-mole}]$$

$$w_R = 0 - RT_2 \ln (P_3/P_2) \qquad \text{(Table 2.5)}$$

$$= -1.34 \times 10^7 \; [\text{J/Kg-mole}]$$

$$\Delta\overline{H} = q_R - w_R = [-0.17 \times 10^7 + 1.34 \times 10^7] \text{ J/Kg-mole}$$

$$= 1.17 \times 10^7 \text{ [J/Kg-mole]}$$

For path B

$$q_R = -RT \ln(P_2/P_1) + C_p (T_3 - T_2)$$

$$= -8,314 \text{ [J/Kg-mole }^\circ K] \times (298) \text{ [}^\circ K] \times \ln (10.1\text{bar}/1.01\text{bar})$$

$$+ 29,100 \text{ [J/Kg-mole }^\circ K] \times (700 - 298) \text{ [}^\circ K]$$

$$= (-5.7 \times 10^6 + 1.17 \times 10^7) \text{ [J/Kg-mole]} = +0.6 \times 10^7 \text{[J/Kg-mole]}$$

$$w_R = -RT \ln P_2/P_1 = -5.7 \times 10^6 \text{ [J/Kg-mole]}$$

$$\Delta\overline{H} = q_R - w_R = (0.6 \times 10^7 + 0.57 \times 10^7) \text{ [J/Kg-mole]} = 1.17 \times 10^6 \text{[J/Kg-mole]}$$

Path	q_R	w_R	$\Delta\overline{H}$
A	$-.17 \times 10^7$	-1.34×10^7	1.17×10^7
B	0.6×10^7	$-.57 \times 10^7$	1.17×10^7

NOTE:

1) ΔH same

2) q_R & w_R differ

3) ΔH independent on path

4) q and w dependent

2.4.3 Evaluation of Entropy

Entropy like enthalpy is a point property. Equation 2.4 showed the total value of $\overline{H}$ is the sum of terms that account for the chemical configuration, the temperature, the phase and the pressure.

$$\overline{H} = \overline{H}_C + \overline{H}_T + \overline{H}_\lambda + \overline{H}_p \qquad (2.4)$$

The value of S is obtained in the same manner.

$$\overline{S} = \overline{S}_C + \overline{S}_T + \overline{S}_\lambda + \overline{S}_p \qquad (2.17)$$

$\overline{S}_C$ is the entropy associated with the chemical configuration and is tabulated in Table 2.7.

$\overline{S}_T$ is the entropy associated with the temperature and
is plotted in Figure 2.8 and tabulated in Appendix A.

$\overline{S}_\lambda$ is the entropy associated with the phase and the
value for water is given on Figure 2.8.

$\overline{S}_p$ is the entropy associated with pressure and is given
in Figure 2.9.

The procedure to determine ΔS over a process is identical
to ΔS and the format presented in Table 2.2 may be used by
replacing H with S.

EXAMPLE 2.14   Evaluate ΔH and ΔS when 1 Kg-mole of CO_2 at 298°K and 10.1 bar
and 2 Kg-mole of N_2 at 2.02 bar and 600°K are mixed to produce
a mixed gas at 1.01 bar and 400°K. No work is done. The ΔH
and $\Delta \overline{S}$ will be defined to be the difference between the enthalpy
and entropy of the "mass flow streams."

SOLUTION 2.14

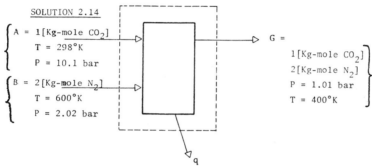

$$A = 1[\text{Kg-mole } CO_2]$$
$$T = 298°K$$
$$P = 10.1 \text{ bar}$$

$$B = 2[\text{Kg-mole } N_2]$$
$$T = 600°K$$
$$P = 2.02 \text{ bar}$$

$$G =$$
$$1[\text{Kg-mole } CO_2]$$
$$2[\text{Kg-mole } N_2]$$
$$P = 1.01 \text{ bar}$$
$$T = 400°K$$

q

Solution to ΔH

Stream Identification	Mass Flow	T	P	$\overline{H}_C$	$\overline{H}_T$	$\overline{H}_\lambda$	$\overline{H}_p$	$\overline{H}$	$H = m\overline{H}$
Units on Terms	[Kg-mole]	[°K]	[bar]	[J/Kg-mole]					[J]
Inlet									
A(CO_2)	1	298	10.1	0	0	0	0	0	
B(N_2)	2	600	20.2	0	8.9×10^6	0	0	8.9×10^6	17.8×10^6
$\sum a_i \overline{H}_{A_i}$ =									17.8×10^6
Exit									
G(CO_2)	1	400	.33	0	4.0×10^6	0	0	4.0×10^6	4.0×10^6
G(N_2)	2	400	.68	0	3.0×10^6	0	0	3.0×10^6	6.0×10^6
$\sum b_i \overline{H}_{B_i}$ =									10.0×10^6
$\Delta H = \sum b_i \overline{H}_{B_i} - \sum a_i \overline{H}_{A_i}$ = 10.0×10^6[J] $- 17.8 \times 10^6$[J] = -7.8×10^6[J]									

TABLE 2.7

Values of $\overline{S}_C$ and $\overline{B}_C$ For Combustion
(Datum: 25°C, 1 atm, $CO_2(g)$, $H_2O(l)$)

Compound		$\overline{B}$ x 10^{-3} (J/Kg-mole)	$-\overline{S}$ x 10^{-3} (J/Kg-mole°K)
Methane	$CH_4(g)$	818,460	243
Ethane	$C_2H_6(g)$	1,468,331	310
Propane	$C_3H_8(g)$	2,109,686	375
n-Butane	$C_4H_{10}(g)$	2,747,976	439
n-Pentane	$C_5H_{12}(g)$	3,388,720	502
n-Hexane	$C_6H_{14}(g)$	4,030,000	560
n-Heptane	$C_7H_{16}(g)$	4,660,000	656
n-Octane	$C_8H_{18}(g)$	5,307,361	690
Ethene	$C_2H_4(g)$	1,332,068	268
Propene	$C_3H_6(g)$	1,958,235	341
1-Butene	$C_4H_8(g)$	2,599,355	401
1-Pentene	$C_5H_{10}(g)$	3,239,136	466
1-Hexene	$C_6H_{12}(g)$	3,879,378	529
Acetylene	C_2H_2	1,235,896	217
Benzene	$C_6H_6(g)$	3,207,450	316
Benzene	$C_6H_6(l)$	3,204,288	219
Ethanol	$C_2H_6O(g)$	1,332,521	260
Ethanol	$C_2H_6O(l)$	1,329,366	130
Methanol	$CH_4O(g)$	707,267	220
Methanol	$CH_4O(l)$	702,871	108
Toluene	$C_7H_8(g)$	3,834,030	390
Toluene	$C_7H_8(l)$	3,825,883	290
Carbon Monoxide	$CO(g)$	257,238	87
Hydrogen Sulfide	$H_2S(g)$	503,500	199
Sulfur Trioxide	$SO_3(g)$	-70,031	-95
Carbon	C	394,560	-29
Hydrogen	H_2	237,304	165
$CO_2(g)$	CO_2	---	---
$H_2O(l)$	$H_2O(l)$	---	---
N_2	N_2	---	---
O_2	O_2	---	---

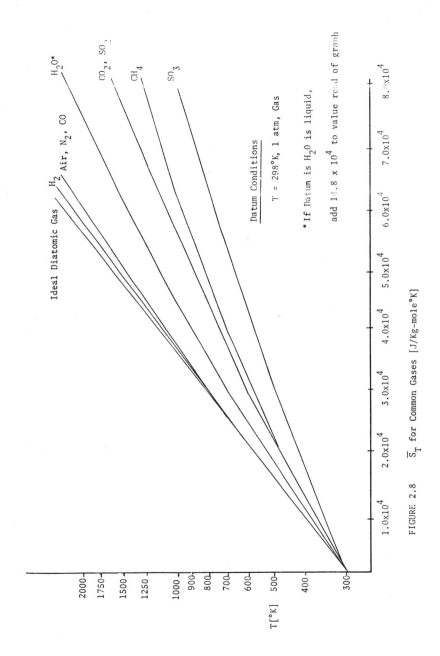

FIGURE 2.8 $\overline{S}_T$ for Common Gases [J/kg-mole°K]

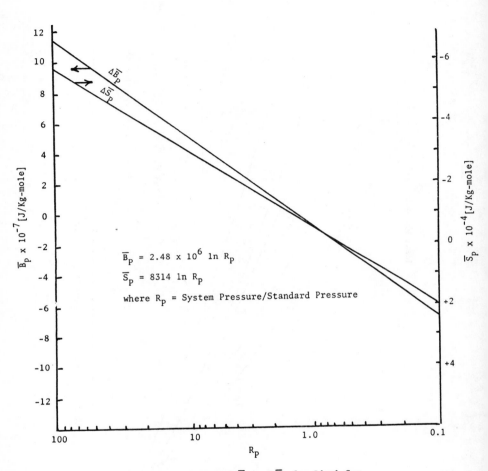

FIGURE 2.9 Values of $\overline{S}_P$ and $\overline{B}_T$ for Ideal Gas
as a Function of Pressure

Solution to ΔS

Stream Identifi-cation	Mass Flow	T	P	$\overline{S}_C$	$\overline{S}_T$	$\overline{S}_\lambda$	$\overline{S}_P$	$\overline{S}$	$S = m\overline{S}$
Units on Terms	[Kg-mole]	[°K]	[bar]			[J/Kg-mole°K]			[J/°K]
Inlet									
A(CO_2)	1	298	10.1	0	0	-	-1.9×10^4	-1.9×10^4	-1.9×10^4
B(N_2)	2	600	2.02	0	2.06×10^4	-	-0.5×10^4	1.49×10^4	2.98×10^4
$\sum a_i \overline{S}_{A_i}$ =									1.09×10^4
Exit									
G(CO_2)	1	400	0.33	0	1.16×10^4	-	0.93×10^4	2.09×10^4	2.08×10^4
G(N_2)	2	400	0.68	0	0.85×10^4	-	0.33×10^4	1.18×10^4	2.36×10^4
$\sum b_i \overline{S}_{B_i}$ =									4.44×10^4
$\Delta S = \sum b_i \overline{S}_{B_i} - \sum a_i \overline{S}_{A_i}$ = 4.44×10^4 $- 1.09 \times 10^4$ = 3.35×10^4 [J/°K]									

NOTE: The pressures used are the partial pressures of the gases.

The next example considers a problem where a chemical reaction takes place.

EXAMPLE 2.15 Determine the value of ΔS for the process shown below.

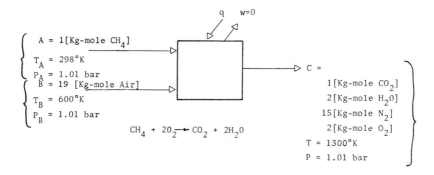

CH$_4$ + 2O$_2$ → CO$_2$ + 2H$_2$O

SOLUTION 2.15

Evaluation of ΔS

Stream Identification	Mass Flow	T	P	$\bar{S}_C$	$\bar{S}_T$	$\bar{S}_\lambda$	$\bar{S}_P$	$\bar{S}$	$S = m\bar{S}$
Units on Terms	[Kg-mole]	[°K]	[bar]			[J/Kg-mole°K]			[J/°K]
Inlet									
A(CH$_4$)	1	298	1.01	24.3x10^4		-	0	24.3x10^4	24.3x10^4
B(air)									
N$_2$	15	600	0.68		2.06x10^4	-	3.29x10^3	2.39x10^4	35.85x10^4
O$_2$	4	600	0.33		2.14x10^4	-	9.30x10^3	3.07x10^4	12.28x10^4
$\Sigma a_i \bar{S}_{A_i}$ =									72.43x10^4
Exit									
C(CO$_2$)	1	1300	.05	0	7.0x10^4		2.5x10^4	9.5x10^4	9.5x10^4
C(H$_2$O)	2	1300	.10	0	5.5x10^4	14.8x10^4	1.9x10^4	22.2x10^4	44.4x10^4
C(N$_2$)	15	1300	.76	0	4.5x10^4	-	2.4x10^3	4.74x10^4	71.1x10^4
C(O$_2$)	2	1300	.10	0	4.6x10^4	-	1.9x10^4	6.5x10^4	13x10^4
$\Sigma b_i \bar{S}_{B_i}$ =									138x10^4
$\Delta S = \Sigma b_i \bar{S}_{B_i} - \Sigma a_i \bar{S}_{A_i}$ = 138 x 10^4 - 72.43 x 10^4 = 65.57 x 10^4 [J/°K]									

Just as in the case of ΔH the procedure for the evaluation of ΔS may be simplified when there is a single mass stream that does not undergo a chemical reaction or phase change. In this case the value of $\bar{S}$ is given by

$$\bar{S} = \bar{S}_T + \bar{S}_P$$

$$\bar{S}_T = \bar{C}_p \ln T/298 \qquad \text{(See Table 2.5)}$$

$$\bar{S}_P = -R \ln P/1.01 \qquad \text{(See Table 2.5)}$$

$\bar{C}_p$ is given in Appendix A
T is in °K
P is in bar

$$\bar{S} = \bar{C}_p \ln(T/298) - R \ln(P/1.01)$$

The value of ΔS is then given by

$$\Delta S = m(\overline{S}_{exit} - \overline{S}_{inlet}) = m[\overline{C}_p \ln(T_{exit}/298) -$$

$$R \ln(P_{exit}/1.01) - \overline{C}_p(T_{inlet}/298) +$$

$$R \ln(P_{inlet}/1.01)] \tag{2.18}$$

If $\overline{C}_p$ is a constant then

$$\Delta S = m[\overline{C}_p \ln(T_{exit}/T_{inlet}) - R \ln(P_{exit}/P_{inlet})] \tag{2.18a}$$

For a diatomic ideal gas

$$\Delta S = m[29,100 \ln(T_{exit}/T_{inlet}) - 8314(P_{exit}/$$

$$P_{inlet})][J/°K] \tag{2.18b}$$

2.5 SECOND LAW OF THERMODYNAMICS

In the previous section there was no question that the processes presented would proceed as shown. However, many processes cannot occur. The First Law of Thermodynamics cannot differentiate between those processes that can take place and those processes that are not possible. This is shown in the next problem.

EXAMPLE 2.16 One Kg of steel at 700°K is plunged into 100 Kg of water at 340°K. Calculate the final temperature of the water.
(a) If the steel temperature decreases to 600°K.
(b) If the steel temperature increases to 800°K.

SOLUTION 2.16

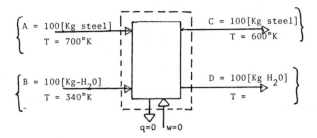

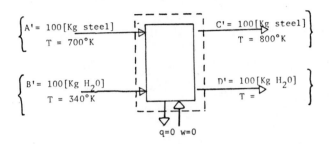

Assume:

$C_p(H_2O) = 4,186$ [J/Kg°K] $C_p(steel) = 1,046$ [J/Kg°K]

Stream	Wt [Kg]	T [°K]	ΔT [T - 298]	$\overline{H}$ [J x 10⁴]
A(Steel)	100	700	402	42.0
B(Water)	100	340	42	17.6
C(Steel)	100	600	302	31.6
D(Water)	100	T_D	ΔT_D	$\overline{H}_D$
A'(Steel)	100	700	402	47.0
B'(Water)	100	340	42	17.6
C'(Steel)	100	800	502	52.5
D'(Water)	100	T_D	ΔT_D	$\overline{H}_D$

$\overline{H} = C_p(T - 298°K)$

First System

$$\Delta H = 0 = 100[Kg\ steel] \times 31.6 \times 10^4 [J/Kg] + 100[Kg\ H_2O] \times \overline{H}_D$$

$$-100[Kg\ steel] \times .42 \times 10^4 [J/Kg] + 100[Kg\ H_2O] \times 17.6 \times 10^4 [J/Kg]$$

$$0 = 100[Kg\ H_2O] \times \overline{H}_D - 2800 \times 10^4$$

$$\overline{H}_D = C_p\ (T_D - 298) = 4{,}186[J/Kg°K]\ [T_D - 298] = 23 \times 10^4 [J/Kg]$$

$$\boxed{T_D = 365°K}$$

Second System

$$\Delta H = 0 = 100[Kg\ steel] \times 52.5 \times 10^4 [J/Kg] + 100[Kg\ H_2O] \times \overline{H}_D$$

$$-100\ [Kg\ steel] \times 42.0 \times 10^4 [J/Kg] + 100\ [Kg\ H_2O] \times 17.6 \times 10^4 [J/Kg]$$

$$0 = 100[Kg\ H_2O] \times \overline{H}_D - 710 \times 10^4$$

$$\overline{H}_D = C_p\ (T_D' - 298) = 4{,}186\ [J/Kg\ °K]\ [T_D' - 298] = 7.1 \times 10^4 [J/Kg]$$

$$\boxed{T'_D = 315°K}$$

In Example 2.16 the First Law of Thermodynamics allowed for a temperature to be evaluated for both cases. However, observation of the world around us helps us to appreciate that when a cold and hot body come together, the hot body temperature decreases and the cold body temperature increases. Under alternative (b) of Example 2.16 the steel got hotter (700°K to 800°K) and the water got colder (340°K to 315°K). This process is not possible. There are other common processes that we feel confident will take place in one direction.

1. Water flows from a higher elevation to a lower elevation. Without assistance it cannot flow from a low elevation to a high elevation.

2. Gas flows from a region of high pressure to one of low pressure. Without assistance it cannot flow

from a low pressure to a high pressure.

3.   Salt will mix with water to form a salt solution.
     Without assistance salt does not separate from
     water.

However, many problems involving energy conversion are more
complex and intuitive judgments are not sufficient to be
sure of what can occur and what cannot occur. It is absolutely
essential that there is a way to screen energy alternatives
to eliminate any that are not possible. The Second Law of
Thermodynamics supplies this screening.

Consider a system of a temperature T_{sys} that is higher
than the surroundings at T_{sur}

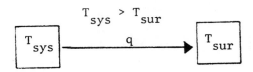

Heat will leave the system at T_{sys} and pass toward the surroundings at T_{sur} (as $T_{sys} > T_{sur}$).

In Section 2.4.2 the expression for q_R was provided

$$\overline{q}_R = \int_{\overline{S}_1}^{\overline{S}_2} T d\overline{S} \qquad (2.16)$$

If the temperature is constant this equation may be easily
integrated and re-arranged to give

$$\Delta \overline{S} = \overline{q}_R / T \qquad (2.19)$$

Writing this equation for both the surroundings and the system
gives

Surroundings $\Delta \overline{S}_{sur} = \overline{q}_R / T_{sur}$

System $\Delta \overline{S}_{sys} = \overline{q}_R / T_{sys}$

The sum of ΔS for the surroundings plus the ΔS of the system will be designated as ΔS of the universe:

$$\Delta S_{universe} = \Delta S_{sur} + \Delta S_{sys} \qquad (2.20)$$

If a given amount of heat is exchanged between the system and the surroundings $|q_R|$ is a constant. The value of ΔS_{sur}, ΔS_{sys} and $\Delta S_{universe}$ may be plotted for various system temperatures (T_{sys} is variable, T_{sur} is constant). These values are plotted in Figure 2.10.

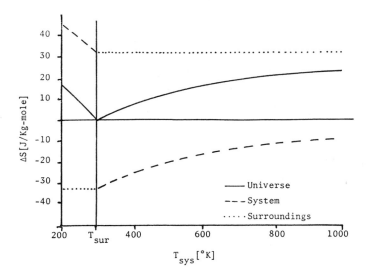

FIGURE 2.10 Plot of ΔS for System, Surroundings, and Universe vs System Temperature (For $|q| = 10,000$ J; $T_{sur} = 300°K$)

When $T_{sys} < T_{sur}$ the heat flows from the surroundings to
the system. ΔS_{sys} is positive, and ΔS_{sur} is negative.
When $T_{sys} > T_{sur}$ the direction that heat flows is reversed
and the signs of ΔS_{sys} and ΔS_{sur} are reversed. The change
in sign occurs at $T_{sys} = T_{sur}$. The sign on $\Delta S_{universe}$ remains positive and becomes zero when the temperature of the
system and the surroundings are the same.

$$\Delta S_{universe} \geq 0 \qquad\qquad\qquad\qquad (2.21)$$

If it were <u>not noted</u> that the direction of heat flow q_R
changed when the temperature of the system became less than
that of the surroundings then a negative $\Delta S_{universe}$ would
have been obtained. The negative value came about because
a mistake was made and an impossible process considered.
This discussion is summarized below.

    (a) For any real process $\Delta S_{universe} \geq 0$

    (b) For any impossible process $\Delta S_{universe} \geq 0$

Equation 2.16 is a definition of entropy. Equation
2.21 is a form of the Second Law of Thermodynamics. It
states that for any change to take place the entropy of the
universe increases. This is a critical criterion for any
energy conversion process proposed. If the Second Law is
applied to any proposed system and the value $\Delta S_{universe}$ is
negative the process is impossible. It is important to
apply the Second Law to any process before spending time,
effort, or money in its development.

EXAMPLE 2.17 Two Kg-moles of a diatomic ideal gas at 3.45 bars and
1,000°K enter a device which operates adiabatically
and produces no work. One Kg-mole leaves at 1.01 bar
and 600°K. The other Kg-mole leaves at 1400°K and 2.02
bar.
(a) Does the First Law of Thermodynamics hold?
(b) Is the process possible?

SOLUTION 2.17

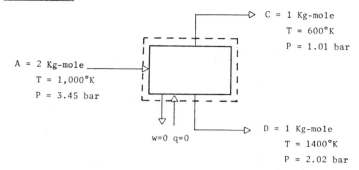

First Law of Thermodynamics $\Delta H = q - w = 0$

Using Equation 2.6

Stream	Kg-moles	T	ΔT	$\overline{H}$
		[°K]	[°K]	[J/Kg x 10^7]
A	2	1,000	702	2.04
C	1	600	302	0.88
D	1	1,400	1,102	3.21

$\overline{H} = C_p (T - 298)[°K]$

$C_p = 29,100[J/Kg\text{-mole °K}]$

$\Delta H = (1)(0.88 \times 10^7) + (1)(3.21 \times 10^7) - (2)(2.0 \times 10^7)$

$= 4.08 \times 10^7 - 4.08 \times 10^7 = 0$

Yes--the first law holds

Second Law of Thermodynamics

Equations Table 2.5

Stream	Kg-moles	T	P	$\overline{S}_T$	$\overline{S}_p$	$\overline{S}$
		[°K]	[bar]	[J/Kg°K]	[J/Kgmole°K]	[J/Kgmole°K]
A	2	1,000	3.45	+35,3000	-10,200	+25,100
C	1	600	1.01	+20,400	0	+20,400
D	1	1,400	2.02	+45,000	-5,760	+39,240

$\Delta S_{sys} = (1)(39,240) + (1)(20,400) - (2)(25,100) = 9,440[J]$

$\Delta S_{universe} = \Delta S_{sys} + \Delta S_{sur}:$

$\Delta S_{sur} = 0[\text{no heat is exchanged}]$

$\therefore \Delta S_{universe} = 9,440[J] + 0$

Process possible ΔS is +

2.5.1 Carnot Engine Efficiency

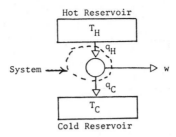

FIGURE 2.11 Engine to Convert Heat to Work

Figure 2.11 shows two energy sources. One is at T_H (Hot Reservoir) and one is at T_C (Cold Reservoir). Clearly, heat would flow between these two reservoirs. The direction of flow would be from the hot reservoir to the cold reservoir. Between these two temperature reservoirs an engine is installed. The purpose of this engine is to produce work. There is no need to know how this engine works to establish the maximum amount of work obtainable. The best that can be attained is when $\Delta S_{universe} = 0$, i.e.,

$$\Delta S_{universe} = \Delta S_{sur} + \Delta S_{sys} = 0 \qquad (2.22)$$

The system shown by the dotted lines is at steady-state and the system does not change, $\Delta S_{sys} = 0$. Substituting this into the previous equation it follows that ΔS_{sur} is also zero for a steady state reversible system (the best that can be done).

If $\Delta S_{sur} = 0$ then the ΔS lost by the hot reservoir must be equal to but opposite in sign from the ΔS gained by the cold reservoir. Entropy, S, travels with heat, and these are the only two places where heat is exchanged between the system and the surroundings.

$$\Delta S(\text{hot reservoir}) + \Delta S(\text{cold reservoir}) = 0$$

$$\Delta S(\text{hot reservoir}) = -q_H/T_H$$

$$\Delta S(\text{cold reservoir}) = q_C/T_C$$

$$-q_H/T_H + q_C/T_C = 0 \quad [q_C/q_H = T_C/T_H] \qquad (2.23)$$

Applying the First Law of Thermodynamics to the system shown in Figure 2.11:

$$\Delta H = q_H - q_C - w = 0 \quad (\text{No mass flow} \quad \Delta H = 0)$$

$$w = q_H - q_C$$

If the efficiency is defined by the relation $\eta = $ work done/heat added

$$\eta = w/q_H = (q_H - q_C)/q_H = 1 - q_C/q_H$$

$$\eta = 1 - T_C/T_H \qquad (2.24)$$

The maximum work depends upon the temperatures between which the engine operates. This is the <u>Carnot Efficiency</u>.

<u>EXAMPLE 2.18</u> Electric home heating is becoming more popular. When resistance heating is used 1 Joule of work provides 1 Joule of heat. As an alternative, a heat pump may be used. The electrical energy is used to take energy from the surroundings and <u>raise the temperature level</u> to heat the home. Assume the outside temperature T_C is 0°C and the hot air for the home is to be 140°F (60°C). Determine the savings in electricity by converting from resistance heat to a heat pump.

SOLUTION 2.18

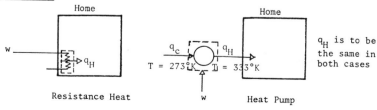

Resistance Heat   w   Heat Pump

q_H is to be the same in both cases

For resistance heating - First Law about Resistance Heater
(Shown by --- line)

$$\Delta H = -q_H + w = 0$$

$$q_H = w$$

For heat pump - First Law about Pump (Shown by --- line)

$$\Delta H = q_C - q_H + w = 0$$

$$w = q_H - q_C$$

$$\Delta S_{universe} = 0 = \Delta S_{sur} = 0 = q_C/T_C - q_H/T_H$$

$$q_C = q_H(T_C/T_H)$$

Ratio of heat delivered to home to work done

$$r = q_H/w = q_H/(q_H - q_C)$$
$$= q_H/[q_H - q_H T_C/T_H] = 1/(1-T_C/T_H) =$$
$$= T_H/(T_H - T_C) = 333/(333-273)$$
$$= 5.55$$
$$q_H = 5.55 w$$

Electric Savings

For electric heat 1 unit of work gave one unit of heat

For heat pump 1 unit of work gives 5.5 units of heat

∴. Heat pump uses $\frac{1}{5.55}$ x 100 = 18% as much electricity

Savings are 82%.

2.5.2 Determination of lw (Lost Work)

The evaluation of the Carnot Efficiency represented a
perfect system where all heat transfer took place reversibly,
and there was no friction in the engine. None of these conditions are achieved in actual systems. Consider a real
engine that has some friction. Some of the work will be
used to overcome this friction. The friction produces heat
that is discharged from the engine. Figure 2.12 compares a
real engine with the ideal Carnot Engine.

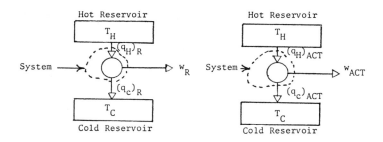

FIGURE 2.12 Comparison of Actual and Carnot Engine

The difference between the ideal work, w_R, and the actual work, w_{ACT}, for engines running between the same two temperatures is termed lost work (lw).

$$w_R - w_{ACT} = lw \text{ or } w_R = w_{ACT} + lw$$

The First Law of Thermodynamics written about each engine gives

Ideal Engine $(q_H)_R = w_R + (q_C)_R$

Actual Engine $(q_H)_{ACT} = w_{ACT} + (q_C)_{ACT}$

Consider the case where the same amount of heat is removed from the hot reservoir or $(q_H)_R = (q_H)_{ACT}$.

In this case

$$w_R + (q_C)_R = w_{ACT} + (q_C)_{ACT}$$

$$w_R - w_{ACT} = lw = (q_C)_{ACT} - (q_C)_R \qquad (i)$$

Writing the Second Law of Thermodynamics over the actual system gives:

$$\Delta S_{universe} = \Delta S_{sur} + \Delta S_{sys} \qquad (2.25)$$

At steady-state $\Delta S_{sys} = 0$, therefore

$$\Delta S_{sur} = \Delta S_{universe} = \Delta S_H + \Delta S_C$$

$$\Delta S_H = -(q_H)_R/T_H \text{ (same as reversible engine)}$$

$$\Delta S_C = (q_C)_{ACT}/T_C = [(q_C)_R + 1w]/T_C \text{ (substituting (i) for } (q_C)_{ACT}$$

$$\Delta S_{universe} = -(q_H)_R/T_H + (q_C)_R/T_C + 1w/T_C$$

For a reversible engine from Equation 2.23

$$(q_C)_R/T_C = (q_H)_R/T_H$$

Substituting this into the previous equation gives

$$\Delta S_{universe} = 1w/T_C$$

$$1w = T_C \Delta S_{universe} \qquad\qquad (2.26)$$

This is an extremely important relationship and will be referred to often. It states that the loss of ability for any process to do work is given by the product of the increase in entropy of the universe and the lowest temperature reservoir available to discharge heat.

A final word on entropy, entropy flows with heat. If there is no heat flow, there is no entropy generation. (Entropy does not "flow with work.")

2.6   POWER CYCLES

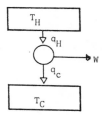

In the previous section the relationship for the efficiency of a Carnot Engine was derived but no suggestion was made as to how it worked.   Before going into various real power cycles,

there are several important points that can be seen by
studying the Carnot Engine: (a) there must be a high temperature energy source, (b) there must be a low temperature
energy sink, (c) not all of the energy that is received from
the high energy source can be converted to work. Several
basic cycles to provide w are discussed in the next few sections.

2.6.1 Closed Gas Turbine--Brayton Cycle

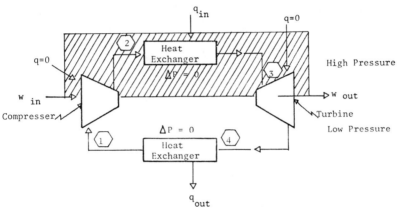

FIGURE 2.13 Gas Turbine (Brayton Cycle)

The material passing through streams $1 \rightarrow 2 \rightarrow 3 \rightarrow 4$ in Figure
2.13 is gas and is termed the working fluid. The Brayton
Gas Cycle consists of two isobaric heat exchanges, one compressor and one turbine. Both compressor and turbine are
adiabatic (q = 0). A gas flows around the closed loop $1 \rightarrow 2 \rightarrow$
$3 \rightarrow 4 \rightarrow 1$. The shaded areas of Figure 2.13 are at a high
pressure (P_H) and the clear area at lower pressure (P_L).
From point ① to point ② the gas is compressed from P_L to
P_H. From point ② to point ③ heat from a high temperature
source is added to raise the gas temperature. From point ③

to point $\langle 4 \rangle$ the gas is expanded to produce work. From point $\langle 4 \rangle$ gas is cooled to temperature of point $\langle 1 \rangle$. The efficiency of the cycle is defined by

$$\eta \quad = \frac{\overline{w}_{net}}{\overline{q}_{in}} \times 100$$

$$\overline{w}_{net} = \overline{w}_{out} - \overline{w}_{in} = (\overline{H}_4 - \overline{H}_3) - (\overline{H}_2 - \overline{H}_1)$$

$$\overline{q}_{in} = \overline{H}_3 - \overline{H}_2$$

When C_p may be considered, $\overline{H}_i = \overline{C}_p(T_i - 298°K)$. Substituting this into the relations for $\overline{w}_{net}$ and $\overline{q}_{in}$ given above yields.

$$\overline{w}_{net} = \overline{C}_p \ (T_4 - T_3 - T_2 + T_1)$$

$$\overline{q}_{in} = \overline{C}_p \ (T_3 - T_2)$$

The cycle efficiency becomes

$$\eta = \frac{(T_4 - T_3 - T_2 + T_1)}{(T_3 - T_2)}$$

Using the relationship for reversible adiabatic compression and expansion given in Table 2.5, the relationship for the Brayton Closed Cycle efficiency becomes

$$\eta = 1 - (P_1/P_2)^{(\gamma-1/\gamma)} \qquad\qquad (2.27)$$

The efficiency increases with pressure ratio.

EXAMPLE 2.19   Determine the efficiency of an ideal Brayton Cycle.
                (Working fluid is an ideal diatomic gas.)
                (a) Using equation 2.27
                (b) Using the definition w_{net}/q_{in}
                Evaluate the value of q_{out}.
                The following conditions apply (the subscript numbers
                refer to Figure 2.13).

                $P_1 = 1.01$ bar; $P_2 = 4.13$ bar; $T_1 = 294°K$; $T_3 = 1089°K$

SOLUTION 2.19

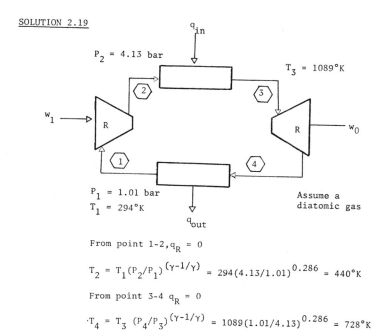

From point 1-2, $q_R = 0$

$$T_2 = T_1 (P_2/P_1)^{(\gamma-1/\gamma)} = 294(4.13/1.01)^{0.286} = 440°K$$

From point 3-4 $q_R = 0$

$$·T_4 = T_3 (P_4/P_3)^{(\gamma-1/\gamma)} = 1089(1.01/4.13)^{0.286} = 728°K$$

Stream	T (°K)	P (bar)	ΔT (°K)	$\overline{H} = \overline{C}_p (\Delta T)$ J/Kg-mole
1	294	1.01	-4°	-1.2×10^5
2	440	4.13	142	41.3×10^5
3	1089	4.13	791	230.2×10^5
4	728	1.01	430	125.1×10^5

$-\overline{w}_{in} = \overline{H}_2 - \overline{H}_1 = [41.3 - (-1.2)] \times 10^5$ [J/Kg-mole] $= +42.5 \times 10^5$ [J/Kg-mole]

$-\overline{w}_{out} = \overline{H}_4 - \overline{H}_3 = (125.1 - 230.2) \times 10^5$ [J/Kg-mole] $= -105.1 \times 10^5$ [J/Kg-mole]

$\overline{q}_{in} = \overline{H}_3 - \overline{H}_2 = (230.2 - 41.4) \times 10^5$ [J/Kg-mole] $= 188.9 \times 10^5$ [J/Kg/mole]

$\overline{q}_{out} - \overline{H}_1 - \overline{H}_4 = (-12. - 125.1) \times 10^5$ [J/Kg-mole] $= -126.3 \times 10^5$ [J/Kg-mole]

From equation 2.27

$$\eta = (1 - P_1/P_2)^{\gamma - 1/\gamma} = (1 - 1.01/4.13)^{0.286} = 0.33$$

From w_{net}/q_{in}

$$\eta = w_{net}/q_{in} = (105.1 - 42.5)/189.9 = 0.33$$

It is more convenient to solve problems involved with power cycles if there is a graphical presentation of the thermodynamic data. The most convenient manner to present the data is in the form of a T-S diagram or a H-S diagram (called a Mollier Diagram).

Figure 2.14 is a $\overline{H}$-$\overline{S}$ diagram for an ideal diatomic gas. In the previous sections given a T the value of $\overline{H}$ could be calculated (Equation 2.6). For each temperature there is a given $\overline{H}$. On Figure 2.14 the right hand scale is the value of T and the corresponding value of $\overline{H}$ is found by going straight over to the left hand scale. The value of $\overline{S}$ depends upon both T and P (see equation prior to 2.18 or Table 2.5). If the value of P is fixed then $\overline{S} = f(T)$. The curve for $\overline{S}$ vs T is plotted for several values of P on Figure 2.14. These curves are labeled with the P values (solid lines). In addition there are curves for the change in S with a constant volume $\overline{V}$. Most power cycles consist of a combination of individual steps that include both constant volume and constant pressure steps. These steps may be followed on Figure 2.14 by following a constant pressure or volume line. For a compression of expansion step that is adiabatic and reversible $\Delta S = 0$, a vertical line on a $\overline{H}$-$\overline{S}$ diagram.

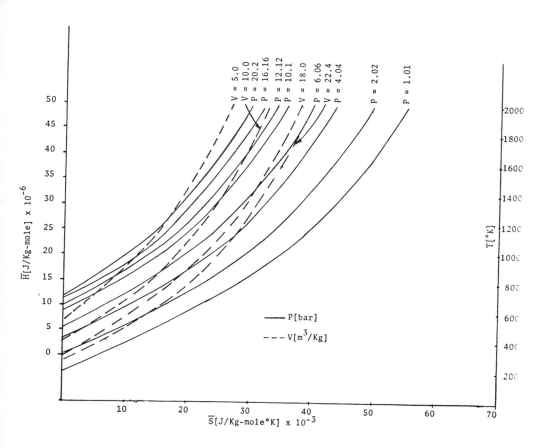

FIGURE 2.14 H-S Diagram for Ideal Diatomic Gas

The ideal turbine cycle shown by Figure 2.13 is sketched on
a T-$\overline{S}$ diagram in Figure 2.15.

The value of the T-$\overline{S}$ diagram is the ease in which power
cycles may be traced. The ideal Brayton Cycle just discussed
consists of two constant pressure steps (follows a constant
pressure curve) and two adiabatic reversible steps ($\Delta \overline{S} = 0$,
$q_R = 0$). The constant $\overline{S}$ steps are vertical lines on the T-$\overline{S}$
diagram. Figure 2.15 is a Brayton Cycle sketched on T-S

coordinates. The numbers refer to those shown in Figure 2.13.

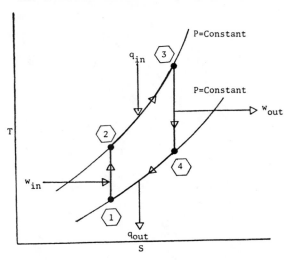

FIGURE 2.15   T-S Diagram for Gas Turbine

EXAMPLE 2.20   Repeat example 2.19 using the T-S diagram for an ideal
diatomic gas to get H values (Figure 2.16).

SOLUTION 2.20

$$\bar{H}_1 = 0.0$$

$$\bar{H}_2 = 4.2 \times 10^6 \ [J/Kg\text{-mole}]$$

$$\bar{H}_3 = 23 \times 10^6 \ [J/Kg\text{-mole}]$$

$$\bar{H}_4 = 13 \times 10^6 \ [J/Kg\text{-mole}]$$

$$\bar{w}_{net} = [(4.2 - 0) - (13 - 23)] \times 10^{+6} \ [J/Kg\text{-mole}]$$

$$\bar{w}_{net} = 5.8 \times 10^6 \ J/Kg\text{-mole}$$

$$\bar{q}_{in} = (23 - 4.2) \times 10^6 \ [J/Kg\text{-mole}]$$

$$\bar{q}_{in} = 18.8 \times 10^6 \ J/Kg\text{-mole}$$

$$\eta = (5.8/18.8) \times 100 = 31\% \ (\text{Error a result of reading graph})$$

For non-ideal gases, the T-$\bar{S}$ diagram is extremely valuable
as the $\bar{H}$ and $\bar{S}$ values cannot be easily evaluated.

2.6.2 Open Cycle Analysis

In the previous section the working fluid circulated
in a closed loop as seen by Figure 2.13. Many power systems
operate as an open-loop flow system. Figure 2.16 compares
the open loop and closed loop gas turbine.

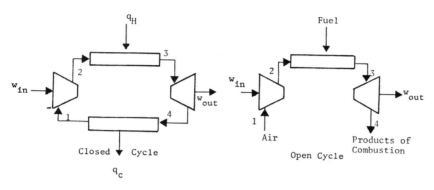

FIGURE 2.16 Comparison of Open and Closed Loop Gas Turbine Cycle

In the open cycle, the compressed air is burned with the
fuel at high pressure. The high temperature combustion products pass through the turbine where they produce work output. It is a common procedure to analyze the open cycle to
consider only the air that enters the combustor to be the
working fluid. It is assumed that the gas passing through
the turbine is air (rather than the combustion products).
Most turbines use large amounts of excess air, and there
is little error in this assumption. It may be further
assumed in this section that air behaves essentially as an
ideal diatomic gas. Figure 2.14 may be used to analyze the
open cycle gas turbine. There are numerous other flow cycles
that can be analyzed in a manner similar to the Brayton Cycle.
A few of the more common ones are discussed below. In all

cases it will be assumed that the working fluid behaves as a
diatomic gas and Figure 2.14 may be used.

The spark ignition internal combustion engine is the
power source for most passenger automobiles. The cycle is
called the Otto Cycle. This Otto Power Cycle is shown on a
T-S diagram below.

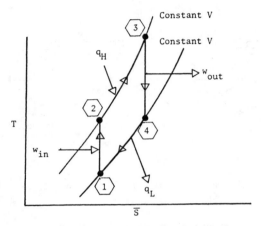

FIGURE 2.17 T-S Diagram for Otto Cycle

The Otto Cycle differs from the Brayton Gas Turbine Cycle
in the heat exchange steps. In the Brayton Cycle they took
place at constant pressure. In the Otto Cycle they occur
at constant volume. The efficiency of the Otto Cycle is
given by

$$\eta = 1 - \frac{T_3}{T_2} = 1 - \left(\frac{V_1}{V_4}\right)^{\gamma-1} \tag{2.28}$$

The diesel cycle is another power cycle alternative.
It differs from the Otto Cycle in that the heat addition
takes place at constant pressure while the heat rejection
occurs at constant volume.

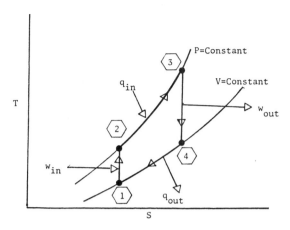

FIGURE 2.18   T-S Diagram for Diesel Cycle

The power cycles described differ from each other in the heat exchange steps. The heat exchange steps are carried out either at constant volume and/or constant pressure To solve problems involving these power cycles using the T-S diagram, it is necessary to follow either constant volume or constant pressure curves. In all cases the compression and expansion steps are adiabatic.

2.6.3 Rankine Cycle (Figure 2.19)

Most modern power plants are based upon the Rankine Cycle with steam as the working fluid. The Rankine Cycle differs from the previous cycles in that the working fluid in the system undergoes vaporization and condensation.

The working fluid flows around the loop 1 → 2 → 3 → 4 → 1, etc. The shaded area is at high pressure and the clear area at low pressure.

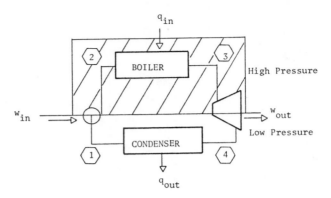

FIGURE 2.19 Basic Rankine Cycle

This cycle may be shown more easily on a T-S diagram.

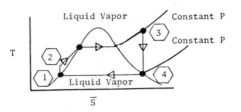

FIGURE 2.20 Diagram for Rankine Cycle

At point ⟨1⟩ the liquid leaves a condenser as a saturated liquid at a low pressure. It passes through a pump where the pressure is increased. This step is adiabatic, and the fluid remains a liquid. The liquid enters a boiler. Heat is added to the liquid, and the water is vaporized and superheated and leaves as a gas at point ⟨3⟩ . This takes place at constant pressure. The steam then enters a turbine where the pressure is reduced to that of point ⟨4⟩ . This step is adiabatic, and the turbine produces work. The low pressure steam leaving the turbine (it can be a mixture of liquid and gas) at point ⟨4⟩ enters a condenser where the working fluid loses heat and condenses to a liquid.

The evaluation of the Rankine Cycle is no different than the procedures used in the other power cycles.

$$\overline{w}_{out} = -(\overline{H}_4 - \overline{H}_3)$$

$$\overline{w}_{in} = -(\overline{H}_2 - \overline{H}_1)$$

$$\overline{q}_{in} = \overline{H}_3 - \overline{H}_2$$

$$\overline{q}_{out} = \overline{H}_1 - \overline{H}_4$$

$$\eta = (\overline{w}_{out} - \overline{w}_{in})/\overline{q}_{in}$$

The only problem is that of locating the $\overline{H}$ values.

The most common way to provide the thermodynamic values is to use a $\overline{H}\text{-}\overline{S}$ diagram. This is the same type of diagram that is shown in Figure 2.14 for an ideal gas. Such a diagram has been given the name "Mollier Diagram." Figure 2.21 is a portion of a Mollier Diagram for steam.

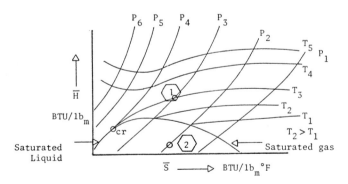

FIGURE 2.21 Portion of T-S Diagram for Steam

The British Units are retained in the discussion of steam cycles as this is likely to be the common set of units in the utility industry for some time. The region in Figure 2.21 above the line labeled saturated gas up to the critical

point (labeled cr.) is the gas phase (or water vapor). Given
the pressure and temperature of the vapor the value of enthalpy
and entropy can be read directly. The point labeled $\langle 1 \rangle$ is
at the intersection of P_3, T_3 which fixes the enthalpy and
entropy of the system. The point labeled $\langle 1 \rangle$ is in the vapor
region above the saturation curve and is "superheated." The
degrees of superheat is determined by the difference between
the actual temperature of the steam and the temperature at
which the steam would begin to condense at the same pressure.
Following the constant pressure line down from point $\langle 1 \rangle$ to
the saturation curve gives the degrees of superheat. In
this case the degrees of superheat is $T_3 - T_2$.

The area under the saturated gas curve is the two phase
region where both the liquid and the gas exist together.
The lines in this region are both constant temperature and
constant pressure lines. The temperature and pressure correspond to the intersection on the saturated gas curve of the
temperature line-pressure line. (For example, see point $\langle 2 \rangle$
is at T_1, P_2).

The curves in the two phase region shown in Figure 2.21
have been cut off, and the lower values of $\overline{H}$ (for the liquid)
are not shown. If the diagram were extended the constant
temperature-constant pressure lines would cross a saturated
liquid curve which would provide the enthalpy and entropy of
the liquid.

An adiabatic reversible process can be followed on this
diagram by following a constant entropy line, a vertical line
on this diagram. The isothermal and isobaric processes follow the constant temperature and constant. pressure lines,
respectively.

EXAMPLE 2.21 A Rankine Cycle operates between the pressure limits of
500 psia and 2 psia. The temperature of steam entering
the turbine is 700°F and the turbine may be considered
to be adiabatic and reversible ($\Delta S = 0$)
(a) Draw the flow sheet. Label all points given. Add
$\overline{H}$ and $\overline{S}$ values for each point.
(b) Evaluate system efficiency.
NOTE: Obtain values from steam tables, Appendix A.

SOLUTION 2.21

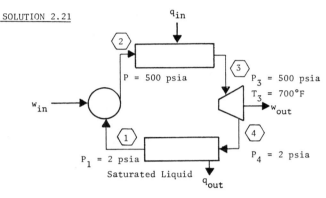

Point 3 Locate the intersection of 700°F and 500 psia

Read $\overline{H}_3$, $\overline{S}_3$

$$\overline{H}_3 = 1357[\text{Btu/lb}], \quad \overline{S}_3 = 1.61[\text{Btu/lb°R}]$$

Point 4 The change from Point 3 to Point 4 was adiabatic and
reversible. This means $\overline{S}_3 = \overline{S}_4$ ($\Delta S = 0$) or $\overline{S}_3 = 1.61$.
From the steam tables in Appendix A, at 2 psia

$\overline{S}$ of liquid = 0.175[Btu/lb°F]

$\overline{S}$ of vapor = 1.92[Btu/lb°F]

Since $S_3 = 1.61[\text{Btu/lb°F}]$ (between 0.175 and 1.92),
the condition at Point 3 must be part liquid and part
vapor. To determine the portion that is liquid, let
x represent the fraction of stream 3 that is a liquid
(1 - x becomes fraction that is vapor)

$$S_3 = 1.61 = 0.175x + 1.92(1 - x)$$

$$x = 0.178$$

Knowing the fraction liquid and vapor $\overline{H}_3$ may be
evaluated. From steam tables in Appendix A at 2 psia

$\overline{H}$ of liquid = 94[Btu/lb]

$\overline{H}$ of vapor = 1116[Btu/lb]

$\overline{H}_3 = 94$ (x) + 1116(1 - x) for x = 0.178

$\overline{H}_3 = 934$

Point 1 This is a saturated liquid at 2 psia.

$$\overline{H}_1 = 94 \quad [\text{Btu/lb}], \ \overline{S}_1 = 0.175[\text{Btu/lb}°R]$$

Point 2 This value must be calculated. The pump is
adiabatic and reversible therefore $\Delta \overline{S} = 0$. For
a non-compressible liquid

$$\Delta \overline{H} = \int \overline{V}dP = \overline{V}(P_2 - P_1)$$

$$= 1/62.4[\text{ft}^3/\text{lb}_m] \ \text{x} \ (500\text{-}2)[\text{lb}_f/\text{in}^2] \ \text{x} \ 144[\text{in}^2/\text{ft}^2]$$

$$\text{x} \ 1/778 \ [\text{Btu/ft-lb}_f] = 1.5 \ [\text{Btu/lb}_m]$$

Writing First Law of Thermodynamics for each unit.

(1-2) $\Delta \overline{H} = \overline{q} - \overline{w}$ (q = 0)
        $-\overline{w} = H_2 - H_1 = (95.4 - 93.9) = 1.5 \ \text{Btu/lb}$

(2-3) $\Delta \overline{H} = \overline{q} - \overline{w}$ (w = 0)
        $\overline{q} = \overline{H}_3 - \overline{H}_2 = (1357 - 95.4) = 1262 \ \text{Btu/lb}$

(3-4) $\Delta \overline{H} = \overline{q} - \overline{w}$ (q = 0)
        $-\overline{w} = \overline{H}_4 - \overline{H}_3 = (934 - 1357) = -423 \ \text{Btu/lb}$

(4-1) $\Delta \overline{H} = \overline{q} - \overline{w}$ (w = 0)
        $\overline{q} = \overline{H}_1 - \overline{H}_4 = (93.9 - 934) = -840 \ \text{Btu/lb}$

        $\eta = \text{Net work/Heat in} = (423\text{-}1.5)/1262 = 0.334$

If the work of the pump were neglected

$$\eta = 422/1262 = 0.334$$

From the example problem just worked it can be seen that
the pump work is small and may be ignored. There are many
modifications to the basic Rankine Cycle to improve the performance. It is important in comparing various energy alternatives to have some meaningful index to measure the comparative merit of a wide variety of process alternatives. Consider the merit of the electric automobile. The energy chain
that starts with the mining of coal and continues through to
the power delivered is depicted in Figure 2.22. The simple
question "What is the efficiency of the automobile?" can
have many answers. It could be 90% if the system boundaries

are drawn about only the transmission (system A), it could
be 65% (.9 x .8 x .9 x 100) if system B were selected for
the evaluation or 22.1% (.65 x .90 x .38 x 100) if the system
boundaries shown by system C that includes the power plant.
If the boundaries of the system were drawn about the total
system shown in Figure 2.22 the efficiency would be only
14.7%. But why not include the loss of coal in the mining
operation or the energy taken up in the mining operation or
the energy that went into the construction of all of the
materials and equipment that went into the mining operation,
the power plant, the steel used, etc.? Where should the calculations of the efficiency of the process begin? There is
no simple answer. The answer is not found in the study of
thermodynamics.

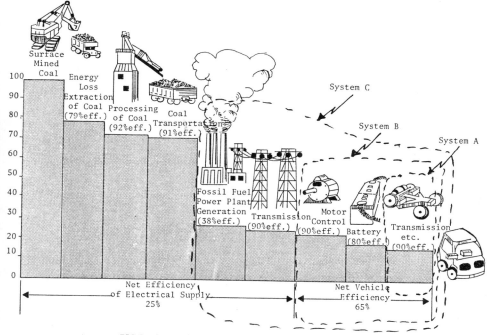

FIGURE 2.22 Efficiencies for Electric Automobile

Adapted from: Living in the Environment - Concepts, Problems, and Alternatives,
                G. Tyler Miller, Jr., Wadsworth Publishing Co., Inc., Belmont, CA.

Only after the system boundaries have been designated can
the thermodynamic analysis begin. Arguments regarding the
establishment of the system boundaries must be resolved if
various systems are to be analyzed and compared on a consistent basis. It is clear that different efficiency values
may be provided by different studies as there is no agreement on a common basis. The statement that it is thermodynam-
ically consistent does not resolve the problems associated
with the choice of boundaries.

Most readers probably have an intuitive feel for what
is meant by the term efficiency. If expressed in equation
form, this intuition would probably take the form of

$$\text{System Efficiency} = \frac{\text{Energy Out}}{\text{Energy In}} \times 100 \qquad (2.29)$$

Applying the First Law of Thermodynamics to this equation
will always give an efficiency of 1.0. This is because the
First Law is a statement that the Total energy into a system
= Total energy out of a system. Upon a reevaluation of the
expression for the system efficiency (Eq. 2.29) it could take
the form

$$\text{System Efficiency} = \frac{\text{Valuable Energy Out}}{\text{Valuable Energy In}} \times 100 \qquad (2.30)$$

This is the type of efficiency definition used in developing
the Carnot efficiency and other cycle efficiencies. The net
work produced was defined as the valuable energy output of
the system, and the valuable energy input was the heat input.
Thermodynamics was used to provide for the values necessary
to calculate the efficiency. However, the ratio is arbitrary
in that the terms to be included were arbitrarily chosen and

did not result from any thermodynamic laws. Unfortunately,
when analyzing a complex system different researchers will
include different terms in the numerator and denominator, and
the same system will provide for a wide variety of calculated
efficiencies. This is shown in the following illustration.

Figure 2.23 shows a simple hypothetical system for converting 1,000 joules per unit mass of fuel material into 600
joules of energy product. This process requires 300 joules
of electrical energy (purchased energy) to operate and it is
assumed this power is generated by consuming 900 joules of
some fossil fuel (raw energy).

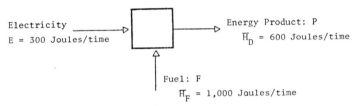

FIGURE 2.23   Hypothetical Energy
Recovery System

The energy recovery efficiency of the hypothetical process may
be defined several ways:

	Valuable Output	Valuable Input	Efficiency
1.	Total Energy in Product 600[J]	Energy in Fuel 1000[J]	60%
2.	Total Energy in Product 600[J]	Total Energy In (1000 + 300)[J]	46.2%
3.	Net Energy Out (600 - 300)[J]	Energy in Fuel 1000[J]	30%
4.	Net Energy Out (600 - 300)[J]	Total Energy Required by Process (1000 + 300)[J]	23.7%
5.	Total Energy in Product 600[J]	Total Energy Required by Process (1000 + 900)[J]	31.6%

6.	Net Energy Out (600 - 300)[J]	Total Energy Required by Process (1000 + 900)[J]	15.8%
7.	Net Energy Generated (600 - 900)[J]	Total Energy Required by Process (1000 + 900)[J]	-15.8%
8.	Net Energy Generated (600 - 900)[J]	Energy in Fuel 1000[J]	-30%

The same set of input and output values can provide a wide range of efficiencies. Reported efficiency values have little importance unless the system boundaries and the definition of efficiency is clearly defined. It is extremely dangerous (but quite common) to compare efficiencies from different studies.

It is unfortunate that the term efficiency is used for so many different situations. The efficiency often provides for the process of the expansion and the compression of gases to be compared with the best possible system.

$$\text{Efficiency of a turbine} = w_{act}/w_{rev} \qquad (2.31)$$

$$\text{Efficiency of a compressor} = w_{rev}/w_{act} \qquad (2.32)$$

These equations compare reversible and actual compressors or turbines operating between the same pressure limits.

Figure 2.24 provides the energy input and output streams for an adiabatic reversible compressor and an actual compressor having an efficiency rating of 85%, operating between 1.01 bar and 5.05 bar pressure.

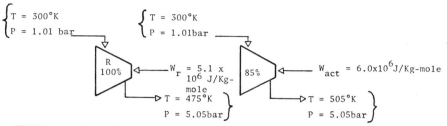

FIGURE 2.24 Comparison of Actual and Reversible Compressor

The actual compressor requires more work input but produces
an exit gas stream of 505°K rather than the 475°K provided
by the reversible compressor. Comparing these two processes,
compares systems that start out at the same starting point
but end up at different conditions. From the Second Law of
Thermodynamics the gas at 505°K is capable of doing more
work at a higher efficiency than the cooler gas at 475°K.

EXAMPLE 2.22   Determine the reversible work that can be obtained from
cooling the gas from the actual compressor to the same
temperature as the reversible compressor. Determine
efficiency = w_{rev}/w_{act} between same conditions of T and
P as a comparable reversible process.

SOLUTION 2.22

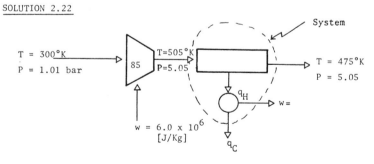

The gas from the 85% efficient turbine is cooled from 505°K
to 475°K. The heat removed runs a Carnot Engine to provide
work.

For the system shown by the --- lines

$$\Delta S_{universe} = 0 = \Delta S_{sys} + \Delta S_{sur}$$

Since system does not change, $\Delta S_{sys} = 0$

$$\Delta S_{sys} + \Delta S_{sur} = 0$$

$$\Delta S_{sys} = 0$$

$$\Delta S_{sur} = 0$$

$$\Delta S_{sur} = \Delta S(\text{for flows stream}) + \Delta S(\text{for } q_c)$$

$$\Delta S(\text{flow stream}) = C_p \ln 475/505$$

$$= 29,100[\text{J/Kg-mole}^\circ\text{K}] \ln 475/505$$

$$= -1.8 \times 10^{+3}[\text{J/Kg-mole}^\circ\text{K}]$$

$$\Delta S(q_c \text{ stream}) = -q_c/T_c = q_c/298^\circ\text{K}$$

$$\Delta S_{sur} = -1.8 \times 10^{+3} - q_c/298 = 0$$

$$q_c = -5.36 \times 10^5[\text{J/Kg-mole}]$$

From the First Law Over the System

$$\Delta H = q_c - w$$

$$\Delta H = C_p(475 - 505) = 29,100[\text{J/Kg-mole}^\circ\text{K}] \times (-30)[^\circ\text{K}]$$

$$= 8.73 \times 10^{+5}[\text{J/Kg-mole}]$$

$$w = q_c - \Delta H = -5.36 \times 10^5 + 8.73 \times 10^{+5}$$

$$= 3.37 \times 10^5[\text{J/Kg-mole}]$$

The net work for the total system

$$w_{net} = (6.0 \times 10^6 - 3.37 \times 10^5)[\text{J/Kg-mole}] = 5.66 \times 10^6[\text{J/Kg-mole}]$$

The efficiency becomes

$$\eta = 5.1 \times 10^6/5.66 \times 10^6 = 0.90$$

The example provided showed that a portion of the additional work required by the real compressor system could "in principle" be recovered from the hotter gas discharged by using the gas to run an engine. For the same process two efficiencies were obtained (85 and 90%).

The definitions of efficiencies used in the discussion above fall into two distinct classifications.

1. The energy content of some output stream or streams are compared to the energy output of some input

stream or streams. The quality of the energy in
the input and output stream is not considered.
One Btu of electrical energy carries the same
weight as one Btu of energy leaving at a low temperature. These efficiencies are referred to as
the First Law efficiencies.

2. The energy requirements for the actual process are
compared to the energy content of an ideal process.
In this case a comparison gives an indication as
to how close a system is to the best possible
system to accomplish a given change. These are
termed Second Law efficiencies.

In an effort to reduce the confusion that goes with the
term efficiency the term Figure of Merit is defined

$$\text{Figure of Merit} = \frac{\text{The maximum amount of work that could be produced by all input streams}}{\text{The maximum amount of work that could be produced by all output streams}} \qquad (2.33)$$

The justification of using this ratio of work terms as a
measure of the merit of a process is provided in the next
paragraph.

The primary need for fossil fuels is in the conversion
of chemical stored energy to work. It is this process that
has lead to the industrial revolution and the way of life
in the United States and other industrialized countries.
Some of the energy is used as a source of heat. Mankind
first learned to harness fossil fuel for heat. He learned
to heat his home, cook his food, and to scare a sabre-toothed
tiger or two; but it was not until he was able to harness
fuel to perform work functions that the great advances were

made. This is the highest use of fossil fuel energy.

Using this definition all streams are included in the
evaluation of the Figure of Merit and there is no need to
arbitrarily select which terms should be included and which
terms should be excluded. To determine the maximum amount
of work that can be obtained from any stream it is necessary
to establish the datum conditions to which all streams must
be compared. The datum temperature of 298°K and a pressure
of 1.01 bar (1 atmosphere) were chosen. These approximate
the average conditions on the earth's surface and are the
conditions chosen in reporting most of thermodynamic data.
In addition to the temperature and pressure the datum conditions for each element must be stated. The choice made
corresponds to the conditions chosen for the evaluation of
$\overline{H}$ and $\overline{S}$.

Carbon as CO_2 (gas)
Hydrogen as H_2O (liquid)
Sulfur as SO_2 (gas)
O_2, N_2 (gas)

Figure 2.25 shows the procedure used to obtain the
Figure of Merit. The energy input and output streams are
a combination of mass flow streams, heat streams, and work
streams. The maximum amount of work that can be obtained
from each stream, if it were diverted from the process
to produce work, must be evaluated. (The maximum amount
of work according to the Second Law is obtained when the
ΔS of the universe is equal to zero). The sum of all of
the work values for the input streams is the denominator
in the Figure of Merit. The same procedure is performed
for the exit streams. The amount of work that could be

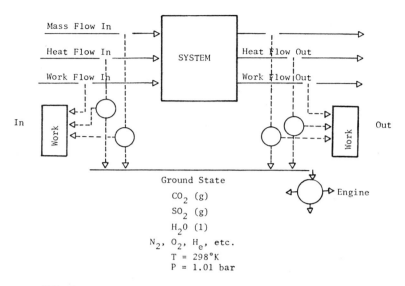

FIGURE 2.25 System for Evaluation of Figure of Merit

obtained if all of the exit streams were converted to work
becomes the numerator. This ratio will always be less than
1.0. The difference between the numerator and denominator
is the lost work term (lw) and is the amount of work that
was lost because the process took place. The ratio defined
as the Figure of Merit is dependent upon the choice of
standard states. The value of lw is obtained from

$$lw = \text{work obtained from input streams} -$$
$$\text{work obtainable from exit streams} \quad (2.34)$$

is independent of the choice of standard states.

In order to evaluate the Figure of Merit it is
necessary to evaluate the maximum amount of work that can
be obtained from three types of energy streams:

1. Work
2. Heat

3. Mass flow streams.

The work streams may be in the form of shaft work or electrical energy. They are already in the form of work, and no
conversion is necessary. The heat streams are all used to
drived a Carnot Engine operating between the temperature
the heat crosses the system boundaries and the datum temperature 298°K. The maximum amount of work can be obtained from
the Carnot Efficiency.

$$w = q(1 - 298/T)$$

For flow streams it is necessary to determine the maximum amount of work that could be extracted in bringing it
to the conditions of the ground state. This is shown in the
following example.

EXAMPLE 2.23 Determine the Figure of Merit for the
systems shown in Figure 2.24.

SOLUTION 2.23

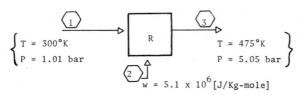

Maximum Work Evaluation

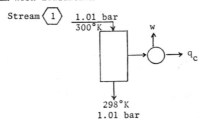

This is identical to the
system used in Example 2.22
and procedure for solution
shown is followed.

$\Delta \overline{S}$(flow stream) = 29,100 ln 298/300 = -195 [J/Kg-mole°K]

$\Delta \overline{S}$(flow stream) = 29,100 ln 298/300 = -195[J/Kg-mole°K]

$\Delta \overline{S}$(q_c stream) = $-q_c/298$

$\Delta \overline{S}_{sur}$ = -195 - $\overline{q}_c/298$ = 0

$\qquad\qquad q_c$ = -58,110[J/Kg-mole]

$\overline{w} = \overline{q}_c - \Delta\overline{H}$ = -58,110[J/Kg-mole] - 29,100(298 - 300)[J/Kg-mole]

$\quad$ = 100[J/Kg-mole]

Stream ⑧②

$\overline{w}$ = 5.1 x 10^6[J/Kg-mole]

Total work available from <u>all</u> input streams

$\qquad w_{in}$ = [5.1 x 10^6 + 100][J/Kg-mole] = 5.1 x 10^6 [J/Kg-mole]

Stream ⟨3⟩ 5.05 bar
475°K

w → q_c

298°K
5.05 bar

→ w

298°K
1.01 bar

ΔS(flow system) = C_p ln T_2/T_1 -

R ln P_2/P_1

= 29,100 ln 298/475 - 8314 ln 1.01/5.05

= -13,568 + 13,380 = -187[J/Kg-mole°K]

$\Delta S(q_c)$ = $-q_c/298$

ΔS_{sur} = -187 - $q_c/298$ = 0

q_c = -5.58 x 10^4[J/Kg-mole]

$\sum\overline{w} = \overline{q}_c - \Delta\overline{H}$ = -5.88 x 10^4 - 29,100[298 - 475]

$\quad$ = 5.1 x 10^6[J/Kg-mole]

Figure of Merit = 5.1 x 10^6[J/Kg-mole]/5.1 x 10^6[J/Kg-mole] = 1.0

lw = [5.1 x 10^6 - 5.1 x 10^6][J/Kg-mole] = 0

(b)$\qquad$ 85% efficient engine -- Follow same procedure as above.

Stream ① Same as part (a) = 100[J/Kg-mole]

$\qquad$② Given 6.0 x 10^6[J/Kg-mole]

$\qquad\qquad$ Total in potential In = 6.0 x 10^6[J/Kg-mole]

Stream ③

$\Delta\overline{S}_{(flow\ stream)}$ = 29,100[J/Kg-mole°K] ln(298/505)

$\qquad\qquad$ -8,314[J/Kg-mole°K] ln(1.01/5.05)

$\qquad\qquad$ = -1919[J/Kg-mole°K]

$$\Delta \overline{S}_{sur} = 0 = -1919[\text{J/Kg-mole}°\text{K}] - q_c/298$$

$$\overline{q}_c = -5.72 \times 10^5 [\text{J/Kg-mole}]$$

$$\sum \overline{w} = -5.72 \times 10^5 - 29,100(298 - 505)$$

$$= 5.45 \times 10^6 [\text{J/Kg-mole}]$$

Figure of Merit = $5.44 \times 10^6 [\text{J/Kg-mole}]/6.0 \times 10^6 [\text{J/Kg-mole}] = 0.91$

$1w = (6.0 - 5.44) \times 10^6 [\text{J/Kg-mole}] = 0.56 \times 10^6 [\text{J/Kg-mole}]$

It can be seen in the example above that the Figure of
Merit is equal to 1.0 for a reversible process and is less
than 1.0 for an actual process. A Figure of Merit value
over 1.0 represents an impossible process. The efficiency
of 0.85 used to determine the work required by the compressor
(defined as the reversible work divided by the actual work to
obtain the same pressure) leads to a figure of Merit of 0.91.
The reason for the higher value is the Figure of Merit took
into account the potential of the output streams from the
actual process to perform more work than the output from
the reversible compressor (at the same pressure). This
additional work available from the actual process may not
be utilized although it is available.

The procedure used in Example 2.23 can be simplified
by defining a new value $\overline{B}$. $\overline{B}$ is the maximum amount of work
that can be extracted from a unit mass in bringing it to
the datum condition of T and P. $\overline{B}$ is defined by the equation

$$\overline{B} = \overline{H} - T_o \overline{S} \qquad [T_o = 298°\text{K}] \tag{2.35}$$

The procedure for evaluating $\overline{H}$ and $\overline{S}$ have been discussed in
previous sections. The value of $\overline{B}$ may be evaluated by
determining both $\overline{H}$ and $\overline{S}$ and substituting into equation 2.35.
If the values of $\overline{B}_C$, $\overline{B}_T$, $\overline{B}_\lambda$, and $\overline{B}_P$ (which correspond to

$\overline{H}_C$, $\overline{H}_T$, $\overline{H}_\lambda$, $\overline{H}_p$ and $\overline{S}_C$, $\overline{S}_T$, $\overline{S}_\lambda$, $\overline{S}_p$ described earlier) are available, then

$$\overline{B} = \overline{B}_C + \overline{B}_T + \overline{B}_p + \overline{B}_\lambda \qquad (2.36)$$

This corresponds to equation 2.4 for $\overline{H}$ and 2.17 for $\overline{S}$.

The value of $B_\lambda \doteq 0$ and need not be included and Equation 2.36 reduces to

$$\overline{B} = \overline{B}_C + \overline{B}_T + \overline{B}_p \qquad (2.37)$$

$\overline{B}_C$ is the maximum amount of work that may be done as
a result of the chemical configuration. Values are
in Table 2.7.

$\overline{B}_T$ is the maximum amount of work that may be done as
a result of temperature. Values are plotted in
Figure 2.25.

$\overline{B}_p$ is the maximum amount of work that may be done as a
result of pressure. Values are plotted in Figure 2.9.

The total amount of work by all entering mass streams
is given by

$$B_{in} = \sum_i a_i \overline{B}_i$$

and by all exiting mass streams by

$$B_{out} = \sum_i b_i \overline{B}_i$$

These values may be evaluated using the same format given
in Table 2.2 for ΔH. To obtain the total amount of work
from all streams the work obtainable from q and w are added
to the $\overline{B}$ terms.

$$w_{in} = \sum a_i \overline{B}_i + \sum w_{in} + \sum q_{in}(1 - 298/T_{in}) \qquad (2.38)$$

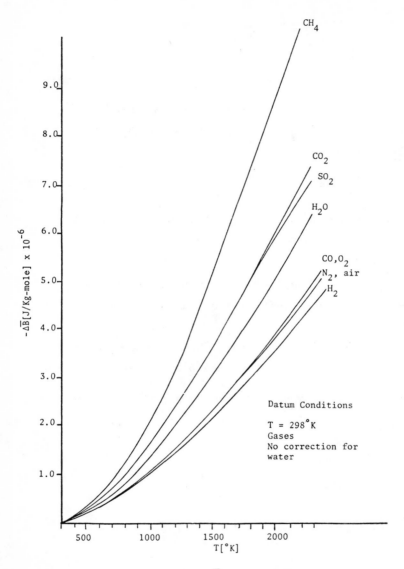

FIGURE 2.25 $\overline{B}_T$ for Common Gases

$$w_{out} = \sum_i b_i \overline{B}_i + \sum w_{out} + \sum q_{out} (1 - 298/T_{out}) \quad (2.39)$$

The lost work, lw, becomes

$$lw = w_{in} - w_{out}$$

EXAMPLE 2.24   It is necessary to heat a stream of 100 Kg-mole
nitrogen available at 5.05 bars pressure and
298°K to a temperature of 600°K and pressure of
1.01 bar. This is done by cooling a stream of CO_2
(50 Kg-mole) available at 1,000°K and 1.01 bars
pressure. 1×10^7[J] of work is extracted.
Determine the Figure of Merit for this process.

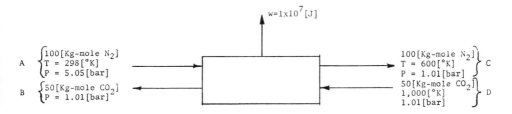

SOLUTION 2.24

It is first necessary to determine T_D. This
is done by applying the First Law of Thermodynamics.

Stream Identification	Mass Flow	T	P	$\overline{H}_C$	$\overline{H}_T$	$\overline{H}_\lambda$	$\overline{H}_P$	$\overline{H}$	H = m$\overline{H}$
Units on Terms	[Kg-mole]	[°K]	[bar]			[J/Kg-mole]			[J]
Inlet									
$A(N_2)$	100	298	5.05	-	0	-	-	0	0
$B(CO_2)$	50	1000	1.01	-	3.33×10^7	-	-	3.33×10^7	1.67×10^9
$\sum_i a_i \overline{H}_{A_i}$ =									1.67×10^9
Exit									
$C(N_2)$	100	600	1.01	-	8.93×10^6	-	-	8.93×10^6	8.93×10^8
$D(CO_2)$	50	$\overline{H}_D$	1.01	-	$\overline{H}_D$	-	-	$\overline{H}_D$	$50\overline{H}_D$
$\sum_i b_i \overline{H}_{B_i}$ =									$8.93 \times 10^8 + 50\overline{H}_D$

$$\Delta H = \sum_i b_i \overline{H}_{B_i} - \sum_i a_i \overline{H}_{A_i} = 8.93 \times 10^8 + 50\overline{H}_D - 1.67 \times 10^9$$

$\Delta H = 50\overline{H}_D - 7.77 \times 10^8$

First Law of Thermodynamics

$\Delta H = q - w$

$50\overline{H}_D - 7.77 \times 10^8 = 0 - 1 \times 10^7$

$\overline{H}_D = (7.87 \times 10^8)/50 = 1.57 \times 10^7 [J/Kg\text{-}mole]$

Extrapolating from values given in

$\overline{T}_D = 650[^\circ K]$

Evaluation of B

Stream Identification	Mass Flow	T	P	$\overline{B}_C$	$\overline{B}_T$	$\overline{B}_\lambda$	$\overline{B}_P$	$\overline{B}$	$B = m\overline{B}$
Units on Terms	[Kg-mole]	[°K]	[bar]			[J/Kg-mole]			[J]
Inlet									
A(N_2)	100	298	5.05	-	0		4×10^6	4×10^6	4×10^8
D(CO_2)	50	1000	1.01	-	1.68×10^7		-	1.68×10^7	8.4×10^8
$\sum a_i \overline{B}_{A_i}$ =									12.4×10^8
Exit									
C(N_2)	100	600	1.01	-	2.78×10^6		-	-	2.78×10^8
D(CO_2)	50	650	1.01	-	5.39×10^6		-	-	2.7×10^8
$\sum b_i \overline{B}_{B_i}$ =									5.48×10^8

$w_{in} = 12.4 \times 10^8 [J]$

$w_{out} = 5.48 \times 10^8 [J] + 1 \times 10^7 [J] = 5.58 \times 10^8 [J]$

Figure of Merit $= 5.58 \times 10^8/12.4 \times 10^8 = 0.45$

$_l\overline{w} = 12.4 \times 10^8 - 5.58 \times 10^8 = 6.8 \times 10^8 [J]$

In the calculations performed in the previous example problems, there were no chemical reactions taking place and there was no heat exchanged with the surroundings.

EXAMPLE 2.25   Consider a home heated with methane. The methane
                is burned in 200% excess air. The stack gases leave
                at 600°K. The furnace provides heat at 120°F (48.6°C).
                a. Evaluate the efficiency of the furnace
                    defined as the ratio of the heat delivered
                    to the heating value of the methane.
                b. Evaluate the Figure of Merit.
                c. Evaluate lost work.

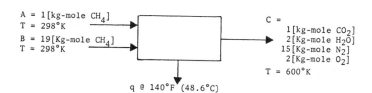

A = 1[kg-mole CH_4]
T = 298°K

B = 19[Kg-mole CH_4]
T = 298°K

C =
    1[kg-mole CO_2]
    2[Kg-mole H_2O]
    15[Kg-mole N_2]
    2[Kg-mole O_2]

T = 600°K

q @ 140°F (48.6°C)

SOLUTION 2.25

Use the First Law of Thermodynamics to evaluate q

Evaluation of q

Stream Identification	Mass Flow	T	P	$\bar{H}_C$	$\bar{H}_T$	$\bar{H}_\lambda$	$\bar{H}_P$	$\bar{H}$	$H = m\bar{H}$
Units on Terms	[Kg-mole]	[°K]	[bar]	[J/Kg-mole]					[J]
Inlet A(CH_4)	1	298	1.01	89.1×10^7	-	-	-	89.1×10^7	89.1×10^7
B(air) N_2	15	298	0.8	-	-	-	-	0	0
O_2	4	298	0.21	-	-	-	-	0	0
$\sum a_i \bar{H}_{A_i}$ =									89.1×10^7
Exit C CO_2	1	600	.05	-	1.29×10^7	-	-	1.29×10^7	1.29×10^7
H_2O	2	600	.10	-	1.05×10^7	4.4×10^7	-	5.45×10^7	10.9×10^7
N_2	15	600	.76	-	0.89×10^7	-	-	0.89×10^7	13.4×10^7
O_2	2	600	.10	-	0.93×10^7	-	-	0.93×10^7	1.9×10^7
$\sum b_i \bar{H}_{B_i}$ =									27.5×10^7
$\Delta H = \sum b_i \bar{H}_{B_i} - \sum a_i \bar{H}_{A_i}$ = 27.5 x 10^7 - 89.1 x 10^7 = -61.6 x 10^7[J]									

First Law of Thermodynamics

$\Delta H = q - w$ $w = 0$

$\Delta H = q = -61.6 \times 10^7 [J]$

Heat removed = $61.6 \times 10^7 [J]$

Furnace Efficiency = 61.6 x 10^7/89.1 x 10^7 = 0.69

Evaluation of B values

Stream Identification	Mass Flow	T	P	$\bar{B}_C$	$\bar{B}_T$	$\bar{B}_\lambda$	$\bar{B}_P$	$\bar{B}$	$B = m\bar{B}$
Units on Terms	[Kg-mole]	[°K]	[bar]	[J/Kg-mole]					[J]
Inlet A(CH$_4$)	1	298	1.01	81.8	0			81.8×10^7	81.8×10^7
B(air) N$_2$	15	298	0.8	0	0		-5.8×10^5	-5.8×10^5	-8.7×10^6
O$_2$	4	298	0.21	0	0		-3.9×10^6	-3.9×10^6	-15.6×10^6
$\sum a_i \bar{B}_{A_i}$ =									79.4×10^7
Exit C CO$_2$	1	600	0.05	0	4.1×10^6		-7.5×10^6	-3.4×10^6	-3.4×10^6
H$_2$O	2	600	0.10	0	3.3×10^6		-5.7×10^6	-2.4×10^6	-4.8×10^6
N$_2$	15	600	0.76	0	2.8×10^6		-0.7×10^6	$+2.1 \times 10^6$	$+31.5 \times 10^6$
O$_2$	2	600	0.10	0	2.9×10^6		-5.7×10^6	-2.8×10^6	-5.6×10^6
$\sum b_i \bar{B}_{B_i}$ =									17.7×10^6

$$w_{in} = \sum_i a_i \bar{B}_{A_i} = 79.4 \times 10^7 [J]$$

$$w_{out} = \sum_i b_i \bar{B}_{B_i} + q(1 - 298/T)$$

$$= 1.77 \times 10^7 + 61.6 \times 10^7 (1 - 298/321.6)$$

$$= 6.29 \times 10^7 [J]$$

Figure of Merit = $6.29 \times 10^7 [J]/79.4 \times 10^7 [J] = 0.08$ (8%)

$lw = (79.4 \times 10^7 - 6.29 \times 10^7)[J] = 73.1 \times 10^7 [J]$

The previous example showed that the efficiency of home heating provided a high value. This would indicate that the heating of the home with methane is a highly effective utilization of the methane. The low Figure of Merit showed that the ability of the methane to perform work has been largely destroyed by converting the energy available in the methane to heat at 120°F. It indicates poor utilization of methane. An extremely high quality energy was used to do a job that can be performed by a low quality energy resource. The use of a high quality energy source when it is not needed is a waste of valuable fuel.

Thermodynamics provides guidance in the consideration of any conversion system. It sets goals; it places limits for any process; it allows the best that can be done to be evaluated. It is a study of the theoretical limits, not the practical limits. It does not help us to design a machine or to carry out an energy conversion. It states that the process can be done; it remains the application of sound engineering to find a way to do the conversion.

The Laws of Thermodynamics are applied about a boundary, but it does not suggest how the boundary should be selected. There are no clear procedures for establishing consistent boundaries for comparison of energy conversion alternatives, and confusion shall exist in comparing values reported.

Problems for Chapter 2.0

1. A furnace is used to provide hot air. The furnace pro-
    vides 6 x 10^7 [Joules/sec].

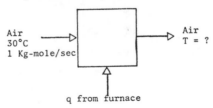

q from furnace

    (a) What is the temperature of the exit air?
    (b) What would the flow-rate be if the temperature
         required is 2000°K and the heat rate from the
         furnace remains constant?

2. The exhaust from a gas turbine may be as high as 800°F
    and may be used to produce steam. Assume that the gas
    behaves as an ideal diatomic gas. Determine the lbs of
    steam that is produced for each 100°F that one lb-mole
    of ideal gas drop in temperature. Assume that water at
    70°F is used to produce steam at 600°F and 150 psia.
    (Note: This is a useful number to remember and may be
    used for many rough calculations.)

3. A power plant produces 350 Mw of electrical energy from
    a nuclear fuel with a plant efficiency of 33%. How
    much cooling water is required for cooling if the tem-
    perature rise of the water is 6°C?

4. Determine the amount of oxygen required for the combus-
    tion of hexylene if 200% excess air is used? What is
    the composition of the exit gas?

5. If the hexylene is used as a fuel what is the amount of
    heat removed from the combustion system if the gases
    exit the system at 400°K (assume air and hexylene enter
    at 298°K). Given: ΔH_c = -935 K cal/kg-mole. What is
    the adiabatic reaction temperature?

6. A fluidized bed furnace has many desirable qualities that
    provide for good combustion of many fuels. The major
    drawback is that the gases after combustion leave the
    furnace at the same temperature that combustion occurs
    (in most cases about 1700°F). These hot gases contain
    significant energy that will be lost. To reduce this
    loss it was decided to install a heat exchanger to use
    these hot gases to heat up the air coming into the furnace.
    (Fuel is C burned in 50% excess air.)

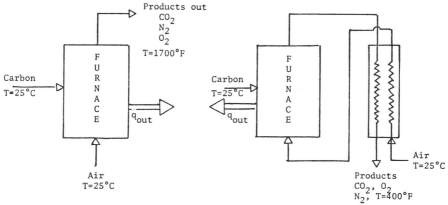

Original System   Revised System

(a) Determine the composition of the product gas.
(b) Determine the temperature of air to the furance.
(c) Determine the increase in q_{out} by adding the heat exchanger.
(d) Steam that could be generated from the heat saved (200 psi sat.)

Selected Values of $\overline{H}$

$\overline{H}$ Values [J/Kg-mole]

T	Air	CO_2	N_2	O_2
25°C	0	0	0	0
1700°F(1225°K)	2.9×10^7	4.6×10^7	2.9×10^7	2.9×10^7
400°F(502°K)	0.60×10^7	0.83×10^7	0.60×10^7	0.61×10^7

Note: Assume $\overline{H}$ linear between 1700°F and 400°F for air.

7. Butane is burned in a furnace to produce steam. The combustion system requires 50% excess air and the product gases leave at 600°F.
   (a) Calculate air flow rate (Kg mole butane).
   (b) Calculate the kg steam generated/kg mole butane.

8. A new water heating system is developed that will heat water as it is used. This reduces the heat loss from the hot water tank used in most systems. The maximum heater power is 10 Kw. What is the maximum flow rate (lb/hr) that can be heated to 150°F?

9. A fossil-fuel power plant may be assumed to operate at about 50% of the Carnot Efficiency. It has been suggested that the air pollution from a plant that operates at

1000°F would contribute 1/2 the pollution (per ton of coal burned) that the present system that operates at about 1500°F. For the same electrical output what is the % difference in the pollution resulting from the combustion?

10. A small steam turbine is designed to operate on 200 psi steam. The turbine operates on a simple Rankine cycle that rejects heat at 90°F. What temperature must the steam be heated to assure that no water condenses in the turbine. Calculate the cycle efficiency.

11. A reversible heat engine reviews energy from a high temperature reservoir at 3000°R and rejects heat at a low temperature reservoir at 600°R.
    (a) Determine the entropy change of the two heat reservoirs when 5000 Btu is added to the engine.
    (b) Determine the work output of the engine.

12.   (a) Determine the efficiency of an ideal-closed Otto cycle with a compression ratio of 8/1. (The compression ratio is the ratio of volume before to after the compression step). The low temperature may be taken as 298°K.
        (b) Determine the heat that may be added to the cycle if the maximum temperature that can be achieved is 1100°K.
        (c) Compare the efficiency to the Carnot engine operating between the same two limits.

13.   (a) Determine the efficiency of an ideal-closed cycle diesel cycle with a compression ratio of 15/1. (The compression ratio is the ratio of the volume before to the volume after the compression step.) The low temperature may be taken as 298°K.
        (b) 5.4×10^7 [J/Kg-mole] is added to the cycle. What is the maximum temperature?
        (c) What is the cycle efficiency?
        (d) Compare the cycle efficiency to the Carnot engine operating between the same two temperature limits.

14. An air standard Brayton closed cycle receives air at 1 atm and 298°K. The upper temperature and pressure limits are 5 atm and 1100°K.

    (a) Determine the cycle efficiency.

    (b) Compare the cycle efficiency with a Carnot engine operating between the same two limits.

15. An open Brayton cycle is operated using methane as a fuel. The conditions are the same as those given in the previous problem.

    (a) Determine the amount of air required/kg-mole of CH_4.

    (b) Determine the cycle efficiency.

    (c) Compare the efficiency with that of the previous problem.

16. In the previous problems the compression and expansion steps were adiabatic and reversible. The text limited the discussion to adiabatic-reversible compression and expansion steps. In a real system the steps are not reversible.

    (a) What would the value of ΔS be for the adiabatic compression and expansion step? +, -, or 0?

    (b) From your answer to (a) draw a T-S diagram for a real Brayton closed cycle where the steps are not reversible but operate between the identical pressures.

    (c) The actual performance of a compressor is given by the ratio of the work required for an ideal reversible cycle to the work required by the actual compressor to compress a gas to the same final pressure, $n_{compressor} = (w_R/w_{ACT})$. For a turbine (expander) the performance is given by $n_{turbine} = (w_{ACT}/w_R)$. Repeat Problem 14 for $n_{compressor}$ and $n_{turbine} = 0.9$. HINT: The value for w given in Table 2.5, i.e., $w = C_p(T_2 - T_1)$ is the same for both reversible and actual systems. The procedure to obtain w_{ACT} is to calculate w_R and use the ratio n to obtain w_{ACT}. w_{ACT} may then be used to determine the temperature following the compression or expansion step by using the equation above.

17. Two Kg-moles of a diatomic ideal gas at 3.45 bars pressure and 1000°K enter a device that operates adiabatically and produces no work. One Kg-mole of gas leaves this device at 1500°K and 3.0 bar pressure and the second leaves at 500°K and 1.0 bar pressure.

    (a) Does the first law hold?

    (b) Does the second law hold?

18. Consider an air preheater that uses a hot gas stream of CO_2 at 1600°K. The CO_2 is cooled to 600°K. The air is heated from 298°K to 800°K. There is no heat loss.
    (a) What is the flow rate of air per Kg-mole CO_2?
    (b) What is the Figure of Merit for the exchanger?
    (c) What is the lost work in the heat exchanger?

19. Consider the combustion of methane in 100% excess air in a furnace. What is the Figure of Merit and lost work for the combustion process. HINT: Determine the adiabatic flame temperature and then determine the work that can be obtained in cooling this gas.

20. Methane is used to heat a home. The methane is burned with 100% excess air and the combustion products leave at 160°F. The furnace nameplate provides the following data: Heat input 110,000 Btu/hr; Heat delivered @ 140°F, 85,000 Btu/hr. Assume ambient air at 70°F. Evaluate:
    (a) Figure of Merit
    (b) Efficiency ≡ heat delivered/heat input.

3.0 CHEMICAL REACTION EQUILIBRIUM AND KINETICS

In Chapter Two the First and Second Laws of Thermodynamics were presented. They allowed for the evaluation
of the energy changes that take place. They also provide
for the determination of processes that are possible and
those that are not possible. There are additional considerations that must be taken into account in the analysis of
fuel conversion systems. Most of these conversions
involve chemical reactions. Chemical reactions always tend
toward a state of chemical equilibrium. How do we determine
what conditions exist at chemical equilibrium? Where
is the reaction going? These questions are answered in
Section 3.1, Chemical Equilibrium.

Once it has been established what the equilibrium
conditions are, it is necessary to understand how fast
the reaction is moving toward this equilibrium state.
These questions are answered in Section 3.2, Chemical Kinetics.

Carbon at room temperature that is exposed

to oxygen is not at equilibrium. Equilibrium considerations
will show that it should react and form CO_2. The rate of
formation is so slow, however, that this reaction is not
detected at normal room temperature.

Chemical equilibrium provides the basic information on what
can occur and is based on thermodynamics. Chemical kinetics
provides the information on how fast reactions actually occur.

3.1   CHEMICAL EQUILIBRIUM

In discussions in Chapter Two, it has been assumed
that a reaction will go to completion

$$aA + bB \longrightarrow rR + sS \tag{3.1}$$

Some reactions under certain conditions will go toward a
state of completion while others do not. What can be
stated is that all chemical reactions move toward a
state of equilibrium. If a process does not go to completion the reaction is written

$$aA + bB \rightleftharpoons rR + sS \tag{3.2}$$

The R formed is limited by an equilibrium value. It is
essential to be able to predict the equilibrium value in
order to assess the potential of producing R from chemicals
A + B. The equilibrium conversion changes with temperature
and pressure. This section provides the essential concepts
required to be able to evaluate the equilibrium conversion
and how it changes with T and P.

This discussion of equilibrium is limited to an ideal

gas or an ideal gas-solid system. For the reaction given by equation 3.2, the equilibrium constant is given as

$$K_p = \frac{pp_R^r \; pp_S^s}{pp_A^a \; pp_B^b} \qquad (3.3)$$

where pp represents the partial pressure <u>at equilibrium</u>. The partial pressure is given by Dalton's Law

$$pp_A = \Pi y_A \qquad (3.4)$$

where Π is the total system pressure and y_A is the mole fraction of A at equilibrium. Substituting equation 3.4 for each chemical component in the gas into equation 3.3 gives

$$K_p = \frac{y_R^r \; y_S^s}{y_A^a \; y_B^b} \; \Pi^{r+s-a-b} \qquad (3.5)$$

Figure 3.1 provides values of K_p as a function of temperature for several reactions. (When a solid is involved in the reaction it may be left out of the equilibrium constant K_p.)

The effect of temperature on the equilibrium constant K_p can be seen in Figure 3.1. It is approximately a straight line when (1/T) is plotted versus Log K_p. For many reactions (where the heat of reaction does not change appreciably with temperature) the following relationship may be used to estimate the equilibrium constant at any temperature if it is known at one temperature.

$$\ln \left[K_p(T_2)/K_p(T_1) \right] = \Delta H_R [1/T_1 - 1/T_2] \ /R \qquad (3.6)$$

where ΔH_R is the heat of reaction, R is the ideal gas constant and T_1 and T_2 the absolute temperatures.

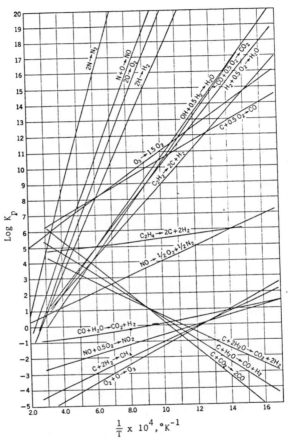

FIGURE 3.1 Equilibrium Constants K_p for Some Common Reactions

Reference: Introduction to Chemical Engineering Thermodynamics, J. M. Smith & H. C. van Ness, McGraw-Hill Book Company, 1959.

EXAMPLE 3.1   Determine the equilibrium composition for the dissociation
of CO_2 to CO and oxygen at 3200°K(1.01 bar)

$$CO_2 \rightleftharpoons CO + 1/2\ O_2$$

SOLUTION 3.1

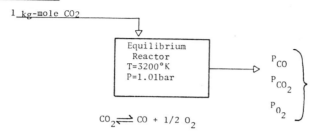

$$CO_2 \rightleftharpoons CO + 1/2\ O_2$$

Let α = moles of CO formed

Component	In +	Generated	- Consumed	= Out	Fraction
CO_2	1	0	$-\quad \alpha$	$= P_{CO_2}$	$\frac{1-\alpha}{1+(1/2)\alpha}$
CO	0	$+\quad \alpha$	$-\quad 0$	$= P_{CO}$	$\frac{\alpha}{1+(1/2)\alpha}$
O_2	0	$+(1/2)\alpha$	$-\quad 0$	$= P_{O_2}$	$\frac{(1/2)\alpha}{1+(1/2)\alpha}$
			$P_{total}\quad = 1+(1/2)\alpha$		

(See Appendix B for set up of this Table)

$$K_p = \frac{\dfrac{\alpha}{1+\alpha/2}\left(\dfrac{\alpha/2}{1+\alpha/2}\right)^{1/2}}{\dfrac{1-\alpha}{1+\alpha/2}} \quad \Pi \quad 1 + 1/2 - 1$$

$$= \frac{(\alpha)(\alpha/2)^{1/2}}{(1-\alpha)(1+\alpha/2)^{1/2}} \quad (1)^{1/2}$$

Figure 3.1 does not show the reaction

$$CO_2 \rightleftharpoons CO + 1/2\ O_2$$

But the reverse reaction

$$CO + 1/2\ O_2 \rightleftharpoons CO_2$$

The K_p for this reaction is $1/K_p$ for the reaction as written for the problem (to see this you may wish to write out K_p for $CO + 1/2\ O_2 \rightleftharpoons CO$ and compare to the expression derived above.)

From Figure 3.1 $\log K_p = 0.2$ (for $CO + 1/2\ O_2 \rightleftharpoons CO_2$)

$$K_p = 1.58$$

$$K_p' = 1/K_p = 0.63 \qquad\qquad \text{(for } CO_2 \rightleftharpoons CO + 1/2O_2\text{)}$$

$$0.63 = \frac{(\alpha)(\alpha/2)^{1/2}}{(1-\alpha)(1+\alpha/2)^{1/2}}$$

Solving for α gives $\alpha = 0.57$

Mole fraction of equilibrium gas

$$CO_2 = \frac{1-\alpha}{1+1/2\alpha} = \frac{0.43}{1.285} = 0.34$$

$$CO = \frac{\alpha}{1+1/2\alpha} = \frac{0.57}{1.285} = 0.44$$

$$O_2 = \frac{\alpha/2}{1+1/2} = \frac{0.285}{1.285} = \frac{0.22}{1.00} = 0.22$$

The equilibrium constant contains the products of the reaction in the numerator and the reactants in the denominator. A high equilibrium constant indicates the equilibrium composition will have high concentrations of products and low concentrations of reactants. The reverse is true low value of K_p.

$K_p \gg 1.0$ reaction very favorable

$K_p = 1.0$ reaction moderately favorable

$K_p \ll 1.0$ reaction not favorable

For an exothermic reaction,* an increase in reaction temperature reduces the amount of products formed at equilibrium.
For an endothermic reaction an increase in reactor temperature
increases the amount of products formed at equilibrium. The
reaction $CO_2 \rightleftharpoons CO + 1/2O_2$ is endothermic and the equili..ium
constant and conversion decreased as the temperature was decreased. This will always be the case with endothermic reactions.

The system pressure does not affect the equilibrium constant, K_p, but does affect the equilibrium conversion. Equation 3.5 may be rearranged to give

$$\frac{y_R^r \, y_S^s}{y_A^a \, y_B^b} = K_p \, \Pi^{(a+b)-(r+s)}$$

If $(a+b)-(r+s)$ is positive (more moles of reactant than products) the term on the left hand side increases as pressure
increases. This increases the yield of products.

The opposite is true if $(a+b)-(r+s)$ is negative.

EXAMPLE 3.2   Calculate the equilibrium conversion for the
               reaction $C(s) + H_2O(g) = CO(g) + H_2(g)$ at 110
               and 34 atm for temperatures of 800°K, 1,000°K,
               and 1500°K. This reaction takes place by pass-
               ing steam through a bed of hot char.

SOLUTION 3.2

$$C(s) + H_2O(g) = CO(g) + H_2(g)$$

---

*Exothermic reaction gives off heat; endothermic reaction
absorbs heat.

Component	In	+	Gen	−	Consumed	=	Out	Mole Fraction
CO	0	+	α	−	0	=	P_{CO}	$\alpha/(1+\alpha)$
H_2O	1	+	0	−	α	=	P_{H_2O}	$(1-\alpha)/(1+\alpha)$
H_2	0	+	α	−	0	=	P_{H_2}	$\alpha/(1+\alpha)$
Totals	1	+	2α	−	α	=	P	

$$P = 1 + \alpha$$

$$K_p = \frac{(pp)_{CO}\,(pp)_{H_2}}{pp_{H_2O}} = \frac{(y)_{CO}\,(y)_{H_2}}{y_{H_2O}}\,\Pi^{1+1-1}$$

$$= \frac{\left(\frac{\alpha}{1+\alpha}\right)\left(\frac{\alpha}{1+\alpha}\right)}{\left(\frac{1-\alpha}{1+\alpha}\right)}\,\Pi = \frac{\alpha^2}{(1-\alpha)\,(1+\alpha)}\,\Pi$$

$$= \frac{\alpha^2}{(1-\alpha)^2}\,\Pi$$

From Figure 3.1

$K_p = 0.044$ @ 800°K; $K_p = 608$ @ 1,500°K; $K_p = 2.6$ @ 1000°K

For P = 1 atm, T = 800°K

$$0.044 = \alpha^2/(1-\alpha)^2$$

$$\alpha^2 = \frac{0.044}{0.956} = 0.046; \quad \alpha = 0.21$$

The values for all combinations are summarized below:

Π	T = 800°K	T = 1,000°K	T = 1,500°K
1	.21	.85	1.00
10	.07	.46	0.99
34	.04	.27	0.97

The results of the example problem provided above shows that the conversion can be increased by increasing the temperature or by lowering the pressure.

This brief discussion on chemical equilibrium acquaints the reader with the evaluation of the equilibrium conversion for a single reaction. In most cases more than one reaction takes place and several equilibrium constraints must be satisfied simultaneously.

3.2 CHEMICAL REACTION KINETICS

Thermodynamics will provide for the evaluation of the energy changes resulting from both physical and chemical changes. It allows for the determination of what can occur in an energy conversion process. It sets limits on what is possible. The maximum amount of work that can be done, the maximum chemical conversion that may be accomplished.

Thermodynamics does not provide answers to whether a reaction will actually occur or how long it will take to occur. The fact that a reaction can occur and that work can be extracted from an energy source provides the incentive to devise a real machine or a reactor system that will allow for the desired conversion to occur. In the design of real systems it is necessary to be able to evaluate how changes can occur.

The subject termed Chemical Reaction Kinetics investigates the rate that a chemical conversion takes place. The relationships provided in this section are not "Laws" as provided in the section on thermodynamics. The relationships interpret experimental data and provide information needed to design a reactor system. They are usually based upon experimental results.

The purpose of the chapter is to introduce some of the basic concepts used in chemical reactor design. It should provide an appreciation of the role played by kinetics, the nomenclature and some ability to utilize the concepts to understand why particular conversion systems are being developed.

The discussion will start by considering the gas phase reaction:

$$A(g) + B(g) \rightarrow P(g) \tag{3.7}$$

A and B are reactants

P is a product

(g) designates gas

The reaction goes essentially to completion (i.e., the reaction will continue until either A or B is consumed).

For a molecule of A to react with a molecule of B it is essential that they come close to each other. To help visualize how reactions occur, consider the following analogy. A billiard table contains black and white balls. The cushions are perfectly elastic. This means that a ball striking the cushion will leave the cushion at the same velocity as it arrived. The balls will continue to move at the same velocity until they hit another ball. When they hit another ball, assume either of two events will take place:

(1) If the balls are of like color, they will rebound and continue to move about the table.

(2) If the balls are of different color, they will stick to each other and will be removed from the table.

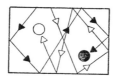

FIGURE 3.2 Billiard Ball Analogy

When two unlike colored balls strike each other and are removed
these will be termed a reaction. The number of balls on
the table will be termed a concentration (balls/unit table).

The reaction rate is then the number of reactions per
unit time (or the number of collisions between black and white
balls) and is represented by the symbol r. For a
given number of black balls on the table the probability
of a reaction increases directly with the number of white
balls present

$$r \propto C_W$$

C_W represents the concentration of white balls; r is the
reactions per unit time. The number of black balls that
react with concentration of the white balls held constant
is written

$$r \propto C_B$$

C_B represents the concentration of black balls. Combining
these equations:

$$r = + k \, C_W \, C_B \qquad (3.8)$$

If the velocity of the balls are increased the chance
of reaction per unit time are increased. The balls travel
greater distance and there is a greater chance for a

collision in a unit time. This increase in velocity affects
the value of k. Increasing the size of the balls increases
the probability of collision. This also affects k

$$k = f \text{ (velocity, size of balls)}$$

Leaving this diversion, consider the chemical reaction

$$B + W \longrightarrow P$$

the rate of reaction of molecules C and W may be written

$$r_B = r_W = -k \, C_B \, C_W \tag{3.9}$$

The rate that B or W (r_B or r_W) react is proportional to the
concentration of B and W molecules. The minus sign appears
because r is defined as the number of molecules produced from
a chemical reaction; B and W are consumed, thus, the minus
sign. The values C_W and C_B are in molecules per unit volume.
The number of molecules is proportional to the number of
gram-moles.

$$1\text{g-mole} = 6.023 \times 10^{23} \text{ molecules (Avagadro's Number)}$$

therefore, $C_B = (n_B/V)A$ where n_B is number of moles and A is
Avagadro's Number. For an ideal gas

$$n_B/V = pp_B/RT$$

where pp_B is the partial pressure of B.

$$C_B = (n_B/V)A = pp_B(A/RT) = k'pp_B: \quad \text{where } k' = A/RT$$
$$C_W = k'pp_W$$

Substituting these into Equation 3.9 gives:

$$r_A = -k' \, pp_W \, pp_B \qquad\qquad (3.10)$$

EXAMPLE 3.3   Consider the effect of tripling the pressure on a reactor system. Assume that the reaction rate r_A was found to be 1.0 kg-mole/m^3 at 298°K for the reaction

$$A(g) + B(g) \longrightarrow R(g) + S(g)$$

$$r_A = -k'' \, pp_A pp_B$$

SOLUTION 3.3   Original Reaction rate at P:

$$(r_A)_P = 1.0 \left[\frac{Kg\text{-mole}}{m^3}\right] = k'' \, pp_A pp_B$$

Reaction rate at 3 P:

$$(r_A)_{3P} = -k''(3pp_A)\,(3pp_B)$$

Ratio

$$\frac{(r_A)_{3P}}{(r_A)_P} = \frac{-k''(3pp_A)(3pp_B)}{-k'' \, pp_A pp_B} = 9$$

$$(r_A)_{3P} = (9) \times 1.0 \,[kg\text{-mole}/m^3] = 9 \,[kg\text{-mole}/m^3]$$

The effect of pressure on a gas phase reaction is to increase the reaction rate. Another way to increase the reaction rate is to increase the constant k". In the simple analogy to billiard balls the rate increased when the velocity increased. The kinetic theory of gases shows that the velocity of molecules increases with the temperature. The change of k" with temperature follows the expression

$$k'' = k''_o e^{-E/RT} \qquad\qquad (3.11)$$

The value E has the units of energy per unit mass, R is the ideal gas constant.

EXAMPLE 3.4 A typical reaction rate constant doubles when the temperature increases by about 15°C at 300°K. For this change, what is the value of E?

SOLUTION 3.4

$$k''(315) = k''_o e^{(-E/315R)}$$

$$k''(300) = k''_o e^{(-E/300R)}$$

$$\frac{k''(315)}{k''(300)} = 2 = \frac{k''_o e^{(-E/315R)}}{k''_o e^{(-E/300R)}} = e^{[-E/R(1/315 - 1/300)]}$$

$$2 = e^{[-E/R[(300-315)/(315 \times 300)]]} = e^{-[(E/R)[-15/(300 \times 315)]]}$$

$$\ln(2) = -(E/R)[-15/(300 \times 315)]$$

$$E/R = 4370$$

$$E = (4370)(8314)[J/Kg\text{-mole}] = 363 \times 10^8 [J/Kg\text{-mole}]$$

Another method to increase the reaction rate is to provide a catalyst. Returning once again to the billiard ball analogy the effect of the catalyst is to increase the effective size of the reacting balls (molecules). This is particularly valuable when there are several competing reactions such as

$$A + B \xrightarrow{k''_1} R \text{ (desired product)} \qquad\qquad (3.12)$$

$$A + B \xrightarrow{k''_2} S \text{ (waste product)} \qquad\qquad (3.13)$$

where R is the desired product and S is a waste product. The distribution ratio is defined as the rate the desired product is generated to the rate the undesired product is generated.

$$\Phi = \frac{r_R}{r_S} = \frac{k''_1 \, pp_A pp_B}{k''_2 \, pp_A pp_B} = \frac{k''_1}{k''_2} \qquad\qquad (3.14)$$

The use of a catalyst will increase k''_1 faster than k''_2 and increase the distribution ratio.

EXAMPLE 3.5   The distribution ratio for the pair of reactions.

$$A + B \xrightarrow{k''_1} R \quad \text{(desired)}$$
$$A + B \xrightarrow{k''_2} S \quad \text{(waste)}$$

is found to be 0.5. Two catalysts are available. Catalyst A has no effect on k''_1, but doubles k''_1. Catalyst B increases k''_1 by 250%, but increases k''_2 by 50%. Which catalyst would be recommended?

SOLUTION 3.5   Without catalyst

$$\Phi = k''_1/k''_2 = 0.5$$

$$k''_1 = 0.5 \ k''_2$$

With catalyst A

$$\Phi = 2 \ k''_1/k''_2 = 2(0.5) = 1.0$$

With catalyst B

$$\Phi = 2.5 \ k''_1/1.5 \ k''_2 = 1.667 \ k''_1/k''_2 = 1.667 \ (0.5) = 0.833$$

Catalyst A preferred as it provides the best product distribution.

3.2.1 Reversible Chemical Reactions

The discussion thus far has been limited to reactions that are irreversible and go to completion. Most reactions do not fall into this class but are reversible and approach some equilibrium value.

$$A + B \underset{k''_2}{\overset{k''_1}{\rightleftharpoons}} R + S$$

This reaction equation indicates that A + B react to form R + S at a rate proportional to k''_1 and that R + S react to form A + B at a rate proportional to k''_2.

$$r_{forward} = k_1'' \, pp_A \, pp_B$$

$$r_{backward} = k_2'' \, pp_R \, pp_S$$

the net rate that the products R (or S) is given by

$$r_R = k_1'' \, pp_A \, pp_B - k_2'' \, pp_R \, pp_S \qquad (3.15)$$

At equilibrium the rate that R forms is equal to the rate that R reacts to form A + B and the net formation is 0 (there is no change in concentration of R)

$$r_R = 0 = k_1'' \, pp_A \, pp_B - k_2'' \, pp_R pp_S$$

$$\frac{pp_R \, pp_S}{pp_A \, pp_B} = \frac{k_1''}{k_2''}$$

The term on the left hand side is the chemical equilibrium constant (see Equation 3.3).

$$K_p = \frac{pp_R \, pp_S}{pp_A \, pp_B} = \frac{k_1''}{k_2''} \, ; \, k_2'' = \frac{k_1''}{K_p} \qquad (3.16)$$

The effect of temperature on the equilibrium constant for an exothermic reaction is shown in Figure 3.1 (for all exothermic reactions the equilibrium constant decreases with temperature). Since the equilibrium constant decreases, the equilibrium conversion of A decreases as the temperature increases. At low temperature complete conversion is possible while at high temperature little conversion is possible. This is shown by the curve labeled "Equilibrium" in Figure 3.3.

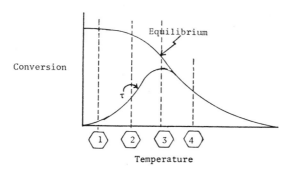

FIGURE 3.3   Conversion for Typical Exothermal Reaction
$A(g) \longrightarrow P(g)$

At low temperature, however, the reaction rate is very slow and little conversion is achieved. As the temperature increases the reaction rate increases. The curve labeled, τ in Figure 3.3 is the actual conversion achieved when the reaction is allowed to take place for a fixed time. At low temperature the equilibrium relation favors complete conversion but the reaction rate is so slow that almost no conversion is achieved (see T_1). At increased temperature the reaction rate increases and significant conversion is achieved (see T_2). At higher temperature the reaction rate is so fast that equilibrium is close to being achieved (see T_4). Unfortunately, at this temperature the equilibrium conversion is low.

For an exothermic reaction a compromise between high equilibrium conversion and rapid reaction rates must be made. For the typical system shown in Figure 3.3 the temperature T_3 gives the maximum conversion for a reaction time.

3.2.2 Heterogeneous Reactions

The discussion above has considered a gas phase reaction. Many common reactions take place between a solid and a gas

$$A(solid) + B(gas) \longrightarrow R(gas) \tag{3.17}$$

This is called a heterogeneous reaction as it occurs between
two distinct phases.

The reaction between two phases is complex and the
discussion that follows will consider the solids to be in
the shape of slabs. They are more often analyzed as spheres
but this adds mathematical complications that are avoided
by considering the slab geometry. The more complex analysis
adds little to the appreciation of most of the factors
to be considered in the analysis of heterogeneous reactions.

Figure 3.4 shows a heterogeneous reaction that occurs
between a gas phase and solid phase as more time for reaction
is allowed. Several time frames are shown.

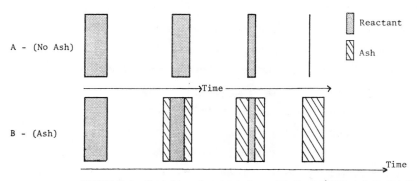

FIGURE 3.4   Shrinking Reactant Core (A) No-Ash (B) Ash Build Up

For the first row of Figure 3.4, no solid layer of
non-reacting product builds upon the surface and the gaseous
reactor has direct access to the solid reactant. For the
second row as the solid reacts a non-reacting solid layer
builds up on the surface. The gaseous reactant must be able
to penetrate this layer in order to react. The ash layer
protects the solid reactant from the reactant gas and retards
the reaction. A familiar example is a charcoal bar-b-que.
The white ash retards the combustion. If the ash is blown

or shaken off the char the reaction rate increases (and more heat is released).

The non-reactive (or ash) layer serves as a resistance to the transfer of the gaseous reactant to the surface of the solid reactant.

The rate of reactant flow can be expressed as

$$\text{Rate} = \frac{\text{Driving Force}}{\text{Resistance}} \text{ or Driving Force x Conductance}$$

For an electric circuit this may be written

$$I(\text{current}) = \frac{V(\text{voltage})}{R(\text{resistance})} \qquad (3.18)$$

where I the current is the electron flow rate, V is the potential for electron flow and R is the resistance. When several resistances are placed in series

Equation 3.8 is written:

$$I = \frac{\Delta V}{\Sigma R} = \frac{V_1 - V_4}{R_1 + R_2 + R_3} \qquad (3.19)$$

Since there is no accumulation of electrons in the wire or resistance

$$I = \frac{V_4 - V_3}{R_3} = \frac{V_3 - V_2}{R_2} = \frac{V_2 - V_1}{R_1} \qquad (3.20)$$

Similar equations may be used for studying a heterogeneous reaction. The driving force is the concentration of the gaseous reactant. There are several resistances to the reaction of this gaseous reactant.

1. Gas film--there is a stagnant gas layer on the surface of the solid. This offers a resistance to the reactant gas.

2. Ash layer--this has been discussed above.

3. Chemical reaction--once the gas has reached the solid surface there is a resistance to the actual reaction.

These resistances to a heterogeneous chemical reaction and an electrical analogy are shown in Figure 3.5.

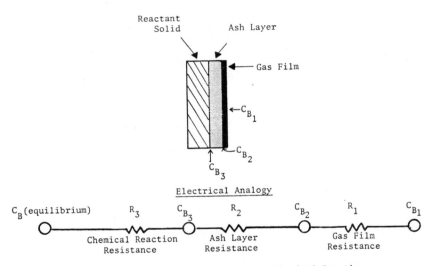

FIGURE 3.5 Resistance to Gas-Solid Chemical Reaction

Electrical current represents the number of electrons flowing through a resister per unit time. By analogy the chemical current represents the molecules flowing through a resister per unit time. This chemical current is given by the product,

$$\text{Total Chemical Current of } B = J_B \times A_C \qquad (3.21)$$

J_B (called flux) is the number of molecules of B passing through a unit area in a unit time [molecules/area x time]. A_C is the cross sectional area through which the flux $[J_B]$ passes.

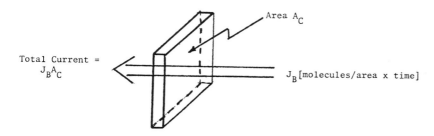

By analogy with electrical current (see Figure 3.5).

Current = Driving Force/Resistance

$$\text{Current} = J_B A_C = [C_{B_1} - C_B(\text{equilibrium})]/(R_1 + R_2 + R_3)$$

$$= [C_{B_1} - C_{B_2}]/R_1 = [C_{B_2} - C_{B_3}]/R_2$$

$$= [C_{B_3} - C_B(\text{equilibrium})]/R_3 \qquad (3.22)$$

As in the electrical system where there is no accumulation of electrons in the resistances there is no accumulation of B molecules in the resistances for the flow of B.

The definition of the reaction rate, r, given in Eq. 3.8 is the molecules produced per unit time per unit volume. For the heterogeneous system, the value of r becomes

$$r_B = -(J_B \times A_C)/V \qquad (3.23)$$

The negative sign results from the fact that B is consumed (see reaction 3.17). The reaction rate is proportional to the flux and to the area for reaction per unit volume (A_C/V). If the value of (A_C/V) is constant, the reaction rate is proportional to the flux, J_B.

Resistance to chemical reaction: The flux, J_B, is proportional to the difference between the concentration at the reacting surface C_{B_3} and the concentration at equilibrium.

$$J_B = k_R [C_{B_3} - C_B (\text{equilibrium})] \qquad (3.24)$$

From Equation 3.22

$$J_B = [C_{B_3} - C_B (\text{equilibrium})]/A_C R_3$$

Solving for R_3

$$R_3 = 1/(A_C k_R) \qquad (3.25)$$

Resistance to ash layer: Fick's Law of Diffusion states that the flux is directly proportional to the concentration difference over the solid layer and inversely proportional to the thickness of this layer. Referring to Figure 3.5

$$J_B = D(C_{B_2} - C_{B_3})/L \qquad (3.26)$$

where D is the proportionality constant (called diffusivity) and L is the ash thickness.

From Equation 3.22

$$J_B A_C = (C_{B_2} - C_{B_3})/R_2$$

Solving for R_2

$$R_2 = L/DA_C \qquad (3.27)$$

Resistance to gas film: The flux through a gas film is directly proportional to the concentration difference across the gas layer.

$$J_B = k_g (C_{B_1} - C_{B_2}) \qquad (3.28)$$

where k_g is a gas film coefficient.

From Equation 3.22

$$J_B A_C = (C_{B_1} - C_{B_2})/R_1$$

Solving for R_1

$$R_1 = 1/(k_g A_C) \tag{3.29}$$

Total resistance: The total resistance

$$R_T = R_1 + R_2 + R_3$$

can be obtained from adding the R values (Eq. 3.25, 3.27, and 3.29).

$$R_T = 1/A_C k_R + L/D A_C + 1/k_g A_C \tag{3.30}$$

and Equation 3.22 becomes

$$\text{Current} = J_B A_C = [C_{B_1} - C_B(\text{equilibrium})]/$$
$$[1/k_R A_C + L/DA_C + 1/k_g A_C] =$$
$$[C_{B_1} - C_{B_2}]/[1/k_g A_C] = [C_{B_2} - C_{B_3}]/[L/DA_C]$$
$$= [C_{B_3} - C_B(\text{equilibrium})]/[1/k_R A_C] \tag{3.31}$$

An important consideration for any chemical reaction is the effect of temperature. A change in temperature has an effect on

a) The value $C_B(\text{equilibrium})$

b) The reaction rate constant, k_R

The effects on D and k_g are small in comparison to the effect on k_R. In most cases the effect on k_R is shown by Equation 3.11.

Figure 3.6 shows the effect of temperature on the resistance to a chemical reaction. In this figure it is assumed that the reaction will go to completion (as predicted by thermodynamics) i.e., $C_B(\text{equilibrium}=0)$. The gas film resistance remains constant and not influenced by temperature. Three ash thicknesses were considered (thicknesses 0, L, 10L).

To understand Figure 3.6 consider the equation for the

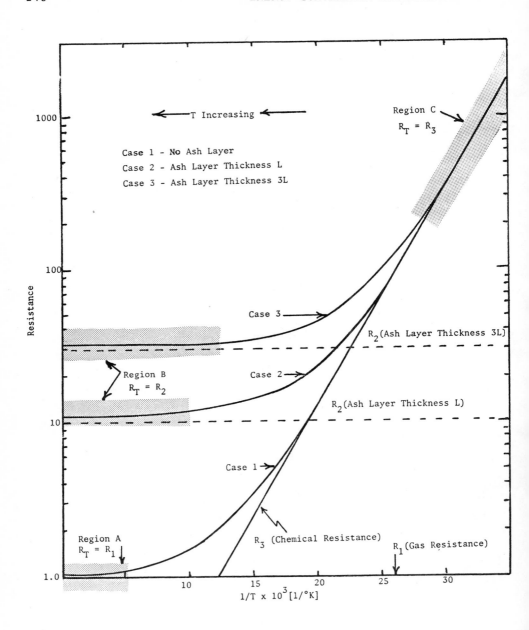

FIGURE 3.6   Effect of Temperature on the
Resistance to Chemical Reaction

total resistance:

$$R_T = R_1 + R_2 +$$
(gas resistance)   (ash resistance)

$$R_3$$
(chemical resistance)

At low temperature as can be seen in Figure 3.6

$$R_T = R_3$$

The total resistance is equal to the chemical resistance.
The chemical resistance $R_3 = 1/(A_C k_R)$ [Eq. 3.25]. The
reaction rate constant, k_R, decreases rapidly with a lowering of temperature (see Eq. 3.11) and therefore R_3 increases.
When $R_T = R_3$ is is said that the "reaction rate is <u>controlled</u>
by chemical reaction." This is region (C) in Figure 3.6.

If there is no ash film (L=0; R_2=0) as the temperature
increases the value of R_3 goes to zero and $R_T = R_1$. The
resistance is due almost entirely to the resistance resulting from the gas film. It is said that the "reaction rate
is <u>controlled</u> by the gas film." This is shown as region (A)
in Figure 3.6.

If there is a thick layer of ash (L=10) the total resistance at some temperatures becomes the major resistance

$$R_T = R_2$$

This then becomes the controlling mechanism for the reaction
and is shown in region (B).

As the ash layer builds up during the course of reaction (see Figure 3.3), the controlling resistance may shift
from chemical reaction or gas layer to ash layer controlled.
When a single mechanism controls the analyses of the reactor
system it is much easier.

3.2.3 Flow Patterns in Chemical Reactors

The amount of chemical reaction that takes place depends
upon both
a.   the rate of chemical reaction, and
b.   the time the reactants remain in the reactor.
The rates of chemical reaction and the effects of some of
the important variables were discussed in the previous
section. The purpose of this section is to determine how
long the reactants are retained in the reactor.

To determine the length of time that would be available
for a chemical reaction a tracer experiment might be performed. In a tracer experiment a small amount of a foreign
material that clearly distinguishes it from the rest of
the stream is introduced into the inlet to the reaction
vessel. The tracer material does not react and all of
the tracer added must appear in the exit stream. A
monitoring station is set up at the exit stream and the
amount of the tracer leaving per unit time as a function
of time is determined. For the experiments that will
be described below the tracer will be added as a pulse.
To help visualize what is happening, picture a pipe through
which water flows. A syringe is loaded with ink. The
point of the syringe is inserted into the water stream at
the inlet pipe to the reactor. The syringe is emptied into the
water stream as rapidly as possible and a stop watch is
started. A monitoring station is placed at the end of the
pipe and color intensity of the water leaving the pipe is
measured. This is shown in Figure 3.7.

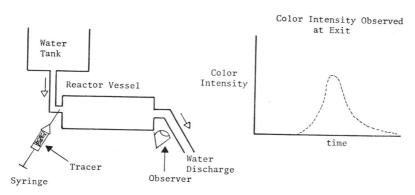

FIGURE 3.7 Typical Tracer Experiment

The output curve is typical of what might have been anticipated by the reader. There is a period of time for the tracer (ink) to be observed at the exit. During this time there is no color in the exit stream. There is some mixing while the stream moves through the pipe and the tracer does not come out "all at once" in the manner it was introduced but comes out over a range of time. This is shown in the output curve.

It is clear that not all of the tracer was retained in the system for the same period of time. Some tracer was retained a far shorter period of time than other portions of the tracer. It might be said that when the tracer was added into the system that the tracer was "born" (introduced into the system). When the tracer left the system in the exit it might be said that the system suffered a death (removed from the system). The time that transpired between the birth and death of the tracer in the system (the adding and removing of the tracer from the system) is defined as the age of the tracer. By measuring the age of the tracer exiting the reactor an exit age distribution can be determined. An analogy to man might be useful to understand the age distribution. The "Exit Age

Distribution" is the distribution that would be obtained by
St. Peter if he collected the age of each person passing
through the Pearly Gates. (Leaving system earth)

There are many flow patterns that are found in chemical
reactors. Two extreme cases will be described in this section.
They represent the two limits and are used most often in
describing reactors.

Consider the flow system shown in Figure 3.8. The
level in the tank does not change and the system is at
constant volume.

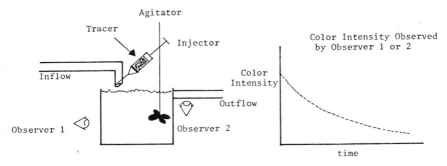

FIGURE 3.8
CONSTANT VOLUME STIRRED TANK REACTOR

The system is well agitated and the contents are completely
mixed. This is referred to as a constant volume stirred
tank reactor (or sometimes just stirred tank or a completely
backmixed reactor). As soon as the tracer enters the system
it becomes mixed with the contents of the reactor. If the
tracer were dark blue ink as soon as it reached the tank
(was born) it became mixed with the contents and became
light blue. An observer (St. Peter) would see some of the
tracer at the exit (Pearly Gates) almost instantly. Some
of the tracer has an exit age of zero. They die immediately
after they are born. As some of the tracer leaves there

is less in the system and the color decreases. Some of the tracer is retained until a very old age. Figure 3.8 shows two observers. One views the exit stream and one the tank contents. The tracer intensity of the water flowing out of the reactor is identical to the intensity of that in the tank. <u>What leaves is identical to that in the vessel.</u>

In contrast to the constant volume stirred tank reactor is the system referred to as plug flow reactor. In this type of reactor no mixing whatsoever occurs. This is shown in Figure 3.9. All tracer leaves the system (death of tracer) at the same time. All the tracer has the same age.

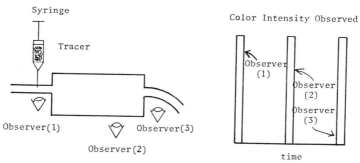

FIGURE 3.9 Plug Flow Reactor

In most all cases the age distributions can be described by a series of plug flow and constant volume stirred tank reactors placed in a parallel and series combination.

<u>EXAMPLE 3.6</u>   A particular reactor may be approximated by the following system (in parallel)

1. 20% of the inlet bypassed the reactor.
2. 60% of the material is completely mixed.
3. 20% of the material undergoes no mixing.

Sketch the exit age distribution. The inlet tracer concentration is given.

SOLUTION 3.6

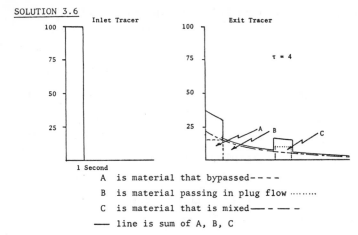

A is material that bypassed-- - -
B is material passing in plug flow ·········
C is material that is mixed—— - —— -
—— line is sum of A, B, C

3.2.4   Reactor Design

The previous sections discussed some of the basic flow
patterns for flow systems and the determination of the rate
constants for chemical reacting systems. They did not cover
how to determine the size of the reactor once the flow pattern and the reaction rate expression have been determined.
This section will develop the design equations for the constant volume stirred tank reactor and the plug flow reactor
(for the isothermal reaction). Consider the reaction $A \longrightarrow P$

$$r_A = -k_1 C_A \qquad\qquad\qquad (3.32)$$

There is no change in volume as the reaction proceeds.
This is representative of a liquid phase reaction or of a gas
reaction where there is no change in the number of moles as
the reaction takes place.

Figure 3.10 shows a constant volume stirred tank reactor.
V is the volume of the reactor, v_o is the volumetric flow
rate through the reactor, r_A is the reaction rate (gram-moles/
unit volume-unit time) and C_{A_o} and C_{A_f} the initial and final
concentration of the reactant (grams-moles/unit volume).

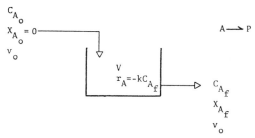

FIGURE 3.10 Constant Volume Stirred Tank Reactor

Writing a material balance for the reactant, A, over the reactor gives:

Input + Generation = Output + Consumption

Input = C_{A_o} (gram-moles/unit volume) x v_o (volume/time)

Output = C_{A_f} (gram-moles/unit volume x v_o (volume/time)

Generation = No A was generated

Consumption = r_A (gram-moles/volume-time) x V(volume))

$$C_{A_o} v_o + 0 = C_{A_f} v_o + (-r_A V) \qquad (3.33)$$

Rearranging Equation 3.33

$$V/v_o = [C_{A_f} - C_{A_o}]/r_A$$

Substituting the expression $r_A = -k_1 C_{A_f}$

$$V/v_o = (C_{A_f} - C_{A_o})/(-k_1 C_{A_f}) = \tau^* \qquad (3.34)$$

EXAMPLE 3.7   For a given system it was determined that for 90% conversion that the residence time was 60 minutes. It is necessary to increase the conversion to 95%. Two methods have been suggested:

1. Build a new reactor, and
2. Add another constant volume stirred tank reactor to the existing system.

The flow rate is the same for a new system with the 95% conversion. Determine the total volume for each reactor system alternative.

$^*\tau$ - defined as V/v_o.

Alternative 1

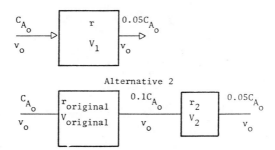

Alternative 2

SOLUTION 3.7

Let C_{A_o} represent the initial concentration in all cases. The concentration after 90% has reacted in 0.1 C_{A_o} and after 95% has reacted is $0.05C_{A_o}$. Let τ_{new} represent the residence time for the new reactor and $\tau_{original}$ represent the residence time of the original system.

$$\frac{\tau_{new}}{\tau_{original}} = \frac{[V/v_o]_{new}}{[V/v_o]_{original}} \quad \frac{[(C_{A_f} - C_{A_o})/+r_A]_{new}}{[(C_{A_f} - C_{A_o})/+r_A]_{original}}$$

$$= \frac{[(0.05-1.0) \, C_{A_o}/ \, (k_1 \, 0.05C_{A_o} \,]_{new}}{[(0.10-1.0) \, C_{A_o}/ \, k_1 0.1C_{A_o} \,]_{original}} = 2.11$$

Since v_o is the same in both cases

$V(new) = 2.11 \, V(original)$

Alternative 2--for the second reactor

$(V/v_o)_2 = [(C_{A_f} - C_{A_o})/-r_A] = (0.05-0.10)/(-k_1 0.05)$

$= 1/k_1$

For the original reactor

$(V/v_o) = (0.1-1)/0.1k_1 = 9/k_1$

$$\frac{V_{(new \, system)}}{V_{(original)}} = \frac{V_1 + V_2}{V_1} = \frac{(V/v_o)_1 + (V/v_o)_2}{(V/v_o)_1}$$

$=(9/k_1 + 1/k_1)/9/k_1 = 10/9 = 1.11$

$V(new) = 1.11 \, V(original)$

The basic disadvantage in using the constant volume stirred tank reactor lies in the fact that the concentration of the feed (C_{A_o}) immediately drops to the concentration of the exit stream. This was shown in the tracer test discussed in the previous section. The rate of reaction in the reactor is proportional to this concentration. For high conversion the concentration in the reactor is low, the reaction rate is low and the reactor becomes large. This can be seen from the previous example problem. When the higher yield was sought by carrying out the reaction in a single constant volume stirred tank reactor, the concentration throughout the whole reactor was cut to half the original concentration and the reactor size more than doubled. In the second case the original reactor was not altered and the original concentration retained. Only in the second reactor was the concentration and, thus, the reaction rate reduced.

The smallest reactor size would be achieved if the reactant was not mixed or diluted with the contents in the reactor. The plug flow reactor system has no mixing. The only reason for a reduction in the concentration of the reactant is that it has been consumed in the reaction. Figure 3.11 provides a schematic flow sheet for a plug flow reactor along with a curve of the typical concentration of the reactant A throughout the reactor. In the constant volume stirred tank reactor the concentration throughout the reactor was constant while in the plug flow reactor the concentration is reduced as the reactant moves through the reactor. To determine the amount of reactant that is consumed, it is necessary to determine the change over a small element of reactor length, and integrate over the length of the reactor.

Writing a material balance over a small element of

length shown in Figure 3.11.

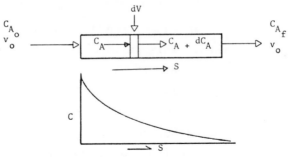

FIGURE 3.11   Plug Flow

where dV (volume) = cross sectional area of
reactor times the distance (AdS)

Input	+	Generation	=	Output	+	Consumpution
$C_A v_o$	+	0	$= v_o(C_{A_o} + dC_A) + r_A dV$			

$$v_o dC_A + r_A dV = 0$$

$$\int_0^V dV/v_o = \int_{C_{A_o}}^{C_{A_f}} dC_A / [-r_A] \qquad (3.35)$$

Defining τ as before

$$\tau = V/v_o = \int_{C_{A_o}}^{C_{A_f}} dC_A/[-r_A] \qquad (3.36)$$

substituting the expression for the reaction rate $-r_A = k_1 \, C_A$
and carrying out the integration

$$\tau = (V/v_o) = - \int_{C_{A_o}}^{C_{A_f}} dC_A/k_1 C_A = [1/k_1] \ln [C_{A_f}/C_{A_o}] \qquad (3.37)$$

This represents the design equation for the plug flow reactor.

EXAMPLE 3.8   For Example 3.7, determine the ratio of $V_{plug\ flow}$ to V_{CSTR} for the reaction given

SOLUTION 3.8

$$\frac{\tau(\text{Plug Flow})}{\tau(\text{CSTR})} = \frac{-\frac{1}{k_1} \ln \frac{C_{A_f}}{C_{A_o}}}{\frac{C_{A_f} - C_{A_o}}{-k_1 C_{A_f}}} = \frac{-\frac{1}{k_1} \ln \frac{0.1C_{A_o}}{1.0C_{A_o}}}{\frac{(0.1 - 1.0)\ (C_{A_o})}{-k_1\ 0.10C_{A_o}}} = .255$$

The size of the plug flow reactor shown in the example given above is only about one quarter the size of the constant volume stirred tank reactor. The difference between these volumes can be seen graphically in Figure 3.12. In this figure $1/-r_A$ is plotted vs. the concentration. For a low concen-

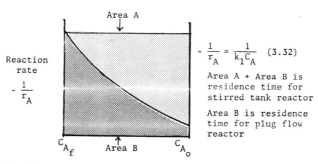

$$-\frac{1}{r_A} = \frac{1}{k_1 C_A} \quad (3.32)$$

Area A + Area B is residence time for stirred tank reactor

Area B is residence time for plug flow reactor

FIGURE 3.12   Time Required to Convert Reactant A from Conversion C_{A_o} to C_{A_f} in Either A Plug Flow or Stirred Tank Reactor System

tration the rate, r_A, is small. From Equation 3.36, the area under the curve between the limits of C_{A_f} and C_{A_o} is the residence time τ (V/v_o). The area within the rectangle with a height of $1/k_1 C_{A_f}$ $(1/[-r_A])$ and a width of C_{A_o} to C_{A_f} (see Equation 3.34) is the τ for the constant volume stirred tank reactor. It is clear that the constant volume stirred tank reactor will always have a larger size, and the ratio of sizes will become larger as the final concentration approaches zero.

Most problems are more complex than the ones presented
in this section. The rate expression is often more complex
and the integration more difficult in the case of the plug
flow reactor. The flow nature of the reactor system is
neither of the two discussed. The fact is that chemical
reactor design is only part science, part art. In most cases
pilot reactors will be operated to provide substantial
evidence of a new systems performance before a commercial
system is considered.

Table 3.1 provides more equations that may be used to
estimate the size of chemical reactors.

3.3   REACTION SYSTEMS (SOLID-GAS)

This section describes many types of solid-gas reactor
systems that are used. Many of the reactions discussed in
the subsequent chapters make use of these reactor systems.
The purpose in this section is to introduce the various
reactor systems and describe some of the unique characteristics of each system. It will develop a familiarization with
the various systems and will develop the nomenclature that
will be used in describing energy conversion systems.

In reading the material to be presented two overriding
needs for most any chemical reactor should be kept in mind
(for a solid-gas reaction).

a.   In order to have the reaction take place the gas
must come into contact with the solid.

b.   As most chemical reactions either release significant
quantities of energy or absorb significant quantities of energy, there must be a means to remove or add energy to the reactor.

The reactor must be a material handling device, a solids-gas
contactor, and a heat exchanger. There are several batch or

TABLE 3.1

Summary of Design Expressions for Single Ideal Reactors
($r_A = -kC_A^n$, A → Products)

Type of Reactor	Capacity Measure	General Design Relationship	Design Equation/w $V = V_0(1 + \varepsilon_A X_A)$	Design Equation for Isothermal nth Order Reaction/w $V = V_0(1 + \varepsilon_A X_A)$
Batch V = Const	t	$= C_{A0} \int \dfrac{dX_A}{-r_A}$		$= \dfrac{1}{kC_{A0}^{n-1}} \int \dfrac{dX_A}{(1-X_A)^n}$
Batch = Const	t	$= N_{A0} \int \dfrac{dX_A}{V(-r_A)}$	$= C_{A0} \int \dfrac{dX_A}{(1 + \varepsilon_A X_A)(-r_A)}$	$= \dfrac{1}{kC_A^{n-1}} \int \dfrac{(1 + \varepsilon_A X_A)^{n-1}}{(1-X_A)^n} dX_A$
Plug Flow	τ	$= C_{A0} \int \dfrac{dX_A}{-r_A}$	$= C_{A0} \int \dfrac{dX_A}{-r_A}$	$= \dfrac{1}{kC_{A0}^{n-1}} \int \dfrac{(1 + \varepsilon_A X_A)^n}{(1-X_A)^n} dX_A$
Backmix reactor	τ	$= \dfrac{C_{A0} X_A}{-r_A}$	$= \dfrac{C_{A0} X_A}{-r_A}$	$= \dfrac{1}{kC_{A0}^{n-1}} \dfrac{X_A(1 + \varepsilon_A X_A)^n}{(1-X_A)^n}$

V - volume; X_A - fraction reacted; C_A - concentration of A;

N_A - moles of A; r_A - reaction rate; ε_A - fractional change in volume
due to reaction; n - reaction order; subscript 0 - initial value.

semicontinuous systems that have been used but the discussion
below is limited to the continuous systems.

3.3.1 Vertical Moving Bed

The reactor is composed of a vertical vessel. Solids
are introduced at the top and move slowly down through the
vessel and are removed. The solids move in plug flow
from the top to the bottom. The reacting gas may flow upward
through the bed of particles or downward (either counter-

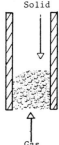

FIGURE 3.13 Vertical Moving
Bed Type

current or parallel to the solid flow). The gas flow is
essentially in plug flow and flows through the spaces
between the solid particles.

The gas flow is distributed evenly over the bottom (or
top) of the bed. Ideally the gas would flow around each
particle (except where it contacted other solid particles) ,
and most of the solid surface would be available for contact
with the reactant gas. Unfortunately, the bed usually forms
channels for the gas. It also forms dead spots where no
gas reaches. The gas becomes poorly distributed and the
solid-gas contacting in a moving bed is likely to be poor.

In the locations where the reaction occurs large heat
effects provide either hot or cold spots in the reactor. The
cold spots are self-regulating. As the temperature falls

the reaction slows down and less heat is absorbed. For exothermic reactions the reaction rate increases as the temperature increases releasing more heat and making the region even hotter. The heat can be carried away from the hot region by the sensible heat of the gas. Gas has a low volumetric heat capacity, and the heat removal potential is poor. The reaction will continue rapidly in these hot spots until the gaseous reactant is consumed. There is almost no possibility of placing any heat transfer surface within the system as it will interfere with the movement of the particles.

The advantage of the moving bed is in its structural simplicity. It can be built to operate at elevated pressures. The disadvantages of the system are those associated with the poor gas distribution and poor temperature control. It is a complex system in terms of reactor design.

3.3.2 Horizontal Moving Bed

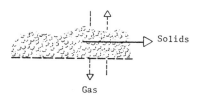

FIGURE 3.14 Horizontal Moving Bed Type

In the horizontal moving bed the solids are placed on a horizontal moving distributor. The solids move through the reactor in plug flow and all solid particles spend the same time within the reactor system. The gas passes either upward or downward through the solid bed. The gas may be considered to flow in plug flow. The gas flow is cross-flow to the solids (in contrast to either parallel flow or countercurrent flow in the vertical moving bed). The bed is much

more shallow than the vertical moving bed, and the gas passing
through the bed has less opportunity to contact the solid.
The residence time of the gas in the reacting region is low.
The problem of channeling exists in the horizontal moving
bed but to a lesser degree as the channeling becomes more
prevalent as the depth increases. The horizontal moving bed
has less pressure drop than the vertical bed.

The gas is distributed over the bottom (or top) of the
bed. The gas that passes through the solid bed toward the
end where the solid enters passes through only the reacting
solids. The gas passing through the bed where the solid exits
passes through little or no reactant solid and little or no
conversion of the gas can occur. The amount of gas that
reacts varies in the direction of flow of the solids.

The advantage of this system lies in that the solid
residence time in the reactor can be easily controlled by
controlling the velocity of the moving grid.

The hot spots are not as severe as in the vertical unit.
In the vertical unit all of the gas passes through all of the
solids. When a hot spot develops there will be a large amount
of gas supplied to the region of the hot spot. In the horizontal bed there is a relatively small amount of gas supplied
over any cross section and the gas supply is the more readily
depleted.

The horizontal moving bed provides for less gas solid
contact. The gas has a limited opportunity to contact the
solid as it passes only a short distance through the solid
bed. The final gas product is a mixture of gases produced
from passing the gas through varying amounts of solid reactant.
The system cannot achieve high gas conversions but can achieve
high solid conversions. Little energy can be added to or
removed from the bed without interfering with the movement
of the solids. The reactor is an extremely good system for

solids handling which is important in some systems.

3.3.3. Flat Hearth Type

FIGURE 3.15 Flat Hearth Type

The flat hearth is composed of a moving bed of solids as in the horizontal moving bed. The gas flow is over the top of the bed. The flow may be parallel, counter or cross flow to the solids. The gas can only contact the solids at the top of the bed. If the solid reaction leaves no ash or if the ash is easily dislodged by the gas flow then new solid surface is continually exposed to the gas for reaction. If an ash is retained then the gas must diffuse downward into the solid bed.

The solid gas contact in the horizontal moving hearth is poor. For most of the solids residence time in the reactor the solid does not have any chance to contact the gas and react.

The flat hearth type reacting system would have a large cross sectional area to provide the necessary surface for good contact between the gaseous and solid reactants.

3.3.4 Rotary Type

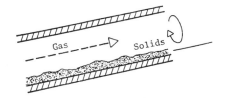

FIGURE 3.16 Rotary Type

The reactor is composed of a cylindrical vessel that is inclined to the horizontal. The cylindrical vessel is slowly rotated. The inside of the vessel contains ridges (called flights) that carry the solid up the walls of the cylinder where they drop off as the ridge reaches the top. The solid rains down through the gas that is passed through the cylinder. The tumbling action assures that the solid is contacted with the gas. The gas may be passed either parallel to or counter to the flow of the solid.

The rotary system provides good gas solid contact. Hot spots to the extent that they will form in the vertical moving bed do not occur in the rotary system. The residence time for the solids in the reactor varies. Some particles may move more rapidly through the system than others. Although the solid is effectively exposed to the gas, all of the gas does not have the opportunity to contact the solid. The vessel is mostly gas with solids tumbling through the gas.

The average residence time of the solids in the reactor can be controlled by changing the speed of rotation or the inclination of the vessel. It would be possible, but difficult to install heat transfer surface in the gas space in the reactor system and to remove or to add heat.

The rotary system is mechanically complex because of the problem of sealing a rotary reactor but does provide a good environment for the solid to react.

3.3.5 Fluidized Bed

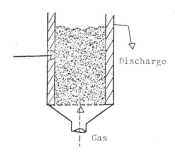

FIGURE 3.17   Fluidized Bed Type

The fluidized bed consists of a vertical vessel. At
the bottom of the vessel is a grid plate. The plate supports
the solid bed and serves to distribute the gas that is introduced below the plate. The velocity of the gas is adjusted
to a velocity that is less than the terminal velocity of the
particles but large enough to keep the particles in suspension.
In the fluidized state each of the particles is surrounded by
a layer of gas. The particles are not static but move about in
the bed. These solid particles may be treated as a completely
mixed system. The solids introduced into the bed are removed
from the bed in either of two manners. For particles where
there is no ash or there is an ash that is easily dislodged, the
solid size is reduced as the reaction proceeds, and some unreacted solid will be carried from the reactor with the
fluidizing gas along with the ash. When this is not the case,
the solids must be removed from the bed. Because the solids
are completely mixed, the exit stream will contain some of the
solids that have just entered the reaction zone as well as some
that have had long residence times. There will always be some
unreacted solids in the exit solid stream under such conditions.
The fluidized bed has many of the characteristics of a liquid:

a. The bed will all be at the same temperature
    throughout. The movement of the solids carries heat
    and stabilizes the temperature. Any energy released
    in a local zone is dissipated because of the particle
    movement.

b. Foreign objects including heat transfer surface
    is easily inserted into the bed.

c. Solids may be easily transported into and out of
    the reactor system.

d. The heat transfer coefficient in the bed is high.
    This means that a large amount of heat can be re-
    moved or added through small heat transfer areas.

3.3.6 <u>Pneumatic Conveying Type</u>

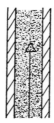

FIGURE 3.18 Pneumatic Conveying Type

In the pneumatic conveying system the solid is suspended in the gas stream. The velocity of the gas exceeds
the terminal velocity of the solid and the solid is carried
along with the gas. The solid particles must be crushed
and reduced in size in order to be carried by the gas stream.
This provides large surface areas. The gas surrounds each
of the particles. This provides an ideal environment for
solids to react. If a solid particle is smaller than the
average size it has a lower terminal velocity and is carried
out of the reactor with a shorter residence time. The time for
complete reaction is less for the small particle. Conversely,
for the larger particles a longer residence time is needed.

Because of the larger size it moves through the reactor slower.

The gas takes up most of the space in the reactor and may not come in contact with the solid. The solid and gas move upward in parallel flow. For a single-sized particle the solid moves in plug flow. Internal heat exchange surface may be placed in the bed to control the temperature.

3.3.7 Suspension Type

Gas

FIGURE 3.19 Suspended Bed

The suspension bed differs from the entrained bed in that the particles fall <u>downward</u> against the moving gas. The larger particles will fall faster than the smaller ones and are likely to reach the bottom before they are completely reacted. There is less area for gas solid contact than in the pneumatic conveying system because the particle size is greater. There is little opportunity to provide for heat removal in the reaction zone. The solid residence time depends upon the particle size.

The gas solid reactor systems described above are only a partial listing of the systems that are available. When additional systems are used in the rest of the text they will be described. The ones given above will be mentioned frequently in future chapters.

Problems for Chapter 3.0

1. The reaction $A_2 \longrightarrow 2A$ takes place at $1000°K$ and pressure
    of 200 atm and at equilibrium 75% of A_2 dissociates.
    The mole table follows.
        (a) Determine K_p.

$$\alpha = \text{moles of } A_2 \text{ reacted}$$

Component	In	+	Gen	-	Cons	=	Out		
A_2	1	+	0	-	α	=	P_{A_2}	x_{A_2} =	$\dfrac{1-\alpha}{1+\alpha}$
A	0	+	2α	-	0	=	P_A	x_A =	$\dfrac{2\alpha}{1+\alpha}$
Total	1	+	2α	-	α		P_T		

For the case where 75% of A_2 reacts $\alpha = 0.75$

        (b) If the pressure is reduced to 100 atm, what
             is the % of A_2 that dissociates?

2. For the reaction $C + 2H_2 \rightleftarrows CH_4$ evaluate the value of H
    by using Equation 3.6 and data from Figure 3.1. Compare
    it to the value using $\overline{H}_C$ from Chapter 2.

3. At $1200°F$ determine the amount of H_2O that must be added
    to a pure CO stream to obtain a mixture of product gas
    where the ratio of H_2/CO is 3/1.

        $CO + H_2O \rightleftarrows CO_2 + H_2$

4. Compare the equilibrium conversion of hydrogen to methane
    at $800°K$ at pressures of 1 and 30 atm. ($K_{800} = 1.41$)

        $C(S) + 2H_2(g) \rightarrow CH_4(g)$

5. Calculate the equilibrium conversion for the reaction

        $C(S) + H_2O(g) \rightarrow CO(g) + H_2(g)$

    at 1.10 and 30 atm and $1000°K$. K_p @ $1000°K = 2.617$.
    (Hint: The solid does not appear in the equation for
    K_p) [Answer = 0.46 @ 100 atm]

6. Calculate the equilibrium composition for the reaction

        $CO + H_2O \rightarrow CO_2 + H_2$

    at $700°K$ for the following steam/CO ratios: (a) 0.75
    (b) 1.00 (c) 1.50 (d) 2.00. Plot the % equilibrium con-
    version as a function of feed ratio. What is the effect of
    pressure on the equilibrium conversion?

7. Determine the temperature required to get 90% conversion for the reaction

$$3H_2 + CO \rightleftarrows CH_4 + H_2O$$

(a) at 1 atm (b) at 10 atm (c) at 100 atm.

Methanation Reaction
Free Energy and Heat of Reaction
$$CO(g) + 3H_2(g) + CH_4(g) + H_2O(g)$$

$T, °K$	$\Delta G°_T, kcal$	$Log_{10}K$	K	$\Delta H°_T, kcal$
298.16	-33.968	+24.896	7.870×10^{24}	-49.271
400	-28.574	+15.611	4.083×10^{15}	-50.352
500	-23.015	+10.059	1.145×10^{10}	-51.280
600	-17.287	+ 6.296	1.977×10^{6}	-52.059
800	-5.513	+ 1.506	3.206×10^{1}	-53.194
1000	+6.522	- 1.425	3.758×10^{-2}	-53.878
1500	+36.902	- 5.376	4.207×10^{-6}	-54.355

8. A tracer experiment is carried out on a reactor system. The output of a pulse test is shown below. Sketch a two reactor system that gives this response.

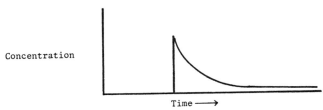

Concentration

Time ⟶

9. An experimental reactor (constant volume stirred tank) is run in the laboratory. For a value of $\tau(V/v_0) = 2$ it was determined that 55% of the feed stream reacted. We have two reactors that are not used. One is 1,000 ft^3 and behaves as a plug flow reactor. It is necessary to obtain a conversion of 93%. Determine:
    (a) The feed flow rate v_0 for the stirred tank reactor.
    (b) The feed flow rate v_0 for the plug flow reactor.

10. A solid fuel reacts without the formation of an ash layer. At 800°K the time for a one inch slab to react is 30 minutes. Under these conditions both the resistance to the gas layer and to chemical reaction are equal.

     A(gas) + S(solid) = 2P(gas)

  (a) The velocity of the gas passing the solid is doubled. This results in reducing the gas film resistance by 50% (1/2 the original value). How long did it take to react a one inch slab?

  (b) The resistance to chemical reaction is cut in half for every 25°K increase in temperature. How long would it take for the slab to react at 900°K?

  (c) Repeat (b) at 1000°K. At about what temperature would you say the gas film became controlling? State reason.

11. Continuation of Problem 10. Assume that the reaction is exothermic and that raising the temperature results in an increase in C_A equilibrium. The figure shown below gives C_A equilibrium as a function of temperature.

  (a) Repeat part b and c in previous problem.

  (b) What is the effect of doubling the system pressure?

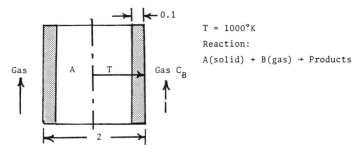

12. Consider the reaction between a solid two units thick.

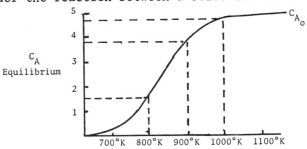

When the ash layer is one unit thick, $dT/dt = 0.1$ and the resistance R_T is given by

$R_T = R_3$(chemical reaction) + R_2(ash layer) +

$\qquad R_1$(gas layer)

$\qquad = 1.0 + 0.5 + 0.3 - 1.8$

(a) Determine the time it takes to consume (i) 50% of solid reactant, (ii) 100% of solid.

(b) Determine the time it takes to consume the solid if the ash layer is continuously re-moved.

(c) Determine (a) at 1100°K if k_R doubles every 25°K.

(d) Assume that increasing T results in an increase in C_B(equilibrium) such that ΔC is 50% of that at 1000°K, Repeat C.
Hint: Determine dT/dt for values of T. Plot

$$\int \frac{dT}{-r}$$

13. Consider two reactions

(a) $A \xrightarrow{k_1} P$ (P - product)

(b) $A \xrightarrow{k_2} U$ (U - undesirable side reaction)

The activation engines may be represented by E_1 and E_2, respectively. Derive an equation for the distribution ratio Φ as a function of T. Assume that the ratio is known at one temperature (T_o).

14. We have on hand two reactors of equal size. One is a plug flow reactor and one a back-mix reactor. Consider three possible ways of connecting these reactors.

(a) Plug-flow followed by back-mix.

(b) Back-mix followed by plug-flow.

(c) Split flow with 50% of feed going to plug flow and 50% to back-mix reactor.

Show graphically which system will provide the largest conversion.

Problem 15

The Figure shown below gives a pollution model where the industrial plant contributes pollutants to a pill shaped control volume surrounding the plant.

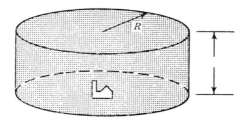

The air movements give rise to a turn-over rate. The turn-over rate is the time it takes for one control volume of outside air to exchange with one control volume. This is the same as the residence time τ (v/V_0) developed in the section on reactor design. The control volume is taken as completely mixed and the equations for the constant volume stirred tank reactor apply. For a control volume of

R = 10 km

H = 2 km

for a 1,000 MW thermal power plant that emits 98 metric tons/day of SO_2 determine the SO_2 concentration $(\mu g/m^3)$

a) For a day where the turn-over rate is 16 hrs.

b) For a day where the turn-over rate is 6 hrs.

c) For a day the same as part a) where measurements show that the actual concentration is 1/2 that predicted. Assume that the SO_2 reacts in the control volume and is thus removed. The reaction rate is given by

$$r_{SO_2} = - k \; C_{SO_2}$$

Evaluate k

d) Repeat b) but account for chemical reaction using value determined in c).

4.0 COMBUSTION OF FOSSIL FUEL FOR HEAT AND POWER

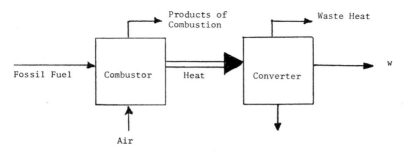

FIGURE 4.1 Conversion of Fossil Fuel to Work

The industrial revolution which lead to the modern industrial society was born from the discovery of how to convert
fossil fuel energy to work. Figure 4.1 provides a generalized
flow sheet for this conversion. In the first unit the chemical

Richard C. Bailie, Energy Conversion Engineering, ISBN 0-201-00840-8

stored energy provided by the fossil fuel is released by
oxidation (chemical reaction). This destroys the fossil
fuel and provides hot gases. The hot gases are then converted
into work (or electricity).

Almost all of the energy needs for heat and power are
presently being provided by the combustion of fossil fuels or
their derivatives. Many of the energy conversion schemes to
be discussed in subsequent chapters provide for the conversion
of some unattractive solid fossil fuels into more attractive
synthetic fuels. These synthetic fuels must then be burned
to produce heat, work, or electricity. Combustion is a major
step in almost all schemes to utilize the world's fossil fuel
resources.

4.1 COMBUSTION OF SOLID FUELS

Solid fossil fuels (coal) has, in the past, been the
most common energy source for large central electrical generating stations. These solid fuels have recently shared the
market with gas and oil. As the supplies of these alternate
fuels decrease it is likely that coal will return to a more
dominate role. Combustion of solid fuels is a complex process. This section discusses some of the major features of
solid fuel combustion.

Ranked in order of increasing age solid fossil fuels
are designated as peat, lignite, subbituminous, bituminous,
and anthracite coal. The information provided in Table 4.1,
Figure 4.2, and Figure 4.3 are given in terms of increasing
age (or rank).

Solid fossil fuels are similar in that they have the
physical state of a solid at room temperature and contain

C, H, S, and O. They were all derived from vegetable matter.
Table 4.1 shows the fuels available as living plants are converted to coal. This table provides a chemical analysis
(called the ultimate analysis) and a proximate analysis. The
proximate analysis is quicker to determine in the laboratory
and provides information needed to understand the behavior
of the fuel in a furnace. In the proximate analysis the
sample is dried and then heated in an inert atmosphere to
950°C for 7 minutes. The loss of weight is termed volatile
matter (V.M.). This is a measure of the degree that a fuel
will vaporize or pyrolyze in a furnace. The vapor will contain some of the carbon. The carbon remaining in the solid
phase after the volatile matter is driven off is termed fixed
carbon. The ash is the residue that remains after solid is
ignited and all combustible matter is oxidized.

TABLE 4.1

COALS OF THE U.S.A. PROGRESSIVE STAGES OF TRANSFORMATION OF VEGETABLE MATTER IN COAL

FUEL CLASSIFICATION	LOCALITY	MOISTURE (As-received)	ANALYSIS ON DRY BASIS								HEATING VALUE J/Kg- × 10⁻⁶ (As-received)
			PROXIMATE			ULTIMATE					
			V.M.	F.C.	Ash	S	H	C	N	O	
Wood		46.9	78.1	20.4	1.5	..	6.0	51.4	0.1	41.0	20.52
Peat	Minnesota	64.3	67.3	22.7	10.0	0.4	5.3	52.2	1.8	30.3	21.03
Lignite	North Dakota	36.0	49.8	38.1	12.1	1.8	4.0	64.7	1.9	15.5	25.63
Lignite	Texas	33.7	44.1	44.9	11.0	0.8	4.6	64.1	1.2	18.3	25.72
Subbituminous C	Wyoming	22.3	40.4	44.7	14.9	3.4	4.1	61.7	1.3	14.6	24.60
Subbituminous B	Wyoming	15.3	39.7	53.6	6.7	2.7	5.2	67.3	1.9	16.2	28.08
Subbituminous A	Wyoming	12.8	39.0	55.2	5.8	0.4	5.2	73.1	0.9	14.6	29.94
Bituminous High Volatile C	Colorado	12.0	38.9	53.9	7.2	0.6	5.0	73.1	1.5	12.6	30.31
Bituminous High Volatile B	Illinois	8.6	35.4	56.2	8.4	1.8	4.8	74.6	1.5	8.9	31.08
Bituminous High Volatile A	Pennsylvania	1.4	34.3	59.2	6.5	1.3	5.2	79.5	1.4	6.1	33.42

TABLE 4.1 (cont.)

COALS OF THE U.S.A. PROGRESSIVE STAGES OF TRANSFORMATION OF VEGETABLE MATTER INTO COAL

FUEL CLASSIFICATION	LOCALITY	MOISTURE (As-received)	ANALYSIS ON DRY BASIS								HEATING VALUE J/Kg- x 10^{-6} (As-recieved)
			PROXIMATE			ULTIMATE					
			V.M.	F.C.	Ash	S	H	C	N	O	
Bituminous Medium Volatile	West Virginia	3.4	22.2	74.9	2.9	0.6	4.9	86.4	1.6	3.6	35.2
Bituminous Low Volatile	West Virginia	3.6	16.0	79.1	4.9	0.8	4.8	85.4	1.5	2.6	34.8
Semi-Anthracite	Arkansas	5.2	11.0	74.2	14.8	2.2	3.4	76.4	0.5	2.7	30.5
Anthracite	Pennsylvania	5.4	7.4	75.9	16.7	0.8	2.6	76.8	0.8	2.3	29.5
Meta-Anthracite	Rhode Island	4.5	3.2	82.4	14.4	0.9	0.5	82.4	0.1	1.7	26.9
Graphite	New York	1.8	1.4	91.2	7.4	0.4	0.2	91.6	-	0.4	34.1

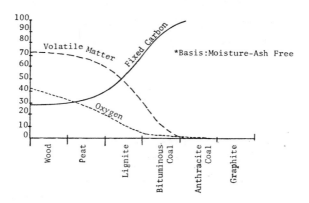

FIGURE 4.2   Graphic representation of the transformation of fuels from wood to anthracite.

Figure 4.2 shows the information given in Table 4.1 in graphical form. This figure shows:

a)   Oxygen decreases as one moves away from living plants.

b)   Volatile matter decreases as one moves away from living plants.

c) Fixed carbon increases as one moves away from living
    plants.

Figure 4.3 shows the comparative heating value for various
solid fuels. It shows the heating value to be the greatest
for semi-bituminous coal. When the heating value of a solid
fuel is not known it may be approximated by Dulong's Formula

$$Btu/lb = 14,544C + 61,500 \left(H - \frac{O}{8}\right) + 4,500S$$

[To convert to J/Kg multiply by 2321]   (4.1)

C = Weight fraction carbon
H = Weight fraction hydrogen
O = Weight fraction oxygen
S = Weight fraction sulfur

EXAMPLE 4.1 Determine the heating value of Wyoming Subbituminous
coal (A). Use the compositions given in Table 4.1.
Compare the value calculated using Dulongs Formula
with the value listed in Table 4.1.

SOLUTION 4.1

$$Btu/lb = 14,544(.617) + 61,500(.041 - \frac{.146}{8})$$
$$+ 4,500(.034)$$
$$= 10,520 \ Btu/lb$$
$$= 24,430,000 \ J/Kg$$

Value in Table 4.1 (24,600,000)

$$\% \ Error = \frac{24,430,000 - 24,600,000}{24,600,000} \times 100$$

$$= 0.7\%$$

Prior to describing the systems that have been developed for
the combustion of solid fuels (particularly coal) it is appropriate to appreciate what occurs when solid fuel is heated.

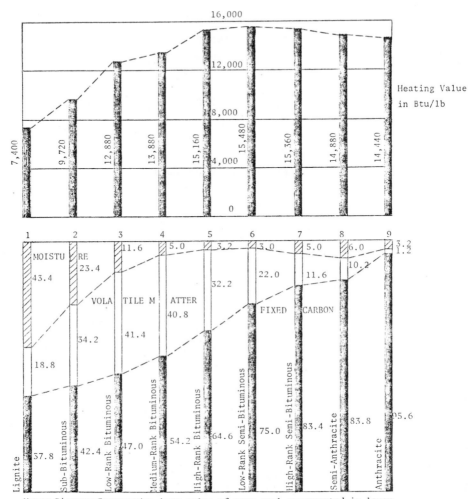

Upper Diagram: Comparative heat value of same coal represented in lower diagram ash-free basis.

Lower Diagram: Variation in fixed carbon, volatile matter, and moisture of coals, from lignite to anthracite, as received, ash-free basis.

FIGURE 4.3 Comparison of the heating values and compositions of various coals.

The characterization of solid fuel according to volatile
matter is important. When coal is exposed to high temperatures, some of the mass is volatilized and leaves the solid
in the form of a gas. This gas contains most of the hydrogen
and some of the carbon. The solid residue that remains is
char and ash. The combustion system designed for a high
volatile coal would differ from that of a low volatile coal.
For a high volatile coal, most of the chemical oxidation
reactions occur in the gas space as a homogeneous reaction,
not on the surface of the solid as a heterogeneous reaction.
For a low volatile coal, most of the reaction occurs between
the solid surface and the oxygen. Furnaces designed to burn
wood or peat, which are high volatile fuels, are significantly
different than a furnace designed to burn anthracite coal.
The volatile fraction is an indicator to how difficult a fuel
is to burn. The anthracite coal, because of the low volatile
fraction, is difficult to burn.

   Another characteristic of coal that must be considered
is the tendency of some coal to swell and cake when heated.
When coal is heated it often goes through a stage where the
coal softens and becomes sticky. As heating continues, the
coal once again becomes a solid in the form of solid coke.
It is difficult to appreciate this characteristic unless one
has experienced working with both types of fuels. Accepting
that swelling and caking does occur, it can be appreciated
that the caking tendency will present problems of distributing
any reacting gas that is to contact the solid. This behavior
may completely destroy the gas distribution in a reactor system.

   The terms furnace and combustion refer to a specific type
of chemical reactor and a chemical reaction. All of the discussion on chemical reactors apply to combustion as well as

other reactor systems. The next few paragraphs will apply
some of the principles provided in the previous chapter to
the combustion of fuels in furnaces.

The discussion of reactor kinetics provided in the previous chapter considered only isothermal (constant temperature)
reactor systems. It did not attempt to describe how the
isothermal conditions were established or maintained.
Figure 4.4 shows several flow systems for combustion of char
with air. Char is essentially solid carbon that contains only
minor amounts of volatile matter (example is home charcoal).

The first system in Figure 4.4 shows a plug flow system
showing both the char and the air entering at 25°C. For this
system the exit from the system would be char and air at 25°C.
No reaction would take place. Char does not react <u>until</u> it
is brought up to its ignition temperature. The reactant must
be brought up to the ignition temperature for the combustion
reaction to become self-sustaining.

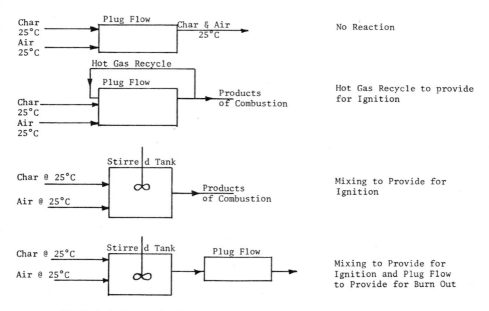

FIGURE 4.4 Generalized Combustion Reactors

A convenient method to heat the reactants would be to recycle some of the hot product gases. This provides the needed energy to heat the inlet reactants. This concept is shown as system 2 in Figure 4.4. The difficulty of this flow pattern is that the oxygen concentration is reduced throughout the furnace.

Another method is shown in system 3 of Figure 4.4. The furnace is built as a stirred tank system. The mixing of the inlet with the combustion products provides for the needed energy to maintain the ignition temperature. The problems associated with a stirred type system have already been discussed. The oxygen concentration would be low and some of the char would be swept out before it had sufficient time to react.

The most effective conversion of the char would occur by the combination of stirred reactor section followed by a plug flow section. The stirred section would be only large enough to assure that the ignition temperature is maintained. The plug flow section would then assure ample residence time to provide complete reaction of the char.

Reference is sometimes made to the "three T's" when speaking of furnace design. These refer to time, temperature, and turbulence. Turbulence provides for mixing in the furnace, temperature provides for a high reaction rate constant and time provides for the needed residence time.

The design of a furnace remains as much an art as a science. The designs are not based upon "basic theory" but more often upon the experience of the designer. Because of experience with the large numbers of combustion systems built and operated, effective furnace designs are usually provided.

4.2 SOLID FUEL FURNACES

Three main types of furnaces are usually used for combustion of solid fuels today. They are:

1. Mechanical stokers
2. Pulverized fuel furnaces
3. Cyclone furnaces

Other types such as multihearth, rotary furnaces are used in some applications. In addition to the three listed, a discussion of the fluidized bed furnace is provided. It is a recent development in furnaces and is receiving a great deal of attention.

4.2.1 Mechanical Stokers

There have been many types of mechanical stokers. The major types are represented diagramatically in Figure 4.5. In (a) the coal is fed to a sloping grate. It is pyrolyzed in the upper part of the furnace. The coke formed travels down the grate. The air flows cross-flow to the coal. This has the flow pattern of a horizontal moving fuel.

The chain grate stoker is shown at (B). The coal is fed on a moving horizontal grate. The coal is pyrolyzed as it enters the furnace. The moving grate carries the coke through the furnace and discharges the ash. The air flows cross ways to the coal movement. In both the overfeed and chain grate stoker furnaces pyrolysis products enter the gas space above the coal bed. For complete combustion additional air (secondary air) is directed toward these regions.

In the underfed stoker (C) fuel is introduced under the bed. The pyrolysis product formed must pass through the hot incandescent coke.

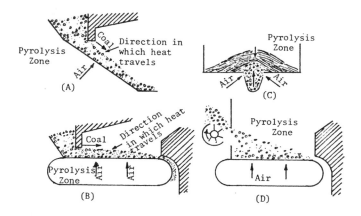

FIGURE 4.5 Principal methods of feeding coal in industrial
furnaces. (A) Overfeed stoker. (B) Chaingrate stoker.
(C) Underfeed stoker. (D) Spreader stoker.

The spreader-stoker is shown in (D). The coal is crushed
and fed into the gas space above a coke bed. The coal pyrolysis
product and small coke burns in the space above the grate.
The larger particles fall upon the grate and form a thin
layer where it burns.

There have been many types of mechanical stokers used
in the past for central generator stations. Their use is
presently confined to industrial or commercial service where
the demand does not exceed 400,000 lbs. of steam/hr.

A wide variety of coals may be used in stokers. The
underfed and spreader stoker systems can fire caking coals.
For underfire systems fines lead to excessive caking and
the size of coal used is (1/4 inch) or greater. For the underfire and overfire air systems the burning rates are in the
range of 35 lbs of coal per hour per square foot of grate
area. The spreader stoker achieves higher rates (60 lbs of
coal per hour per square foot).

It is likely that stokers will continue to play a role

for smaller systems and the spreader stoker will be the major
system to be used. The air is controlled in zones along the
bed. The higher rates obtained by the spreader stoker is
caused by the increase in surface area of the coal as a result of crushing and the thin bed on the grates. This results
in all of the solid fuel being exposed to the air.

4.2.2 Pulverized Coal Furnaces

From the discussion of heterogeneous reaction kinetics
in Chapter 3 the rate of chemical reaction for heterogeneous
chemical reactions would be increased by increasing the
available surface area of the solid. This immediately suggests that the solid fuel should be **crushed** or reduced in size.
This is done in furnace systems referred to as pulverized coal
furnaces. Based upon the theories provided in Chapter 3, the
rate of reaction would be controlled by the resistance to mass
transfer surface, the resistance to any ash and the resistance
to chemical reaction. When the temperature is high the resistance to chemical reaction may be neglected. For small particles the ash layer is always small and may be neglected.
The rate that the particle in a pulverized coal furnace burns
is controlled by the diffusion of the oxygen through the gas
film. Under these conditions the time required to burn a
particle is given by

$$t_c = Kd^2/[pp_{O_2}] \tag{4.2}$$

where pp_{O_2} is the partial pressure of the oxygen, d is the
diameter of the particle and K is known as the burning constant.

Pulverized furnaces are designed using empirical relationships and judgements based upon experience. However, considerable insight may be obtained by some simple theoretical models.
To obtain complete reaction of a particle in a plug flow reactor system, the residence time is reduced by 1/4 if the
diameter is cut in half (Equation 4.2). This is true for
single sized particles. If particles with a wide size distribution are used, the amount of unburned solids leaving
the furnace may increase when average size is reduced. This
may be seen by considering what will happen if some fines
are added to a furnace operating on a fuel crushed to a single size.

Figure 4.6 compares the reaction of the same mass of
char for

1) Single sized particles
2) Two sized particles; 75% mass of the particles sized
    the same as for Case 1, and 25% mass crushed to 1/10
    of the particle size for Case 1 (100 times the area
    per unit mass).

The curves labeled "Mass-Single Sized" and "Mass-Two Sized"
show how the mass of unreacted char changes with time for
both cases. The case where some of the char had been crushed
reacts much faster at the beginning and the amount of char
decreases more rapidly. At time labeled by (1) the same amount
of char remains for either case. For times longer than this,
the effect of crushing some of the char would result in less
char consumption, not more. Why this happens is discussed
in the next paragraph.

In addition to the total mass of char remaining at
any time, Figure 4.6 also shows the reaction-rate for each
case. Consider the initial reaction rates. The reaction

rates are proportional to the available surface area and the
concentration of oxygen concentration. For both cases at the
beginning pp_{O_2} = 0.21 If the reaction rate for Case 1 is
taken as 1.0, then the rate of the large and small particles
for Case 2 are:

a) Large particles = 0.75 (75% of the particles are
    the same size).

b) Small particles = 0.25 x 100 = 2.5 (the 25%
    gives 100 times the area of larger particles).

Thus, the initial rate is 2.5 + .75 = 3.25.

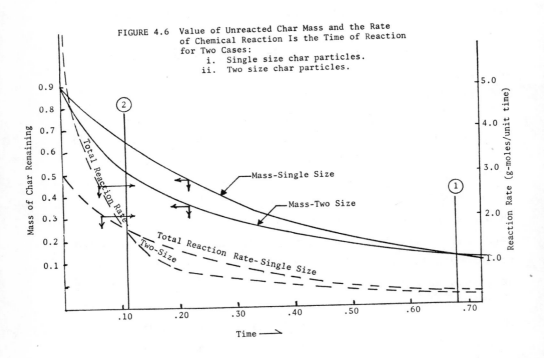

FIGURE 4.6   Value of Unreacted Char Mass and the Rate
             of Chemical Reaction Is the Time of Reaction
             for Two Cases:
                i. Single size char particles.
                ii. Two size char particles.

The higher reaction rate is a result of the high surface
area of the smaller particles. The oxygen is being consumed

by these small sized particles. The small sized particles
are soon consumed while the larger particles undergo little
conversion. The large particles must continue to react in
an atmosphere that is depleted in O_2. Most of the reaction
for the large particles takes place at reduced oxygen levels
and takes longer to react than for single sized particles.
For this reason, the amount of unreacted carbon leaving becomes greater as a result of adding the small particles (at
times longer than t_1 in Figure 4.6).

It is important that the size distribution be considered,
and it is not sufficient to know solely the average size to
predict the burn-out fraction of a fuel. It is often best
to remove fine material and feed a closely sized fuel to a
furnace.

Figure 4.7 shows a typical pulverized coal combustion
system.

Several advantages are attributed to pulverized coal
furnaces. They are:

1. Can be built in much larger sizes than stoker boilers.
2. Better response to load changes.
3. Capable of burning all ranks of coal from anthracite to lignite.
4. Ease of combination firing with fluid fuels.
5. Increased thermal efficiency because of low excess
   air requirements (15-20%) and low carbon loss.
6. Low labor requirements.
7. High volumetric heat release rates (12,000-14,000
   Btu/ft^3).

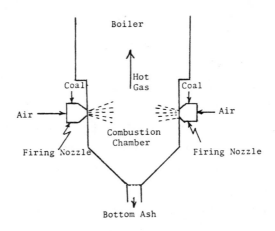

FIGURE 4.7 Pulverized Coal Furnace

Many of these advantages are a direct result of the
reduction in coal size. The pulverized coal behaves more
like a gas than a solid. When the pulverized coal enters
the furnace, the particles heat up and some of the coal
pyrolyzes. The pyrolysis gas burns and the large solid surface of the remaining char provides for contact with oxygen
and combustion is rapid.

The pulverized coal furnace has a few disadvantages.

1. Costs associated with pulverizing.

2. Energy required for crushing.

3. Generation of fly ash.

The development of pulverized coal systems have resulted
in the emission of finely sized particles (fly ash) that was
not a problem when the ash from a coal system was primarily
a coarse bottom ash. A 1,000 MW electrical power plant burning 7.0% ash coal will produce approximately 234 tons/day of
fly ash. The pulverized coal furnaces can be large. Most
newer units generate 4,000,000 to 4,500,000 lbs of steam/hr.

4.2.3 Cyclone Furnaces

The cyclone furnace is shown in Figure 4.8. It is a horizontal inclined, water cooled, tubular unit. Crushed (not pulverized) coal is burned in air entering the furnace tangentially. The temperatures within the furnace reach temperatures of up to 2,000°K. The ash becomes molten and covers the walls. Coal fines burn in suspension; larger pieces are captured by the molten slag and are retained until they burn.

Volumetric heat release rates are as much as 50 times that of pulverized furnaces. An advantage of the cyclone furnace is the low dust emission because of the wet walls that collect and melt the ash that impinges against it.

The air is added at a high velocity tangent to the walls of the furnace. This requires significant energy consumption that offsets that energy that is used to pulverize the coal in a pulverized coal furnace.

The high temperature in a cyclone furnace leads to high slag formation.

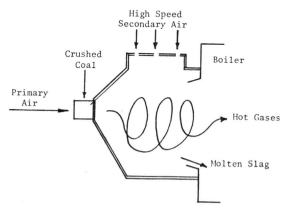

FIGURE 4.8 A Cyclone Furnace

4.2.4 Fluidized Bed Combustion

The combustion of coal in a fluid bed is a recent development. The advantages cited for the fluid bed combustion system include:

a) Low quality, high sulfur coal can be burned without danger of slagging because of low combustion temperatures.

b) The SO_2 formed is removed by the solid within the bed.

c) The heat release rate and the high heat transfer rates reduce the boiler size, weight and costs.

d) The cost of electricity from a fluid bed boiler will be lower than a conventional coal fuel boiler with required flue gas clean-up system.

e) The overall efficiency is predicted to be 39% as compared to 37% for a conventional boiler with a stack clean-up system.

f) The formation of NO_x is low because of low combustor bed temperature.

g) Fluidized beds can be easily built to operate under pressure.

The basic fluid bed system is shown in Figure 4.9. The bed is composed of lime, limestone or dolomite rock (sorbant). Coal and sorbant is fed into the bed and air introduced at the bottom. The temperature is $1200°K$ to $1310°K$.

The sulfur in the coal oxidizes to SO_2 in the bed. The SO_2 reacts with the limestone to form $CaSO_4$ by the reaction

$$CaCO_3 + SO_2 + 1/2 \ O_2 \longrightarrow CaSO_4 + CO_2$$

The sulfur is captured by the bed and retained in the sorbant and removed from the bed. The sulfur is removed from the gas and the gases do not have to undergo further gas clean-up for SO_2.

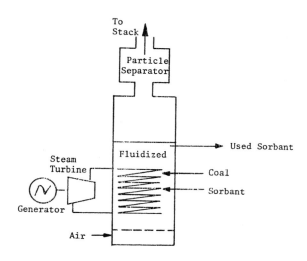

FIGURE 4.9 Fluidized Bed Combustion

EXAMPLE 4.2   Determine the minimum amount of limestone that would be required to capture all of the SO_2 formed in a 1000 megawatt fuel bed facility firing 2% sulfur coal. The energy conversion efficiency is 37%.

SOLUTION 4.2

$$1000 \times 10^6 \text{ watts} \times \frac{3.413 \text{ Btu}}{\text{Kw-hr}} \times \frac{1 \text{ lb Coal}}{13,000 \text{ Btu}}$$

$$\times \frac{2 \text{ lb S}}{100 \text{ lb Coal}} \times \frac{1 \text{ lb-mole S}}{32 \text{ lb S}}$$

$$\times \frac{1 \text{ lb-mole } SO_2}{1 \text{ lb-mole S}} \times \frac{1 \text{ lb-mole } CaSO_4}{1 \text{ lb-mole } SO_2}$$

$$\times \frac{136 \text{ lbs } CaSO_4}{1.0 \text{ lbs-mole } CaSO_4} \times \frac{1 \text{ ton } CaSO_4}{2000 \text{ lbs } CaSO_4}$$

$$= 11.1 [\text{tons/hr}] @ 100\% \text{ efficiency}$$

$$= 30 [\text{tons/hr}] @ 37\% \text{ efficiency}$$

Major programs are being undertaken to develop the fluid
bed combustor. The first commercial system is anticipated
to be on-line in Rivesville, WV, sometine in late 1977. It
will have a capacity of 30 MW. Further discussion on the
fluid bed is given in the chapter on the environment.

4.3 LIQUID FUELS

Fuel oil is presently the most common fuel used in boilers
for steam generation or power plants. Petroleum is sometimes
burned in crude form. This product contains low boiling gasoline constituents that present a fire hazard potential. By
a simple distillation this lighter gasoline may be removed
and a "safe" oil for combustion may be produced. The term
fuel oil is applied to a wide variety of fuels such as
crude petroleum, kerosene, gasoline, and heavy residue.
Each fuel contains a multiplicity of chemical compounds.
Table 4.2 provides typical analysis of liquid fuel oils.
Among the advantages of oil fuel are:
1) Weighs 30% less, occupies 50% less space than coal
   for same amount of energy.
2) No deterioration during storage.
3) No trouble with spontaneous combustion.
4) Oil storage may be located a distance from the
   furnace.
5) May be stored and removed with practically no labor.
6) High combustion rates per unit volume.
7) Has a high turn down ratio.
8) No labor required to remove ash.
9) Little smoke.
10) Less pressure drop (since there is no fuel bed).

TABLE 4.2

TYPICAL ANALYSES AND PROPERTIES OF FUEL OILS*

Grade	No. 1 Fuel Oil	No. 2 Fuel Oil	No. 4 Fuel Oil	No. 5 Fuel Oil	No. 6 Fuel Oil
Type	Distillate (Kerosene)	Distillate	Very Light Residual	Light Residual	Residual
Color	Light	Amber	Black	Black	Black
API gravity, 60F	40	32	21	17	12
Specific gravity, 60/60F	0.8251	0.8654	0.92729	0.9529	0.9861
LB per ·U.S. gallon, 60F	6.870	7.206	7.727	7.935	8.212
Viscos., Centistokes, 100F	1.6	2.68	15.0	50.0	360.0
Viscos., Saybolt Univ., 100F	31	25	77	232	-
Viscos., Saybolt Furol, 122F	-	-	-	-	170
Pour point, F	Below Zero	Below Zero	10	30	65
Temp. for pumping, F	Atomspheric	Atmospheric	15 min.	35 min.	100
Temp. for atomizing, F	Atomspheric	Atmospheric	25 min.	130	200
Carbon residue, per cent	Trace	Trace	2.5	5.0	12.0
Sulfur, per cent	0.1	0.4-0.7	0.4-1.5	2.0 max.	2.8 max.
Oxygen and nitrogen, percent	0.2	0.2	0.48	0.70	0.92
Hydrogen, per cent	13.2	12.7	11.9	11.7	10.5
Carbon, percent	86.5	86.4	86.10	85.55	85.70
Sediment and water, percent	Trace	Trace	0.5 max.	1.0 max.	2.0 max.
Ash, per cent	Trace	Trace	0.02	0.05	0.08
Btu per gallon (1)	137,000	141,000	146,000	148,000	150,000

(1) Multiply by 277.5 to get J/liter

TABLE 4.2 (cont.)

No 1. - A distillate oil intended for vaporizing pot type
burners and other burners requiring this grade of
fuel. Sulfur content less than 0.5%.

No 2. - A distillate oil for general purpose domestic heating
for use in burners not requiring No 1 fuel oil. Sulfur content less than 1.0%.

No 4. - An oil for burner installations not equipped with preheating facilities.

No 5. - A residual type oil for burner installations equipped
with preheating facilities.

No 6. - An oil for use in burners equipped with preheaters
permitting high viscosity fuels. Sometimes referred
to as Bunker C.

11) High efficiencies because of low excess air require-
    ments.

12) Less SO_2 in stack.

The fundamental requirements for efficient boilers is
to break the oil into a fine mist of small droplets (atomize).
The feed is atomized by either steam or air. The small drops
vaporize rapidly to become gaseous and the combustion takes
place in the gas space.

Almost any coal fueled furnace may be converted to oil
firing. The coal handling equipment is removed and replaced
by oil storage tanks and oil atomizers. The retrofit required for pulverized coal burners is usually not prohibitive.
The pulverized coal nozzles are replaced by atomized oil
nozzles. This conversion has been done in many instances to
meet environmental regulations. The conversion of furnaces
designed for oil to coal is more difficult and not always
possible. In anticipation of oil shortage some newer units
provide for switching from oil to coal or even co-firing
these fuels.

4.4 GASEOUS FUELS

The use of natural gas as a fuel for steam and power
generation is preferred because of the low atmospheric pollution levels. It is because of this preference that serious
consideration is being given to the manufacture of synthetic
natural gas from coal to continue to provide this fuel after
our natural gas supplies become depleted.

With the shortage of natural gas, renewed interest is
being shown in the lower quality gaseous fuels that were
produced in the past and can be produced wherever the demand

exists. Table 4.3 reviews the properties of several gases.

Several important observations can be made from this table.

a) The theoretical flame temperature ranged between 2350-2570°K for all gases with a heating value greater than 5 x 10^6 J/m^3 (136 Btu/ft^3).

b) The heating value of a stoichiometric mixture of gas and air ranged between 94.5 to 99.6 Btu/SCF (3.5 x 10^6 to 3.7 x 10^6 [J/m^3]) for gas with a heating value greater than 150 Btu/SCF.

c) The ft^3 of stack gas per 10^6 Btu of fuel released ranges from 9,000-14,000.

The advantage of using gas for an energy source include

a) Low pollution potential.

b) High turndown ratio.

c) Low excess air.

d) Little or no smoke.

To obtain good combustion of the gas, several conditions must be met.

a) The fuel and oxygen must be intermittently mixed.

b) The mixture must be heated to the ignition temperature.

c) The air fuel ratio must be within the flammability limits.

Furnaces designed to burn natural gas can be retrofitted at relatively small cost to burn gases of 10^7 [J/m^3] (270 Btu/ft^3) or more and should operate at little reduction in efficiency. When low energy gas is substituted for natural gas, the furnace output is reduced. The low energy gas requires more excess oxygen and gives higher gas volumes. This would decrease residence time in the furnace as a result of greater

TABLE 4.3

Composition and Combustion Constants of Representative Manufactured and Natural Gases (a)

	Methane (CH$_4$), per cent	Ethane (C$_2$H$_6$), per cent	Propane (C$_3$H$_8$), per cent	Ethylene (C$_2$H$_4$), per cent	Carbon monoxide (CO) per cent	Carbon dioxide (CO$_2$) per cent	Hydrogen (H$_2$), per cent	Oxygen (O$_2$), per cent	Nitrogen (N$_2$), per cent	Btu per cu.ft., high Hg,satd.,60°F.,30in. (gross) (a)	Btu per cu.ft., low (net) 60°F.,30in. Hg,satd.with H$_2$O (a)	Cubic feet air required per cu.ft. gas (b)	Cubic feet CO$_2$ per cu.ft. of gas burned (b)	Cubic feet H$_2$O per cu.ft. of gas burned (b)	Cubic feet N$_2$ per cu.ft. of gas burned (b)	High (gross) Btu per cu.ft. gas-air mixture	Theoretical flame temperature, degrees K	Ft^3 of POC / 10^6 Btu (c)
Coal gas	34.0			6.6	9.0	1.1	47.0		2.3	634	560	5.50	0.573	1.282	4.36	97.5	2428	9,800
Coke-oven gas	33.9			5.2	6.1	2.6	47.9	0.6	3.7	600	538	5.28	0.529	1.260	4.21	95.6	2400	10,000
Syn gas					43.4	3.5	51.8		1.3	310	285	2.28	0.469	0.518	1.81	94.5	2571	9,000
Carburetted water gas	14.8			12.8	33.9	1.5	35.2		1.8	578	529	4.85	0.758	0.904	3.85	98.8	2528	9,500
Oil gas	27.0			2.7	10.6	2.8	53.5		3.4	516	461	4.25	0.458	1.129	3.39	98.3	2461	9,206
Low energy gas	2.6			0.4	22.0	5.7	10.5		58.8	136	128	1.08	0.311	0.165	1.44	65.3	1950	14,106
Low energy gas					26.2	13.0	3.2		57.6	93	91.6	0.70	0.392	0.032	0.55	54.7	1730	10,400
Natural gas, Follansbee, WV		31.8	77.7						0.5	2469	2268	23.80	2.967	4.062	18.81	99.6	2389	10,400
Natural gas, Follansbee WV		79.4	20.0						0.6	1868	1711	17.97	2.188	3.182	14.21	98.5	2383	9,600
Natural gas, McKean County, Pennsylvania	32.3	67.0							0.7	1482	1350	14.25	1.663	2.756	11.29	97.3	2350	10,700
Natural gas, Sandusky, Ohio	83.5	12.5				0.2			3.8	1047	946	10.04	1.087	2.045	7.98	94.7	2333	10,600

Notes: 1. Products of combustion and theoretical flame temperatures figured with theoretical air (21 per cent O$_2$, 79 per cent N$_2$).
2. Theoretical flame temperatures not corrected for dissociation.
(a) Multiply by 3.71×10^4 to obtain J/m^3.
(b) Also m^3.

velocity and increased pressure drops throughout the system.

A gas furnace has no provision to remove ash, almost no gas clean-up equipment, and no storage for liquids or solids. For this reason it is costly to consider the conversion of a gas fired system to either liquids or solid fuels. The easiest conversion is from natural gas to a lower Btu gas.

4.5   THERMODYNAMICS OF COMBUSTION

The discussion above provided a brief description of some of the fuels and systems to burn fuels. All of the systems converted chemical stored energy available in the fossil fuel into thermal energy. Once the chemical energy has been released as thermal energy, a large fraction of the energy can no longer be converted to work. The maximum efficiency to which thermal energy can be converted to work is given by the Carnot efficiency,

$$w = q(1 + T_H/T_C)$$

The conversion of chemical energy to heat results in considerable lost work.

The amount of work that is lost when a fuel is burned in air may be calculated by the methods described in Chapter 3. Figure 4.10 shows the combustion of a fuel (methane is used for this disucssion) in a two step process. In the first unit the methane is burned with the oxygen from the air in a system that exchanges neither work nor heat with the surroundings. All of the chemical energy released in the reactor goes to heat up the reaction products. The temperature reached, i.e., T_C, is termed the adiabatic reaction temperature (or the

adiabatic flame temperature in the special case of combustion).
The hot products of the reaction provide heat to run a Carnot
Engine.

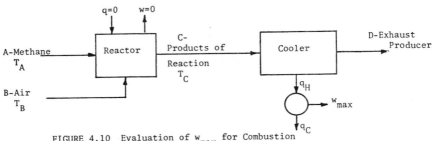

FIGURE 4.10 Evaluation of w_{max} for Combustion

The amount of work that is lost in the combustion unit
may be obtained by evaluating the maximum work of all input
streams and subtracting the maximum work of all exit streams.
This is shown in the next example.

EXAMPLE 4.3   Determine the amount of lost work resulting from
the combustion of 1 Kg-mole of methane with 9.52
Kg-mole of air. Determine the adiabatic reaction
temperature and the Figure of Merit for this process. Both air and methane enter at 298°K.

SOLUTION 4.3

$A = 1$[Kg-mole CH_4]
$T_A = 298$[°K]

Reactor

$T_P =$

$P_{CO_2} =$
$P_{H_2O} =$
$P_{N_2} =$
$P_{O_2} =$

$B = 9.52$[Kg-mole Air]
$T_B = 298$[°K]

$$CH_4 + 2O_2 \rightarrow CO_2 + 2H_2O$$

Material Balance About Both Units (α = mole CO_2 formed)

	IN	+ GEN	− CONSUMED	= OUT
CH_4	1	+ 0	− α	= 0
H_2O	0	+ 2α	− 0	= P_{H_2O}
CO_2	0	+ α	− 0	= P_{CO_2}
O_2	9.52×0.21	+ 0	− 2	= P_{O_2}
N_2	9.52×.79	+ 0	− 0	= P_{N_2}

$$\alpha = 1$$
$$P_{N_2} = 7.52$$
$$P_{CO_2} = 1.0$$
$$P_{O_2} = 0$$
$$P_{H_2O} = 2.0$$

To evaluate H_p values T_p must first be known. The procedure followed is to assume T_p and evaluate q. The temperature leaving the system that results in q = 0 is the desired value.

Evaluation of T_p

Stream Identification	Mass Flow	T	P	$\overline{H}_C$	$\overline{H}_T$	$\overline{H}_\lambda$	$\overline{H}_p$	$\overline{H}$	$m\overline{H}$
Units on Terms	[Kg-mole]	[°K]	[bar]	[J/Kg-mole]					[J]
Inlet									
A(CH_4)	1	298	1.01	891×10⁶	0	0	0	891×10⁶	891×10⁶
B(N_2)	7.52	298	0.80	0	0	0	0		
B(O_2)	2.00	298	0.21	0	0	0	0		
$\sum a_i\overline{H}_{A_i}$ = 891 × 10⁶									

Assume T_p = 2000°K

Exit									
P(CO_2)	1	2000	0.095	0	90.2×10⁶	0	0	90.2×10⁶	90.2×10⁶
P(H_2O)	2	2000	0.190	0	73×10⁶	44×10⁶	0	117×10⁶	234×10⁶
P(N_2)	7.52	2000	0.715	0	56.6×10⁶	0	0	56.6×10⁶	426×10⁶
P(O_2)	0	2000	0	0	58.3×10⁶	0	0	58.3×10⁶	0
$\sum b_i\overline{H}_{B_i}$ = 750×2 × 10⁶									
$\Delta H = \sum b_i\overline{H}_{B_i} - \sum a_i\overline{H}_{A_i}$ = 750.2 × 10⁶ − 891 × 10⁶ = −140.8 × 10⁶									

Assume $T_p = 2500°K$

Exit									
$P(CO_2)$	1	2500	0.095	0	115×10^6	0	0	115×10^6	115×10^6
$P(H_2O)$	2	2500	0.190	0	101×10^6	44×10^6	0	145×10^6	290×10^6
$P(N_2)$	7.52	2500	0.715	0	75.9×10^6	0	0	75.9×10^6	571×10^6
$P(O_2)$	0	2500	0	0	75.4×10^6	0	0	75.4×10^6	0
$\sum b_i \overline{H}_{B_i}$ = 976 x 10^6									
$\Delta H = \sum b_i \overline{H}_{B_i} - \sum a_i \overline{H}_{A_i}$ = 976 x 10^6 - 891 x 10^6 = 85 x 10^6									

For this system $\Delta H = q$

$T[°K]$	$q[J]$
2000	-140.8×10^6
T_p	0
2500	$+85 \times 10^6$

Assuming linear relationship

$$\frac{2500 - T_p}{2500 - 2000} = \frac{85 \times 10^6 - 0}{85 \times 10^6 + 140.8 \times 10^6} \longrightarrow 2312°K \doteq 2300°K$$

Evaluation of ΔB

Stream Identification	Mass Flow	T	P	$\overline{B}_C$	$\overline{B}_T$	$\overline{B}_\lambda$	$\overline{B}_P$	$\overline{B}$	$m\overline{B}$
Units on Terms	[Kg-mole]	[°K]	[bar]	[J/Kg-mole]					[J]
Inlet									
$A(CH_4)$	1	298	1.01	818×10^6	0	0	0	818×10^6	818×10^6
$B(N_2)$	7.52	298	0.80	0	0	0	$-.58 \times 10^6$	$-.58 \times 10^6$	-4.36×10^6
$B(O_2)$	2.00	298	0.21	0	0	0	3.89×10^6	3.89×10^6	-7.78×10^6
$\sum a_i \overline{B}_{A_i}$ = 805.86 x 10^6									
Exit									
$P(CO_2)$	1	2300	0.095	0	75.3×10^6	0	-5.86×10^6	69.4×10^6	69.4×10^6
$P(H_2O)$	2	2300	0.190	0	64.4×10^6	0	-4.14×10^6	60.3×10^6	120.6×10^6
$P(N_2)$	7.52	2300	0.715	0	48.4×10^6	0	0.86×10^6	47.7×10^6	358.7×10^6
$P(O_2)$	0	2300	0	0	0	0			
$\sum b_i \overline{B}_{B_i}$ = 548.7 x 10^6									
$\Delta B = \sum b_i \overline{B}_{B_i} - \sum a_i \overline{B}_{A_i}$ = 548.7 x 10^6 - 805.9 x 10^6 = -257 x 10^6 [J]									

$_1w = -\Delta B = 257 \times 10^6 [J]$

Figure of Merit = $548.7 \times 10^6 / 805.9 \times 10^6 = 0.68$

The temperature reached in the adiabatic reactor depends upon the amount of excess reactant. Not only are the products of the reactant heated up by the chemical energy released, but also all excess reactant. In example problem 4.3 the theoretical amount of air was used. When more air is introduced, the adiabatic reaction temperature is lowered. The total amount of energy available in the combustion product stream is identical. Moreover, there are more moles of gas at a lower temperature. This lower temperature provides for lower efficiency for conversion of this energy to work. (See Problem 4.4.)

EXAMPLE 4.4   Assume that the products resulting from combustion of methane in excess air leave at 1000°K. What is the maximum amount of work attainable? The flow rates resulting in 1000°K adiabatic temperature are those given in the table below.

SOLUTION 4.4

Evaluation of $\sum b_i \overline{B}_{B_i}$ at 1000°K

Stream Identification	Mass Flow	T	P	$\overline{B}_C$	$\overline{B}_T$	$\overline{B}_\lambda$	$\overline{B}_P$	$\overline{B}$	$m\overline{B}$
Units on Terms	[Kg-mole]	[°K]	[bar]			[J/Kg-mole]			[J]
Exit									
P(CO_2)	1	1000	0.026	0	16.8×10^6	-	-9.1×10^6	7.7×10^6	7.7×10^6
P(H_2O)	2	1000	0.053	0	13.0×10^6	-	-7.3×10^6	5.7×10^6	11.4×10^6
P(N_2)	29.4	1000	0.777	0	10.6×10^6	-	$-.7 \times 10^6$	9.9×10^6	291.1×10^6
P(O_2)	5.8	1000	0.153	0	11.2×10^6	-	-4.7×10^6	6.5×10^6	37.7×10^6
$\sum b_i \overline{B}_{B_i}$ = 347.9 × 10⁶[J]									

$\Delta B = 347.9 \times 10^6 - 805.9 \times 10^6 = -458 \times 10^6 [J]$

$lw = 458 \times 10^6$

Figure of Merit = $347.9 \times 10^6 [J]/805.9 \times 10^6 [J] = 0.43$

4.6   CONVERSION OF HEAT TO POWER

Following the combustion of fossil fuel the energy released must be converted to power. The process most used for this conversion follows the Rankine Cycle. The features of the Rankine Cycle were introduced in Chapter 3 along with other cycles. The Rankine Cycle used in power plants uses water-steam as the working fluid. These cycles are presently limited to a pressure of about 5,000 psi and temperature of 1500°F. These constraints limit the maximum conversion of heat received by the working fluid to provide work.

EXAMPLE 4.5   Determine the maximum amount of heat that can be converted into work from steam at 5,000 psia and 1500°F.

SOLUTION 4.5

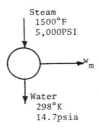

```
        Steam
        1500°F
        5,000PSI

              ────► w_m


        Water
        298°K
        14.7psia
```

From Steam Tables

$@$ 1500°F, 5000 psi $\bar{H}$ = 1729.9 $\bar{S}$ = 1.6169

$@$ 77°F (298°K) $\bar{H}$ = 45 $\bar{S}$ = 0.0874

$\Delta\bar{H}$ = 45 - 1729.9 = -1684.9[Btu/lb]

$\Delta\bar{S}$ = 0.0874 - 1.6169 = -1.5295[Btu/lb°R]

$\bar{w}_{m} = -\Delta\bar{H} + T_{o}\Delta\bar{S}$ = 1684.9[Btu/lb] - 1.5290[Btu/lb°R]

   x 537°R = 863.8[Btu/lb]

Total energy available above 77°F = 1729.9 - 45

   = 1684.9[Btu/lb]

Ratio of $[\bar{w}_{max}/\Delta\bar{H}_{available}]$ = (863.8/1684.9) = 0.51

The previous example shows that the amount of work that could be obtained because of the limitation of temperature and pressure for a steam system was only about 51% of the energy initially available in the steam. The problem was worked in English Units rather than the international units as this is still the practice in the U.S. power industry. The refined Rankine Cycle achieves energy conversions of up to 42%. This would indicate that power utility plants are approaching the maximum level that could be achieved by a steam cycle.

It is important to be able to analyze a process system to determine where the losses in the ability to do work occur. This will help to direct any efforts to improve the effectiveness of the process toward those areas where the maximum improvement can be achieved. This is illustrated in the following example.

EXAMPLE 4.6   Consider a simple power plant cycle generating steam to run a steam turbine under the following conditions: The fuel to be burned is pure methane burned with 100% excess air. Steam will be generated at 500 psia and superheated to 900°F. The condensate is saturated. The work of the pump may be neglected. The stack gases leave the furnace at 500°F. The turbine has an efficiency of 75% compared to an isotropic turbine and will exhaust at 1 psia. Cooling water is available at 77°F and water discharges at 85°F.

a. Draw a flow sheet of the process

b. Sketch a T-S diagram of the power cycle

c. Determine the flow rate thru the power cycle (Kg of water/Kg-mole of CH_4)

d. Determine the amount of work lost in each unit in the flow sheet.

SOLUTION 4.6

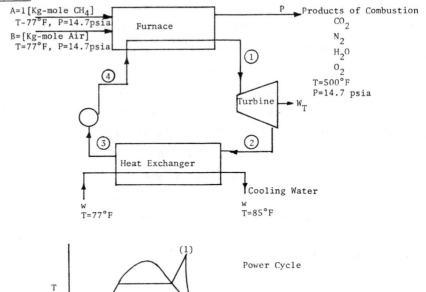

Power Cycle

Values obtained from steam tables

Point	State	T [°F]	P [psi]	$\bar{H}$ [Btu/lb]	$\bar{S}$ [Btu/lb°R]	$\bar{B}*$ [Btu/lb]
1	Superheat vapor	900	500	1466.0	1.6982	554.1
2	L + V					
	$X_v = 0.973$	101.7	1	1105.9	1.9284	70.3
3	Sat. Liquid	101.7	1	69.7	0.1326	-1.5
4	Sat. Liquid	101.7	1	69.7	0.1326	-1.5
Water @ 85°F		85	1	53	0.1061	3.4

$*\bar{B} = \bar{H} - T_o\bar{S}$

$\bar{B} = \bar{H} - 537\bar{S}$

Analysis of Turbine ($q = 0$)

$$\bar{w}_{act} = -\Delta H = -(H_2 - H_1) = -(1105.9 - 1466) = 360.9[Btu/lb]$$

$$\bar{w}_{max} = -\Delta B = -(70.3 - 554.1) = 483.8[Btu/lb]$$

$$l\bar{w} = 483.8 - 360.9 = 122.9[Btu/lb]$$

Analysis of Condensor ($w = 0$)

$$w_{act} = 0$$

$$w_{max} = -(-1.5 - 70.3) = 71.8[Btu/lb]$$

$$lw = 71.8[Btu/lb]$$

Analysis of Furnace

Evaluation of Products of Combustion (See Table 2.3)

1. Fuel Composition: CH_4: $\alpha = 1$, $\beta = 4$
2. Molecular Weight: $12(1) + 1(4) = 16$
3. Balanced Chemical Reaction: $CH_4 + (1 + 4/4)O_2 = CO_2 + 2 H_2O$
4. Theoretical Air: $(1 + 1)/0.21 = 9.52$
5. Actual Air: $9.52(1 + 1)/1 = 19.05$
6. Exit gas:

Exit gas	Moles	Mole Fraction
CO_2	1.0	0.05
H_2O	2.0	0.10
N_2	15.0	0.75
O_2	2.0	0.10
	20.0	

The Flow Rate of Working Fluid

This calculation is not provided but comes from the application of the First Law of Thermodynamics over the furnace.

1 Kg-mole CH_4 ⇌ 452 lbs of water in steam
  burned   power cycle

Stream Identification	Mass Flow	T	P	$\bar{B}_C$	$\bar{B}_T$	$\bar{B}_\lambda$	$\bar{B}_P$	$\bar{B}$	$m\bar{B}$
Units on Terms	[Kg-mole]	[°K]	[bar]	[J/Kg-mole]					[J]
Inlet									
A(CH_4)	1	298	1.01	818×10^6	0	0	0	818×10^6	818×10^6
B(N_2)	15	298	0.8	0	0	0	-0.58×10^6	-0.58×10^6	-8.9×10^6
B(O_2)	4.0	298	0.21	0	0	0	-3.89×10^6	-3.89×10^6	-15.6×10^6
Steam	452 lbs.	101.7°F	1 psi		See Table Above			-1.5[Btu/lb]	-678[Btu]
$\sum a_i \bar{B}_{A_i}$ = 793.5×10^6[J] - 678[Btu]x1055[J/Btu] = 7.93×10^8[J]									
Exit									
P(CO_2)	1	533	0.05	0	2.69×10^6	0	-7.4×10^6	-4.75×10^6	-4.75×10^6
P(N_2)	15	533	0.76	0	1.87×10^6	0	-0.7×10^6	1.17×10^6	17.55×10^6
P(H_2O)	2	533	0.10	0	2.20×10^6	0	-5.73×10^6	-3.53×10^6	-7.06×10^6
P(O_2)	2	533	0.10	0	1.94×10^6	0	-5.73×10^6	-3.79×10^6	-7.58×10^6
Steam	452 lbs.	900°F	500 psi	0	See Table Above			554.1[Btu/lb]	2.51×10^5[Btu]
$\sum b_i \bar{B}_{B_i}$ = -1.84×10^6[J] + 2.51×10^5[Btu] x 1055[J/Btu] = 2.63×10^8[J]									

$1w = 793.5 \times 10^6[J] - 263 \times 10^6[J] = 530.5 \times 10^6[J]$

Summary of 1w term

Turbine	$122.9[Btu/lb] \times 452[lb] \times 1055[J/Btu]$ =	$58.6 \times 10^6[J]$
Condensor	$71.8[Btu/lb] \times 452[lb] \times 1055[J/Btu]$ =	$34.2 \times 10^6[J]$
Furnace		= $530.5 \times 10^6[J]$

Furnace represents the greatest loss.

Example 4.6 clearly shows that it is not the turbine inefficiency or the heat discharged in the condensor that resulted in the lost ability to convert all of the energy to work. The major problem is associated with the furnace - the combustion system.

An alternative to the Rankine Cycle is the gas turbine or Brayton Cycle discussed in Chapter 3. To achieve high energy conversion, it is necessary to operate at high temperatures. However, the turbine is a high speed rotary device and this leads to severe mechanical problems associated with materials of construction. Until recently the temperatures

were limited to 1200°K to 1260°K. Recent work devoted toward
increasing this temperature includes:

1) New materials - metal alloys.

2) Ceramic coatings.

3) Turbine blade cooling
   a) Transpiration
   b) Sweating
   c) Convective.

It is anticipated that in the future temperatures as high as
1922°K may be reached.

The fuel burned in the furnace section of the gas turbine
may be any of the gaseous fuels described earlier. The hot
gas from combustion must not contain any solid particles over
a few microns or materials that might be a vapor at the furnace
temperature and would condense out in the cooler turbine.
The turbine is extremely sensitive to particulate matter that
rapidly erode the blades. For this reason, in spite of many
efforts, solid fuels have not been used successfully to fire
turbines. The problem is one of removing the particulate
material from a hot gas stream.

An attractive approach is to gasify the solid fuel to
provide a low Btu gas. This gas is much lower in volume than
the combustion products. It may be cleaned-up hot or cooled-
cleaned and then combusted in the turbine burner. The
hot clean-up system is preferred because the sensible heat is
not lost during clean-up.

4.6.1 Combined Cycles

There are several systems for generation of electricity
that combine power cycles. The efficiencies of these combined
cycles are higher than for a single cycle energy conversion

process. Figure 4.11 shows a typical combination gas turbine-
Rankine Cycle. The gas leaving the gas turbine, stream ⟨4⟩ ,
contains significant amounts of sensible heat. This sensible
heat is recovered to produce steam for a Rankine Steam Cycle
(shaded area).

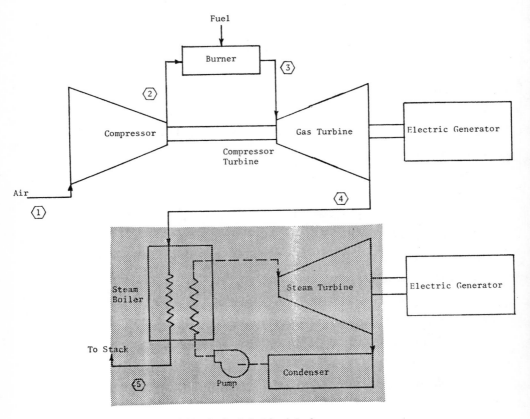

FIGURE 4.11 Typical Combined Cycle

Table 4.4 shows some predicted parameters for firing a
low energy gas in a combined cycle. The I, II, and III refer
to present technology, technology under development and future

technology, respectively. Notice the increase in the temperature
predicted. The efficiencies (work out/chemical energy in) are
higher than the single Rankine Cycle. The increased efficiency
results from development of gas turbines that may operate at
higher temperatures.

TABLE 4.4
Proposed Power Systems

Generation	I	II	III
Number of gas turbines	3	2	2
Turbine inlet temperature, °K	1478	1822	1978
Compressor pressure ratio	8	12	20
Percent of airflow bled for cooling	4.7	8.5	9.0
Turbine exhaust temperature, °K	976	1097	1081
Single steam turbine, of size	431 MW	381 MW	312 MW
Stack temperature, °K	430	378	389
System efficiency, w/ΔH_R	47.0	54.5	57.7

EXAMPLE 4.7   The power plant given in Example 4.6 will be modified
as follows. The furnace shown will be replaced by a
pressurized furnace that operates at 10.1 bar. The
gas and air will be compressed up to 10.1 bar
(adiabatically and reversibly). The fuel and air
will be burned. The turbine temperature cannot exceed
1900°K and the amount of air used is selected so that
this temperature is not exceeded. The product of combustion will leave at 500°F as before. The gas and
compressors are 100% efficient. Determine:

a) The work from the gas cycle (net).
b) The work from the steam cycle.
c) Compare the amount of work from the combined
cycle to the work from the Rankine Cycle (Problem 4.6).
d) Assume all gas behaves as ideal diatomic gases
to evaluate temperatures.

SOLUTION 4.7

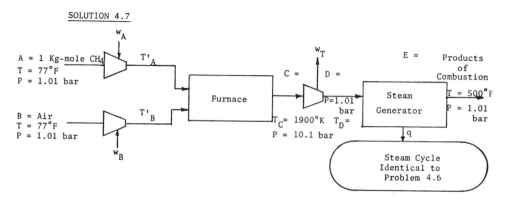

Analysis of Turbine Section

Evaluation of $\overline{w}_A$

$$T'_A = T_A(P'_A/P_A)^{((\gamma-1)/\gamma)} = 298°K(10)^{((1.4-1)/1.4)} = 576°K$$

$$\overline{w}_A = \overline{H}_{576°K} - \overline{H}_{298°K} = [12.1 \times 10^6 - 0][J]$$

Evaluation of $\overline{w}_B$

$$T'_B = T'_A = 576°K$$

$$\overline{w}_B = \overline{H}_{576°K} - \overline{H}_{298°K} = [8.27 \times 10^6 - 0][J]$$

Evaluation of Air Flow Rate - The air is adjusted until the exit from the furnace equals 1900°K for q and w = 0. This is a trial and error solution.

For 100% Excess Air

Stream Identification	Mass Flow	T	P	$\overline{H}_C$	$\overline{H}_T$	$\overline{H}_\lambda$	$\overline{H}_P$	$\overline{H}$	$m\overline{H}$
Units on Terms	[Kg-mole]	[°K]	[bar]	[J/Kg-mole]					[J]
Inlet $A(CH_4)$	1	576	10.1	890.9×10^6	12.1×10^6	0	0	903×10^6	903×10^6
$B(N_2)$	15	576	7.98	0	8.21×10^6	0	0	8.21×10^6	123×10^6
$B(O_2)$	4	576	212	0	8.85×10^6	0	0	8.85×10^6	35×10^6
$\sum a_i \overline{H}_{A_i} = 1061 \times 10^6$									
Exit $P(CO_2)$	1	1900	0.05	0	84.7×10^6	0	0	84.7×10^6	84.7×10^6
$P(H_2O)$	2	1900	0.10	0	67.8×10^6	44×10^6	0	111.8×10^6	223.6×10^6
$P(N_2)$	15	1900	0.75	0	52.9×10^6	0	0	52.9×10^6	793.5×10^6
$P(O_2)$	2	1900	0.10	0	54.8×10^6	0	0	54.8×10^6	109.6×10^6
$\sum b_i \overline{H}_{B_i} = 1211 \times 10^6$									
$\Delta H = \sum b_i \overline{H}_{B_i} - \sum a_i \overline{H}_{A_i} = 1211 \times 10^6 - 1061 \times 10^6 = 150 \times 10^6$									

For 50% Excess Air

Stream Identification	Mass Flow	T	P	$\overline{H}_C$	$\overline{H}_T$	$\overline{H}_\lambda$	$\overline{H}_p$	$\overline{H}$	$m\overline{H}$
Units on Terms	[Kg-mole]	[°K]	[bar]			[J/Kg-mole]			[J]
Inlet									
A(CH_4)	1	576		890.9×10^6	12.1×10^6	0	0	903×10^6	903×10^6
B(N_2)	11.25	576		0	8.21×10^6	0	0		92.4×10^6
B(O_2)	3	576		0	8.85×10^6	0	0	11.5×10^6	26.6×10^6
$\sum a_i \overline{H}_{A_i} = 1022 \times 10^6 \,[J]$									
Exit									
CO_2	1	1900		0	84.7×10^6	0	0	84.7×10^6	84.7×10^6
H_2O	2	1900		0	67.8×10^6	44×10^6	0	111.8×10^6	223.6×10^6
N_2	11.25	1900		0	52.9×10^6	0	0	52.9×10^6	595.1×10^6
O_2	1	1900		0	54.8×10^6	0	0	54.8×10^6	54.8×10^6
$\sum b_i \overline{H}_{B_i} = 958.2 \times 10^6 \,[J]$									
$\Delta H = \sum b_i \overline{H}_{B_i} - \sum a_i \overline{H}_{A_i} = 958.2 \times 10^6 - 1022 \times 10^6 = -63.8 \times 10^6$									

By linear extrapolation between these two excess air values gives:

Let X = % excess air at q = 0

$$\frac{X - 50}{100 - 50} = \frac{0 - (-63.8)}{150 - (-63.8)} = 29.5$$

X = 50 + 29.5 = 79.5%

For 79.5% excess air:

    1. Fuel composition: $\alpha = 1$, $\beta = 4$

    2. Molecular weight: 16

    3. Balanced chemical reaction: $CH_4 + 2O_2 \longrightarrow CO_2 + 2H_2O$

    4. Theoretical air = 9.52

    5. Actual air = 9.52(1.795) = 17.088

    6. Exit Gas

Exit Gas	Moles	Mole Fraction	
CO_2	1	0.06	
H_2O	2	0.11	This is the gas to the turbine and steam generator.
O_2	1.59	0.09	
N_2	13.5	0.75	
	18.1	1.01	

Temperature of exit gas from turbine:

    For a reversible turbine

$$T_D = 1900(1.01/10.1)^{0.286} = 979 \,[°K] \text{ (For adiabatic reversible change)}$$

Evaluation of Gas Turbine

Stream Identification	Mass Flow	T	P	$\bar{H}_C$	$\bar{H}_T$	$\bar{H}_\lambda$	$\bar{H}_p$	$\bar{H}$	$m\bar{H}$
Units on Terms	[Kg-mole]	[°K]	[bar]	[J-Kg-mole]					[J]
Inlet									
$C(CO_2)$	1	1900		0	84.7×10^6	0	0	84.7×10^6	84.7×10^6
$C(H_2O)$	2	1900		0	67.8×10^6	44×10^6	0	111.8×10^6	223.6×10^6
$C(O_2)$	1.59	1900		0	54.8×10^6	0	0	54.8×10^6	87.1×10^6
$C(N_2)$	13.5	1900		0	52.9×10^6	0	0	52.9×10^6	714.2×10^6
$\sum a_i \bar{H}_{A_i} = 1109.6 \times 10^6$									
Exit									
$D(CO_2)$	1	979		0	32.1×10^6	0	0	32.1×10^6	32.1×10^6
$D(H_2O)$	2	979		0	25.2×10^6	44×10^6	0	69.2×10^6	138.4×10^6
$D(O_2)$	1.59	979		0	21.9×10^6	0	0	21.9×10^6	34.8×10^6
$D(N_2)$	13.5	979		0	20.8×10^6	0	0	20.8×10^6	280.8×10^6
$\sum b_i \bar{H}_{B_i} = 486.1 \times 10^6$									
$\Delta H = \sum b_i \bar{H}_{B_i} - \sum a_i \bar{H}_{A_i} = 486.1 \times 10^6 - 1109.6 \times 10^6 = -623.5 \times 10^6$									

$$\Delta H(\text{Turbine}) = q - w \quad (q = 0)$$

$$w_T = -\Delta H = 623.9 \times 10^6 [J]$$

$$w_{net} = w_T - A\bar{w}_A - B\bar{w}_B$$

$$= 623.9 \times 10^6 - (1)(12.1 \times 10^6) - (17.1)(8.27 \times 10^6) = 470 \times 10^6 [J]$$

The amount of heat transferred to the steam is obtained from a balance about the steam generator.

Stream Identification	Mass Flow	T	P	$\bar{H}_C$	$\bar{H}_T$	$\bar{H}_\lambda$	$\bar{H}_p$	$\bar{H}$	$m\bar{H}$
Units on Terms	[Kg-mole]	[°K]	[bar]	[J/Kg-mole]					[J]
$\sum a_i \bar{H}_{A_i} = 632.7 \times 10^6 [J]$ - Same as output from turbine									
Exit									
CO_2	1	533		0	9.8×10^6	0		9.8×10^6	9.8×10^6
H_2O	2	533		0	8.1×10^6	44×10^6		52.1×10^6	104.2×10^6
O_2	1.59	533		0	7.1×10^6	0		7.1×10^6	11.3×10^6
N_2	13.5	533		0	6.9×10^6	0		6.9×10^6	93.2×10^6
$\sum b_i \bar{H}_{B_i} = 218.5 \times 10^6 [J]$									
$\Delta H = \sum b_i \bar{H}_{B_i} - \sum a_i \bar{H}_{A_i} = 218.5 \times 10^6 - 632.7 \times 10^6 = -414.2 \times 10^6 [J]$									

$$\Delta H = q = -414.2 \times 10^6 [J]$$

Analysis of Steam Section

Balance Over Steam Generator

Stream Identification	Mass Flow	T	P	$\bar{H}_C$	$\bar{H}_T$	$\bar{H}_\lambda$	$\bar{H}_p$	$\bar{H}$	$m\bar{H}$
Units on Terms	[Kg-mole]	[°K]	[bar]		[J/Kg-mole]				[J]
Inlet									
$D(CO_2)$	1	979		0	32.1×10^6	0	0	32.1×10^6	32.1×10^6
$D(H_2O)$	2	979		0	25.2×10^6	44×10^6	0	69.2×10^6	138.4×10^6
$D(O_2)$	1.59	979		0	21.9×10^6	0	0	21.9×10^6	34.8×10^6
$D(N_2)$	13.5	979		0	20.8×10^6	0	0	20.8×10^6	280.8×10^6
S(Steam)	S(lbs)	–	See	Point 4	Problem	4.6		69.7Btu/lb	69.7S Btu
$\sum a_i \bar{H}_{A_i}$ = 486.1 + 69.7S									
Exit									
$E(CO_2)$	1	533		0	9.8×10^6	0	0	9.8×10^6	9.8×10^6
$E(H_2O)$	2	533		0	8.1×10^6	44×10^6	0	52.1×10^6	104.2×10^6
$E(O_2)$	1.59	533		0	7.1×10^6	0	0	7.1×10^6	11.3×10^6
$E(N_2)$	13.5	533		0	6.9×10^6	0	0	6.9×10^6	93.2×10^6
S(Steam)	S(lbs)	–	See	Point 1	Problem	4.6		1466Btu/lb	1466S
$\sum b_i \bar{H}_{B_i}$ = 218.5×10^6[J] + 1466S[Btu]									
$\Delta H = \sum b_i \bar{H}_{B_i} - \sum a_i \bar{H}_{A_i}$ = 218.5×10^6[J] + 1466S[Btu] – 486.1×10^6[J] – 69.7S[Btu]									

$\Delta H = -267.1 \times 10^6$[J] $- 1396.3$S[Btu] $\times 1055$[J/Btu] $= 0$

$S = (267.1 \times 10^6)/(1396.3 \times 1055) = 181$[lbs steam]

Work from Steam Turbine (See Problem 4.6)

181[lbs steam] x (1466 - 1105.9)[Btu work/lb steam]

　　x 1055[J/Btu] = 69×10^6[J]

a) Work from Gas Turbine = 470×10^6[J]
b) Work from Steam Turbine = 69×10^6[J]
c) Combined Work = 539.0×10^6[J]
d) Work from Rankine Cycle = in Problem 4.6 = 452[lbs steam] x (1466 - 1105.9)

　　[Btu work/lb steam] x 1055[J/Btu] = 172×10^6[J]

This example shows the combined cycle in the most favorable light. The temperature used was high in terms of future technology and the turbine and compressors were 100% efficient. For more realistic cases, the work from gas cycle is significantly less and the work from the steam cycle is significantly more.

4.6.2 Magnetohydrodynamics (MHD)

The magnetohydrodynamics system relies on the basic phenomena of electromagnetics that states that a voltage will
develop along an electric conductor moved through a magnetic
field. For a conventional generator, the conductor is a copper
wire but in a MHD unit the conductor is a fluid. Figure 4.12
shows a schematic diagram of a MHD electric generator.

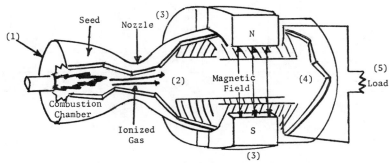

Powdered coal burns in a combustion chamber (1). Ionized
gas passing through a nozzle (2) at supersonic speed.
Ionized gas stream passes between magnetic field (3);
current taken off by electrodes (4) through external load (5).

FIGURE 4.12 Schematic of an MHD Generator

In Section 1 the combustion gas reaches high temperatures in
the range of 2000 to 5000°K. This gas is "seeded" with elements such as cesium that became ionized (lose orbital electrons).
These free electrons makes it possible for the gas to conduct
electricity. The gas is accelerated through a nozzle and sent
to a strong magnetic field. This magnetic field deflects the
electrons to the side causing a voltage difference. This causes
the electrons to flow through the external circuit.

The high velocities are obtained by compressing the gas
to be passed through the channel. The gas is expanded through

a nozzle which provides the high velocity. The MHD converter
takes the place of the turbine in a power cycle. Figure 4.13
shows a MHD power cycle and the Brayton Power Cycle. They
both require a high temperature source, a low temperature sink
and compressor to operate. The difference is in the power
generating unit. The T-S diagram for both units is also given
in Figure 4.13. It is clear that the MHD cycle is a thermal
power cycle and the cycle thermal efficiency is written as

$$\frac{w}{q} = 1 - \frac{T_C}{T_H}$$

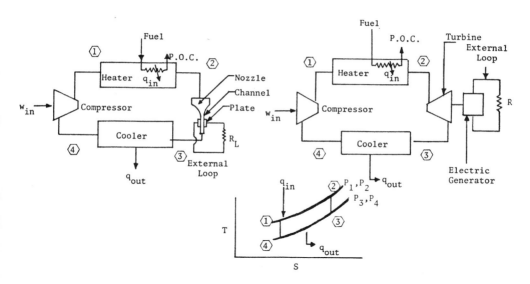

FIGURE 4.13 Comparison of MHD and Brayton Gas Cycle

The advantage of the MHD cycle results from the high
temperature operation. At the high temperatures of the MHD
converter the Carnot Efficiency is high when the temperature
of the surroundings can be used as the low temperature sink.

Unfortunately, the MHD power unit can only operate at high
temperatures and the gas loses its ionizing properties at
lower temperatures (2500 to 2000°K). When the gas can no
longer conduct electricity, there is no longer any output from
the MHD unit. For practical purposes the lowest temperature
from the MHD unit is on the order of 2000°K. For this reason
the MHD power cycle by itself will not have a high efficiency
(electricity out/heat in).

As in the case of the combined cycle power system, the
MHD cycle may be combined with the Rankine Cycle to improve
the performance. The MHD cycle has no moving parts and can
achieve temperatures much higher than those of the gas turbine.
This results in a higher Carnot Efficiency. Both the gas turbine and the MHD power cycle may be termed topping cycles as
they operate "on top of" the basic Rankine Cycle. The heat
output of the topping cycle feeds the Rankine Cycle.

4.6.3 Fuel Cells

Fuel cells are out of place in this chapter on combustion.
The combustion reaction does not take place (a combination of
reactions when added together result in the same products as
combustion). Chemical energy is never converted to thermal
energy. The Carnot Cycle limitation does not apply to the
fuel cell.

The fuel cell is similar to a battery. Two electrodes
are placed in an electrolyte solution. This is shown in
Figure 4.14. The hydrogen and oxygen source are under pressure
and diffuse through the electrodes to the electrolyte.

At the surface of the anode the hydrogen combines with
the hydroxal ions according to the reaction.

$$H_2 + 2\overline{OH} \longrightarrow 2H_2O + 2\overline{e}$$

The electrons flow through the external circuit to the cathode where they react

$$1/2\ O_2 + H_2O + 2\overline{e} \longrightarrow 2\overline{OH}$$

The overall reaction becomes

$$H_2 + 2\ \overline{OH} \longrightarrow 2H_2O + 2\overline{e}$$

$$\underline{1/2\ O_2 + H_2O + 2\overline{e} \longrightarrow 2\overline{OH}}$$

$$H_2 + 1/2\ O_2 \longrightarrow H_2O$$

which is the same reaction as obtained from the combustion of hydrogen.

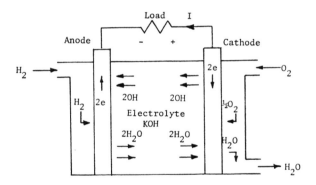

FIGURE 4.14   Schematic Representation of a
Hydrogen-Oxygen Fuel Cell

EXAMPLE 4.8   Determine the efficiency of the fuel cell. The
efficiency is defined as the maximum work out over
the heat of combustion of the hydrogen.

SOLUTION 4.8

The maximum work from the cell may be obtained from the difference between the maximum work of the output streams and the maximum work of the input streams. Assuming that the pressure of oxygen on hydrogen required is 20atm. (H_2 + 1/2 O_2 H_2O)

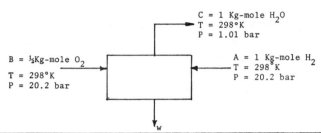

$$C = 1 \text{ Kg-mole } H_2O$$
$$T = 298°K$$
$$P = 1.01 \text{ bar}$$

$$B = \tfrac{1}{2}\text{Kg-mole } O_2$$
$$T = 298°K$$
$$P = 20.2 \text{ bar}$$

$$A = 1 \text{ Kg-mole } H_2$$
$$T = 298°K$$
$$P = 20.2 \text{ bar}$$

Stream	Kg-moles	T [°K]	P [bar]	$\overline{B}_R$ [J/Kg-mole]	$\overline{B}_T$ [J/Kg-mole]	$\overline{B}_P$ [J/Kg-mole]	B [J]
A	1	298	20.2	0	0	7.4×10^6	7.4×10^6
B	1/2	298	20.2	237×10^6	0	7.4×10^6	122×10^6
C	1	298	1.01	0	0	0	0

$$w_{max} = -\Delta B = 122 \times 10^6 [J] + 7.4 \times 10^6 [J] - 0 = 129.4 \times 10^6 [J]$$
$$H_{reactants} = 0.5 [\text{Kg-mole } H_2] \times 286 \times 10^6 [J/\text{Kg-mole}] = 143 \times 10^6 [J]$$
$$\eta = 129.4 \times 10^6 / 143 \times 10^6 = 0.91$$

It is apparent that the fuel cell system is far more effective in converting chemical stored energy to work than a combustion system.

The work that is available in the H_2 and O_2 because of the elevated pressure is used to force the gas through the electrode and is not recovered. It is a potential source of work that is not realized. Eliminating these from the work term

$$\eta = 118.5 \times 10^6 / 143 \times 10^6 = 0.83$$

The fuel cell is a very effective conversion system if the high theoretical conversions could be achieved. The cells have not come close to achieving this potential.

4.6.4   Co-Generation Power Systems

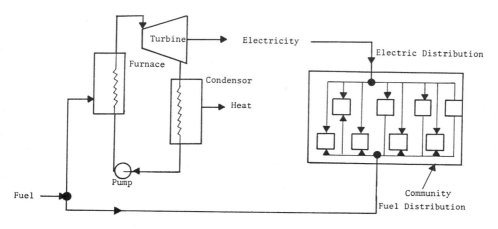

A - Typical Power Utility in U.S.

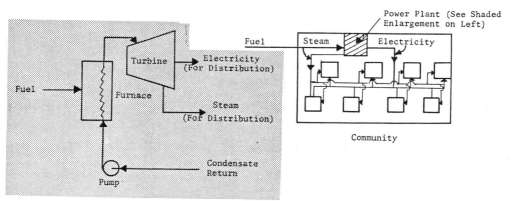

B - Co-Generation System

FIGURE 4.15   Schematic Diagram For (A) Large Power
Station and (B) Co-Generation System

Figure 4.15 shows two systems to provide for both heat and electricity to a community. The top shows the system used in the U.S. The electricity is generated in large generating facilities located remote from the consumer. These large

facilities achieve economies because of the large scale operations and the electrical conversion efficiency is high.
They generate large amounts of waste heat in the condensor.
This heat has no value and is dumped. To provide heat for
the community, separate furnaces are used in each community
unit.

In contrast to providing large remote units, it is the
practice in Europe to build smaller plants that are dispersed.
The heat from the condensor is used to provide heat to the
community. This is shown in the lower system of Figure 4.15.
The power plant is smaller and there is a lower conversion
efficiency to electricity (electricity produced/heating value
of fuel) than in the larger facility. Since the steam leaving
the turbine is needed at a higher temperature for heating,
the exit pressure is much higher. This reduces the cycle
efficiency in terms of the electricity generated. The energy
available in the steam is not dumped as in the remote system,
but is used as a heating energy source eliminating the need
for individual furnaces burning fuel at each community unit.
The amount of energy that is required to furnish both electricity and heat to the community is less for the second system.

Problems for Chapter 4.0

1. Estimate the heating value of a medium volatile West
   Virginia bituminous coal and compare it to the value
   shown in Table 4.1 and Figure 4.3.

2. The developing interest in the fluidized bed combustion
   system can be seen in the attendance at two "International
   Conference on Fluid Bed Combustion." The first conference
   was held in 1967 in Hueston Woods, Ohio. It was sponsored
   by the Environmental Protection Agency and attracted only
   40 participants. The fifth conference in the series was
   held ten years later in Washington, D.C. Over 500 participants from 13 countries were in attendance. Sponsor-
   ship was the U.S. Department of Energy. The theme was the
   same. The fluid bed has the potential to burn high sulfur
   eastern coals and meet the environmental restrictions
   at a lower cost than other alternatives. The most recent
   data suggest that about two times the theoretical amount
   of $CaCO_3$ would be required to remove 90% of the SO_2 formed.
   Evaluate the ratio of $Kg(CaCO_3)/Kg(coal)$ for a 3% coal.

3. Rather than reduce the pressure to 2 psia in the power
   plant given in Example 4.6, the pressure will be reduced
   to 20 psi. Rather than condense this steam directly at
   the power plant, the steam is distributed for district
   heating.
     (a) Compare the amount of electricity generated
         to the amount generated in Example 4.6.
     (b) How much energy is made available for district
         heating? (Assume that the heat of vaporization
         is recovered as heat.)
     (c) Compare the amount of fuel required to provide
         a separate power station to produce electricity
         and heating system for district heating to the
         fuel used to provide the same amount of power
         and district heating from a co-generation sys-
         tem.
   For part (c) assume that 80% of the fuel value goes into
   heat for district heating in a separate unit.

4. A fluidized bed combustion unit operates at a temperature
   of less than 1300°K. One of the advantages of a fluid
   bed is a simple construction that allows it to operate at
   elevated pressure. Compare two fluidized bed systems.
     1. The 1300°K temperature is maintained by adding
        sufficient excess air to the system.

2. The 1300°K temperature is maintained by adding
    heat transfer area to the bed and generating
    steam.

Determine:
    (a) The net amount of energy produced in each system.
    (b) The energy mix (i.e., fraction from Rankine
        Cycle and Gas Turbine) in each case.

Assume:
    1. Operating pressure is 10.1 bar.
    2. 20% excess air is necessary for complete com-
        bustion.
    3. All steam is used in a Rankine Cycle similar
        to Example 4.6.
    4. All turbines and compressors are reversible.
    5. Pure char is the fuel source.

5. The reactivity of a char is often more than 100 times
    that of a hard coal (value of K in Equation 4.2). Com-
    pare the use of char and hard coal under the following
    conditions.
    (a) Both solid fuels are to be burned in 5% O_2
         in the exit stack. The reactor is well mixed.
         Both fuels are the same size. Compare the
         reaction times.
    (b) Both fuels are to be burned in 5% O_2 in the
         exit stack. The reactor is well mixed. The
         reaction time for each fuel is to be the same.
         Compare the particle size required.
    (c) For the hard coal the amount of O_2 is allowed
         to increase to 10%. The reactor is well mixed.
         The reactor size is to be the same. Compare
         the reactor size to obtain complete consumption
         of fuel.

6. A report to EPRI (Electric Power Research Institute) states
    that the efficiency of a medium energy gas is greater than
    that of methane. They define efficiency as $q_{out}/\Delta H$ combustior.
    Assume 25% excess air and an exhaust temperature of 400°F.
    Evaluate the efficiency of CH_4 and compare it to a 50:50
    mixture of H_2:CO. Check the accuracy of this statement.

7. Example 4.3 evaluated the adiabatic flame temperature,
    lost work and Figure of Merit. It was stated that the
    addition of excess air reduced both the excess air and the
    Figure of Merit and reduces the lost work. To check this

out, repeat Example 4.4 using 100% excess air for combustion.

8. Uniformly sized carbon char is fed to an isothermal
    reactor. A 12-gram sample (1 g-mole) provides 12-1
    gram samples (spherical). The char is burned in a plug
    flow furnace with 10% excess air. The reactivity con-
    stant is 1.0[g-mole/(cm^2-unit time)]. In a second case
    three grams (25%) of the original char is sent to a crusher
    where they are crushed into 300 equal sized spherical
    particles (char density is 0.7 g-cc). Determine for
    both cases:
    (a) Mass remaining as a function of time in reactor.
    (b) Reaction rate as a function of time in the
         reactor.
    (c) Time when the reaction rate is the same.
    (d) Time when the mass remaining is the same.

9. General Electric in a report "Integrated Gasification-
    Gas Turbine Cycle Performance," No. 75CRDC 21, March
    1975, for a gas turbine, they gave the following condi-
    tions:

    Hypothetical Gas Turbine Characteristics

Turbine inlet temperature	T = 1800°F
Compressor ratio	X = 10/1
Compressor efficiency	87.5
Turbine efficiency	87.5
Combustion efficiency	99%
Total pressure loss $\Delta P/P$	5%
Leakage loss	3%
Generator loss	2%

    Evaluate:
    (a) Temperature exiting turbine.
    (b) Cycle efficiency
    (c) Draw flow-sheet.

10. A single fuel cell can provide 100-200 milliamps per cm^2
    of electrode surface at a one volt potential. By com-
    bining the output in series and providing sufficient
    area, most any size output can be constructed. For a
    demand of 250 Kw @ 100V:
    (a) Determine the total surface area of electrode.
    (b) Number of cells in series.

11. A fluidized bed furnace has many desirable qualities that
    provide for good combustion of many fuels. The major
    drawback is that the gases after combustion leave the
    furnace at the same temperature that combustion occurs
    (in most cases about 1700°F). These hot gases contain
    significant energy that will be lost. To reduce this
    loss, it was decided to install a heat exchanger to use
    these hot gases to heat up the air coming into the furn-
    ace. (Fuel is C burned in 50% excess air.)

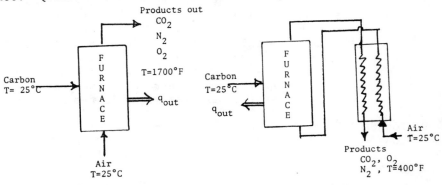

Original System   Revised System

    (a) Determine the composition of the product gas.
    (b) Determine the temperature of air to the furnace.
    (c) Determine the increase in q_{out} by adding the
        heat exchanger.
    (d) Steam that could be generated from the heat
        saved (200 psi sat.).

12. An empirical formula for the organic portion of municipal
    waste is given by

    $$C_{30} H_{48} O_{19} S_{0.05}$$

    Determine:
    (a) The heating value (Btu/lb).
    (b) The gas composition if burned with 100% excess
        air.

    NOTE: Must convert the equation above to weight %.

13. Butane is burned in a furnace to produce steam. The
    combustion system requires 50% excess air and the product
    gases leave at 600°F.

(a) Calculate air flow rate (Kg-mole butane).

(b) Calculate the kg steam generated/kg-mole butane.
Steam conditions are the same as Example 4.6.

14. In Example 4.3 the adiabatic flame temperature of methane
was calculated to be 2312°K. From Table 4.3 the flame
temperature for natural gas streams ranged between 2333
to 2383°K. These gases were not 100% methane but contained ethane and propane. This observation suggests
that ethane and propane have a higher flame temperature.

(a) Determine the flame temperature of ethane.

(b) Assume that the flame temperature of a mixture
of ethane and methane can be obtained from

$$T_{mix} = x_{ethane} \, T_{ethane} + x_{methane} \, T_{ethane}$$

Determine the adiabatic flame temperature for
natural gas from Sandusky, Ohio and McKean,
Pennsylvania. Compare to tabulated value.

(c) Compare the Figure of Merit of ethane to that
of methane.

15. Assume that methane and air are compressed to 10.1 bar
adiabatically and reversibly. This gas is then burned
and used in a Magnetohydrodynamic Generater. The temperature exits at 2000°K because the gas is no longer ionized
at this temperature.

(a) What is the temperature of the inlet gas?

(b) What is the maximum amount of work that may be
extracted from this generator?

16. Consider a community with the following energy demands:

1. Electricity   100 energy units
(Efficiency of Conversion = 40%)

2. Steam   100 energy units
(Efficiency of Conversion = 80%)

3. Heat   50 energy units
(Efficiency of Conversion = 80%)

(a) How much fuel energy is consumed?

(b) Devise a co-generation system to provide all
three needs. Assume smaller electric generator
is only 35% efficient.

5.0 COAL GASIFICATION AND LIQUEFACTION

It has been suggested that there are only two things wrong
with coal today: You cannot mine it, and you cannot burn it. It
would be more appropriate to state that you cannot mine or
burn it in an environmentally acceptable manner; or you cannot
mine it or burn it in the manner that has been practiced in
the past. This chapter will discuss the conversion of coal to
a gas or liquid that can be burned in an environmentally acceptable manner. The necessity of using the nation's coal resources
to replace oil and natural gas supplies was discussed in Chapter
1. This chapter discusses conversion of coal into substitute
gas and liquid fuels.

5.1 GASIFICATION

Figure 5.1 shows a generalized process flow sheet for the
conversion of coal into a substitute natural gas (S.N.G.).

Richard C. Bailie, Energy Conversion Engineering, ISBN 0-201-00840-8

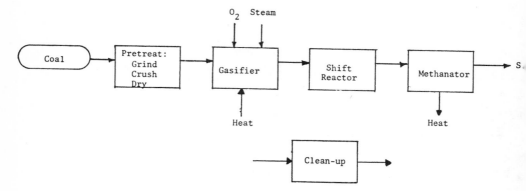

FIGURE 5.1   Process Flow Sheet - Coal to S.N.G.

In order to study the conversion of coal to S.N.G. it
is necessary to appreciate the chemistry of the major reactions
that occur. The important reactions to be considered are
presented in Table 5.1 along with some facts associated with
each reaction. The information provided in Table 5.1 is

Column 1 - The name given to the reaction.

Column 2 - The balanced chemical reaction.

Column 3 - The heat of reaction and a qualitative statement suggesting the heat effects.

Column 4 - The temperature range where the equilibrium
constant is 1, or more, i.e., the reaction
may proceed to a significant level of conversion.

Column 5 - Qualitative statement regarding the reaction
kinetics.

Column 6 - Classification of the reaction as catalytic
or non-catalytic.

This information will be used in the discussion of reaction
schemes.

TABLE 5.1

Basic Chemistry for Coal Gasification

Name	Reaction	Heat of Reaction $J/Kg\text{-}mole$	Temperature $K_p > 1$	Reaction Kinetics	Catalytic
Carbon Steam	$C + H_2O \rightleftharpoons CO + H_2$	$+1.359 \times 10^8$ Highly Endothermic	$\downarrow 947°K$	Fast above 1200°K	No
Pyrolysis	Volatile Coal $\rightarrow$ C + Tar + CH_4 + CO + H_2 + C_2H_4 etc.	Neutral	-	Fast above 980°K	No
Water Gas Shift	$H_2O + CO \rightleftharpoons H_2 + CO_2$	-3.07×10^7 Slightly Exothermic	$\downarrow 1100°K$	Fast above 760°K	Yes
Hydrogasification	$C + 2H_2 \rightleftharpoons CH_4$	-9.148×10^7 Highly Exothermic	$\downarrow 819°K$	Fast above 980°K	No
Methanation	$CO + 3H_2 \rightleftharpoons CH_4 + H_2O$	2.503×10^8 Highly Exothermic	$\downarrow 900°K$		Yes
No Name	$CO_2 + C \rightleftharpoons 2CO$	Moderately Endothermic	$1070°K$	Fast above 1260°K	No
Combustion a) Complete	$C + O_2 \rightleftharpoons CO_2$	-3.937×10^8 Highly Exothermic		Fast above 1000°K	No
b) Partial	$C + 1/2 O_2 \rightarrow CO$	-1.105×10^8 Moderately Exothermic		Fast above 1000°K	No
Column Number:	(1)	(2)	(3)	(4)	(5)	(6)

5.1.1 <u>Generalized Process Flow Sheet for Generation of S.N.G.</u>

Figure 5.1 shows a generalized flow sheet for the production of synthetic natural gas from coal. Almost all of the processes under development include these steps. In some systems each reaction is carried out in a separate reactor and in some systems more than a single reaction is carried out in one reactor. Each of the basic steps is discussed briefly.

1. Fuel Preparation - This may include size reduction pretreatment and/or mild oxidation and/or pyrolysis. When coal is exposed to a high temperature environment, it undergoes a series of chemical and physical changes. Physically the coal softens and swells. This physical change causes serious problems in a reacting system: In a fluid bed the fluidization may be destroyed; in a packed bed a crust can form that inhibits the flow of gas. It is essential that the coal be treated to remove this characteristic prior to any gasification. When the coal is heated, the volatile matter in the coal distills or pyrolyzes off. This may be burned for energy or added to the product from the gasifier. This leaves a solid non-volatile char as the material to undergo the conversion to gas.

2. Gasification (Reaction 1) - This is the critical step in the conversion of coal to synthetic natural gas. It is in this step that the solid is transformed into gas. The subsequent steps consist of gas phase reactions. There are many different ways to carry out this gasification step. The primary difference between different systems is usually found in the manner that this step is carried out.

The major output from this step is CO and H_2.
Depending on the process, some methane is produced in
this step.

3. Waste-Gas Shift (Reaction 3) - This step is carried
   out in a catalyst bed and is common to many systems.
   It modifies the ratio of H_2/CO to three to one. These
   are the conditions necessary for the methanation step.

4. Methanation (Reaction 5) - The H_2, CO mixture is
   converted in the presence of a catalyst to methane
   which serves as the synthetic natural gas.

5. Clean-up - The objective of the conversion of coal
   to gas was to resolve the environmental problem.
   Nowhere in the general process shown in Figure 5.1
   were any of the impurities, primarily sulfur and ash,
   that ultimately contribute to the abuse of the environment removed. Somewhere in the process these impurities must be removed. Where this is done in the
   overall scheme varies with each process. It is
   sometimes carried out at several locations. For this
   reason the clean-up step is shown in the figure but
   has not been located in the generalized process flow
   sheet.

5.1.2 Feed Preparation

Depending upon the process the coal must be sized. For
moving bed systems the gasifier required lumps. Any fines
must be briquetted. Fluidized beds require relatively small
particles, and entrained bed systems require small sized
particles. In these cases the coal must be crushed.

If the coal to be gasified is the type that will cake
or agglomerate in the gasifier, a pretreatment step is required.
This usually consists of a thermal treatment where the particle
is partially reacted to give a nonsticking surface. This may
be done by reacting with steam, oxygen, or by pyrolysis. In
almost all systems the gas leaving the pretreatment step contains some pyrolysis products (Reaction 2).

5.1.3 Gasification Reactor

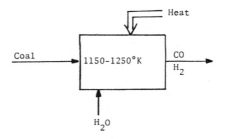

The purpose of the gasification reactor is to convert
the char to a gaseous state. The primary reaction is the
carbon-steam reaction,

$$C + H_2O \rightleftharpoons CO + H_2 \qquad\qquad \text{Reaction 1}$$

It can be seen from Table 5.1 that this reaction absorbs large
amounts of heat and takes place at a temperature about 1200°K.
In order to maintain this temperature there must be some
mechanism to provide this heat. In almost all cases this
heat is obtained from the combustion reaction

$$C + O_2 \rightleftharpoons CO_2 \qquad\qquad \text{Reaction 2}$$

This reaction may take place in a separate reactor and the heat transferred to the gasification reactor or may take place within the gasification reactor.

EXAMPLE 5.1   Determine the composition of the gas produced by the gasification of char (reaction 1) in a system where both the gasification reaction and combustion occurs in a single reactor. The char enters the reactor at 25°C and the steam enters at 1 atm and 600°F. The oxygen is supplied pure at 298°K, 25% excess steam over that required for complete reaction of C with steam is added. The gas leaves the reactor at 1200°K. Determine the Figure of Merit and lw. Determine the heating value of the gas in Btu/ft^3 (dry basis). NOTE: It is assumed that the reactions take place and the system is not at equilibrium.

SOLUTION 5.1

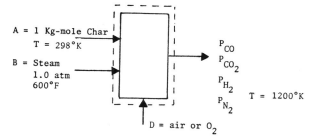

Reactions

(1) $C + H_2O \rightarrow CO + H_2$

(2) $C + O_2 \rightarrow CO_2$

Reaction 2 furnishes energy for reaction 1

Let α be the amount of CO_2 formed and β the amount of CO formed.

Basis: 1 Kg-mole of C

Assume all the oxygen added to reactor reacts according to reaction (2).

Component	Input	+ Generation	- Consumed	= Output
C	1	+ 0	- $(\alpha+\beta)$	= 0
H_2O	1x1.25	+ 0	- β	= P_{H_2O}
O_2	α	+ 0	- α	= P_{O_2}
CO_2	0	+ α	- 0	= P_{CO_2}
CO	0	+ β	- 0	= P_{CO}
H_2	0	+ β	- 0	= P_{H_2}

Also $\alpha + \beta = 1.0 \longrightarrow \alpha = 1 - \beta$

$$P_{H_2O} = 1.25 - \beta$$
$$P_{O_2} = 0$$
$$P_{CO_2} = \alpha = 1 - \beta$$
$$P_{CO} = \beta$$
$$P_{H_2} = \beta$$

Stream Identification	Mass Flow	T	P	$\bar{H}_C$	$\bar{H}_T$	$\bar{H}_\lambda$	$\bar{H}_P$	$\bar{H}$	$m\bar{H}$
Units on Terms	[Kg-mole]	[°K]	[bar]	[J/Kg-mole]					[J]
Inlet									
A(C)	1	298	1.01	394×10^6	0	0	0	394×10^6	394×10^6
B(H_2O)	1.25	589	1.01	0	10.1×10^6	44×10^6	0	54.1×10^6	67.6×10^6
D(O_2)	$1-\beta$	298	1.01	0	0	0	0	0	0
$\sum a_i \bar{H}_{A_i} = 461.6 \times 10^6$									
Exit									
P(CO_2)	$1-\beta$	1200		0	44.5×10^6	0	0	44.5×10^6	$(1-\beta)(44.5 \times 10^6)$
P(CO)	β	1200		283×10^6	28.5×10^6	0	0	311.5×10^6	$(\beta)(311.5 \times 10^6)$
P(H_2)	β	1200		286×10^6	26.8×10^6	0	0	312.8×10^6	$(\beta)(312.8 \times 10^6)$
P(H_2O)	$1.25-\beta$	1200		0	34.5×10^6	44×10^6	0	78.5×10^6	$(1.25-\beta)(78.5 \times 10^6)$
$\sum b_i \bar{H}_{B_i} = 482.2 \times 10^6 \beta + 132.1 \times 10^6$									
$\Delta H = \sum b_i \bar{H}_{B_i} - \sum a_i \bar{H}_{A_i} = 482.2 \times 10^6 \beta + 132.1 \times 10^6 - 461.6 \times 10^6$									

$\Delta H = q - w = 0$

Therefore, $\beta = 0.684$, $\alpha = 0.316$

Using these values the final gas composition is:

Component	Moles	%(Wet)	%(Dry)
P_{H_2O}	0.566	0.25	-
P_{CO}	0.684	0.30	0.41
P_{CO_2}	0.316	0.14	0.19
P_{H_2}	0.684	0.30	0.41
Total	2.25	0.99	1.01

Heating value (heat of combustion of dry product gas)

$\Delta H = 0.41\overline{H}_C(CO) + 0.19\overline{H}_C(CO_2) + 0.41\overline{H}_C(H_2)$

$\quad = 0.41 \times 283 \times 10^6 [J/Kg\text{-mole}] + 0.19 \times 0$

$\qquad + 0.41 \times 286 \times 10^6 [J/Kg\text{-mole}]$

$\quad = 233.3 \times 10^6 [J/Kg\text{-mole}]$

$\Delta H = 233.3 \times 10^6 [J/Kg\text{-mole}] \times 1/2.2 [Kg\text{-mole}/lb\text{-mole}]$

$\qquad \times 1/1055 [Btu/J] = 1 \times 10^5 [Btu/lb\text{-mole}]$

At standard temperature and pressure (298°K, 1.01 bar) 1 lb-mole occupies 359 ft³.

$1 \times 10^5 [Btu/lb\text{-mole}] \times 1/359 [lb\text{-mole}/ft^3] = 280 [Btu/ft^3]$

Evaluation of Figure of Merit and lw

Stream Identification	Mass Flow	T	P	$\overline{B}_C$	$\overline{B}_T$	$\overline{B}_\lambda$	$\overline{B}_P$	$\overline{B}$	$m\overline{B}$
Units on Terms	[Kg-mole]	[°K]	[bar]	[J/Kg-mole]					[J]
Inlet									
A(C)	1	298	1.01	394x10⁶	0	0	0	394x10⁶	394x10⁶
B(H₂O)	1.25	589	1.01	0	3.12x10⁶	0	0	3.12x10⁶	3.9x10⁶
D(O₂)	.316	298	1.01	0	0	0	0		
$\sum a_i \overline{B}_{A_i}$ = 397.9x10⁶ [J/Kg-mole]									
Exit									
P(H₂O)	0.566	1200	.25	0	19.1x10⁶	0	-3.4x10⁶	15.7x10⁶	8.9x10⁶
P(CO)	0.684	1200	.31	257x10⁶	15.6x10⁶	0	-2.9x10⁶	269.7x10⁶	184.4x10⁶
P(H₂)	0.684	1200	.31	327x10⁶	14.6x10⁶	0	-2.9x10⁶	248.7x10⁶	170.1x10⁶
P(CO₂)	0.316	1200	.14	25x10⁶	25x10⁶	0	-4.9x10⁶	20.1x10⁶	6.4x10⁶
$\sum b_i \overline{B}_{B_i}$ = 369.8x10⁶ [J/Kg-mole]									

$w_m(\text{Products}) = 369.8 \times 10^6 [J]$

$w_m(\text{Reactants}) = 397.9 \times 10^6 [J]$

$lw = [397.8 \times 10^6 - 369.8 \times 10^6] = 27.9 \times 10^6 [J]$

Figure of Merit = $369.8 \times 10^6 / 397.9 \times 10^6 = 0.93$

There is little work lost in conversion of one chemical stored energy form to another.

It is necessary to balance the heat given up by the combustion reaction with the heat needed by the carbon steam reaction. Several schemes have been used. These may be divided into two classifications.

    a. Combustion and carbon steam reactions are carried out in the same reactor.

    b. The combustion and carbon steam reactions are carried out in separate reactors.

These two schemes are shown in Figure 5.2.

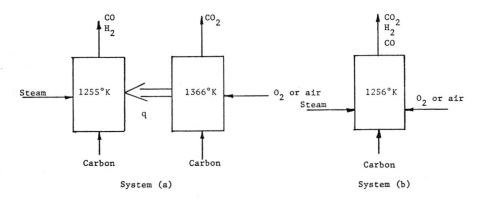

FIGURE 5.2 General Systems for Gasification

EXAMPLE 5.2   Determine the sand flow rate between the two beds
                shown in Figure 5.2 in Kg sand/Kg char gasified.
                The gasifier in this case operates at 1200°K and
                the combustor at 1300°K. C_p(sand) is 0.26 cal/g°K.

SOLUTION 5.2

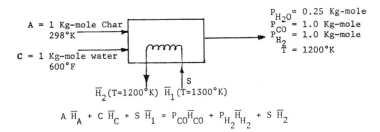

$$A \bar{H}_A + C \bar{H}_C + S \bar{H}_1 = P_{CO}\bar{H}_{CO} + P_{H_2}\bar{H}_{H_2} + S \bar{H}_2$$

Stream Identification	Mass Flow	T	P	$\bar{H}_C$	$\bar{H}_T$	$\bar{H}_\lambda$	$\bar{H}_P$	$\bar{H}$	$m\bar{H}$
Units on Terms	[Kg-mole]	[°K]	[bar]	[J/Kg-mole]					[J]
Inlet									
A(Char)	1	298	1.01	394×10^6	0		0	394×10^6	394×10^6
C(Steam)	1.25	589	1.01	0	10.1×10^6	44×10^6	0	54.1×10^6	67.6×10^6
S(Sand)	S	1300	1.01	0	$1.1\times10^6*$		0	1.1×10^6	1.1×10^6S
$\sum a_i\bar{H}_{A_i} = (461.6\times10^6 + 1.1\times10^6 S)$									
Exit									
P(H_2O)	0.25	1200		0	34.5×10^6	44×10^6	0	78.5×10^6	19.6×10^6
P(CO)	1.00	1200		283×10^6	28.5×10^6	0	0	311.5×10^6	311.5×10^6
P(H_2)	1.00	1200		286×10^6	26.8×10^6	0	0	312.8×10^6	312.8×10^6
S(Sand)	S	1200		0	$0.98\times10^6*$	0	0	$.98\times10^6$	$.98\times10^6S$
$\sum b_i\bar{H}_{B_i} = 643.9\times10^6 + 0.98\times10^6 S$									

$$* \quad \bar{H}_S(T) = 0.26(T - 298)\,[\text{cal/g}] \times 4.18\,[\text{J/cal}] \times 1000\ \text{g/Kg}$$
$$= 1086.8(T - 298)\,[\text{J/Kg}]$$

$$\Delta H = 0 = 643.9\times10^6 + 0.98\times10^6 S - 461.6\times10^6 - 1.1\times10^6 S$$

$$0.12\times10^6 S = 182.3\times10^6$$

$$S = 1519\,[\text{Kg Sand/Kg-mole C}]$$

or $1519\,[\text{Kg Sand/Kg mole C}] \times 1/12\,[\text{Kg mole C/Kg C}] = 126\,[\text{Kg Sand/Kg C}]$

The two-reactor system separates the gas produced from the combustion reaction from the gas resulting from the carbon-steam reaction. The product gas does not contain the CO_2 produced by combustion or the nitrogen if air is used as the source of oxygen. The inclusion of CO_2 is not critical and can be easily removed. Because of the nitrogen in the gas for air blown single reactor systems, they are not used when the gas is to be upgraded to S.N.G.

The gas coming from the air blown unit may be used to fire boilers if the boilers are close to the gasifier. The low energy gas cannot be transferred long distances without a substantial cost for pumping. The gases that are not diluted

with N_2 have a variety of uses other than upgrading the gas to
S.N.G. They may be used to:

1. Fuel a boiler

2. Fuel a gas turbine

3. Fuel a fuel cell

The use of low and medium energy gases for combustion were discussed in Chapter 2. This section emphasizes the upgrading
of the gas to S.N.G.

A common system for gasification is the moving bed
reactor. This is shown in Figure 5.3.

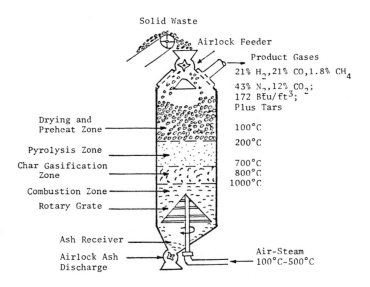

FIGURE 5.3 Moving Bed Gasification Reactor

Figure 5.3 represents a moving bed gasifier. The coal
moves downward through several zones. The following reactions
occur in each zone:

1. Drying and Pyrolysis - The wet coal is dried and
    then pyrolyzed to remove the volatile matter. The

solid remaining is char (Reaction 2).
2. Gasification - The char moves down into a gasification zone where it reacts with steam to produce CO
and H_2 (Reaction 1).
3. Combustion - The remaining char descends into a
combustion zone where it is burned to ash (Reaction 7).
The moving grate discharges the ash at the bottom.

A mixture of steam and air (or oxygen) flows upward
counter-current to the solid. In the bottom of the bed the
oxygen reacts releasing heat which is utilized in the subsequent
reaction of steam with carbon.

The gas leaving the top is a mixture of gases resulting
from pyrolysis, steam-carbon reaction and combustion reaction.
If air is used as the source of oxygen, it contains nitrogen
as well.

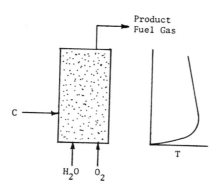

FIGURE 5.4 Entrained Bed

Figure 5.4 shows an entrained bed reactor system. In
this reactor system the flow of gas and solid is co-current
in contrast to the packed moving bed. The bottom combustion

zone reaches extremely high temperatures and the ash in this region forms a slag. The extremely high temperatures reached at the point of coal feed completely breaks down any products from pyrolysis, and higher hydrocarbons above methane are not found with the product gas. Either air or oxygen may be used in this single reactor system.

The remaining systems described do not mix combustion products with product gas. Figure 5.5 shows a system that uses two separate reactors. In the combustion reactor some of the gas produced is burned in a ball heater to heat ceramic balls up to high temperatures. The balls leave this system and pass to the gasifier where they lose heat to the carbon-steam reaction. The sensible heat loss provides the heat for the gasification reaction. The combustion products do not mix with the gas product stream.

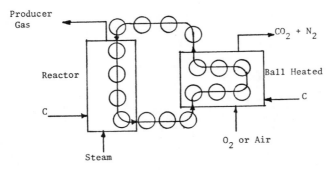

FIGURE 5.5   Two Reactor System - Heat Carried by Ceramic Balls

Another system that has received widespread support makes use of the liquid-like characteristics of the fluid bed which allows the solid from one reactor to be conveyed to another reactor. Solid is circulated between beds. This circulation

carries heat between the beds. This is shown in Figure 5.6.
The solid from the gasification bed contains some char that
circulates to the combustion reactor where it burns and provides
the necessary heat.

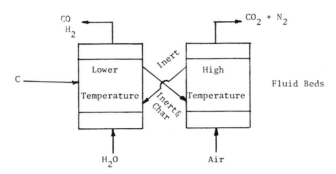

FIGURE 5.6   Two Reactor System - Heat Carried by Bed Material

Figure 5.7 presents a system where two separate reactors
are housed in a single shell. Each of the reactors contain a
molten salt solution that circulates in the pattern shown.

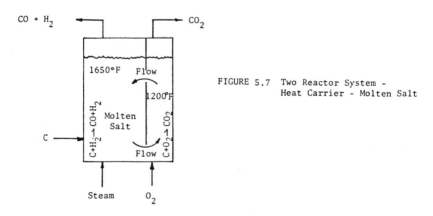

FIGURE 5.7   Two Reactor System -
Heat Carrier - Molten Salt

In the reaction system on the left, coal and steam are introduced. They react, absorb heat, and cool the molten salt
solution. In the right hand side the molten salt containing

char is reacted with oxygen or air. This gives off heat and
heats up the molten salt. The molten salt flows from this
region and provides the heat for the gasification reaction.

In the final system discussed electricity is the source
of heat. This is shown in Figure 5.8. This is an example
of a two reactor system. The combustion reaction occurs in
a power station and the gasification and pyrolysis reaction
in a fluid bed electrically heated. The products of combustion do not mix with this gas, and the gas is free of nitrogen.

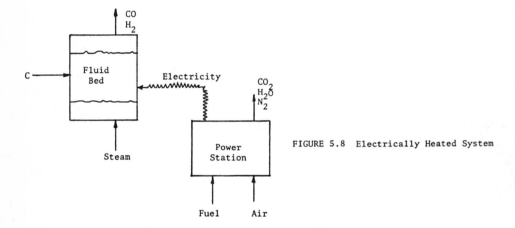

FIGURE 5.8 Electrically Heated System

There are a number of schemes that have been devised to separate
the reactions and provide the energy where it is needed for
the gasification reaction. A few of them that appear to be
the primary candidates for conversion of coal to a synthesis
gas have been described. The discussion above assumed that
only three reactions need be considered: Pyrolysis, Carbon-
Steam, and Combustion. As long as there is little excess
steam and the temperatures shown are selected, this assumption
is reasonable.

The gas produced from all systems described may be used directly for heating, electric power generation, fuel for a turbine or internal combustion engine. The gas produced from systems that do not contain nitrogen have the additional potential as a chemical feed stock or raw material for generation of synthetic natural gas.

EXAMPLE 5.3   Determine the Figure of Merit and lost work for the gasification reactor shown in Figure 5.8. Compare to example problem 5.2.

SOLUTION 5.3

There is insufficient information given in the problem to obtain a solution. If it is assumed that the gasifier conditions given in problem 5.2 apply and the values in example 4.6 represent a typical power plant the values may be estimated.

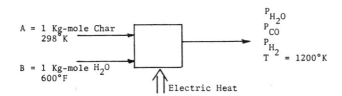

From Example 5.2 the amount of heat needed by the reaction may be determined. In this problem the sand will be replaced by electricity.

$$q(\text{Needed}) = S(\overline{H}_2 - \overline{H}_1) \qquad \text{(See Example 5.2)}$$

$$= 1519[1.1 \times 10^6 - 0.98 \times 10^6] = 1.82 \times 10^8 [\text{J/Kg mole C}]$$

Using this value and a power plant system similar to the one in Example 4.6, the amount of fuel burned in the power plant may be estimated.

$$1 \text{ Kg-mole } CH_4 \times 452[\text{lbs } H_2O/\text{Kg-mole } CH_4]$$
$$\times 360[\text{Btu/lb}] \times 1055[\text{J/Btu}] = 1.72 \times 10^8 [\text{J/Kg-mole}]$$

$$1.82 \times 10^8 [\text{J/Kg-mole C}] \times (1/1.72 \times 10^8)[\text{Kg-mole } CH_4/\text{J}]$$
$$= 1.06[\text{Kg-mole } CH_4] \text{ at power plant.}$$

From the values of lw given in Example 4.6:

$$lw(\text{electric station}) = (58.6 + 34.2 + 530.5) \times 10^6 [J] \times 1.06$$
$$= 6.6 \times 10^8 [J]$$

$$lw(\text{gasification}) = 28 \times 10^6 [\text{estimate from Example 5.1}]$$

$$\text{Total } lw = 6.6 \times 10^8 + 28 \times 10^6 = 6.9 \times 10^8$$

$$B_{in} = 7.93 \times 10^8 (1.06) + 3.98 \times 10^8 = 12.38 \times 10^8 [J]$$

$$B_{out} = B_{in} - lw = 12.38 \times 10^8 - 6.9 \times 10^8 = 5.48 \times 10^8$$

$$\text{Figure of Merit} = (5.48 \times 10^8)/(12.38 \times 10^8) = 0.44$$

This calculation is only an estimate of the correct value. It shows, however, that the use of electricity for a heat source results in an undesirable loss of work.

5.1.4 Water-Gas-Shift

$$CO + H_2O \rightleftharpoons H_2 + CO_2$$

The water-gas-shift reaction is used to provide the proper H_2/CO ratio for the conversion to methane. The ratio of 3/1 is usually used. This is a rapid reaction and may be assumed to approach equilibrium. The reaction is only slightly exothermic, and there is little problem in controlling the heat effects. The proper ratio of H_2/CO is obtained by adjusting the temperature and the amount of excess steam.

EXAMPLE 5.4   Determine the equilibrium temperature that is necessary to convert an equal molar mixture of CO and H_2 from a gasifier to a 3/1 : H_2/CO ratio needed for methanation. Assume 100% excess steam is used.

SOLUTION 5.4

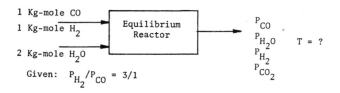

$$H_2O + CO \longrightarrow H_2 + CO_2 \quad [\text{Let } \alpha = CO_2 \text{ Gen.}]$$

Component	Input	+	Generation	-	Consumed	=	Out
CO	1			-	α	=	P_{CO}
H_2	1	+	α			=	P_{H_2}
CO_2	0	+	α			=	P_{CO_2}
H_2O	2			-	α	=	P_{H_2O}

$$K_p = (P_{H_2})(P_{CO_2})/(P_{H_2O})(P_{CO})$$
$$= (1+\alpha)(\alpha)/(2-\alpha)(1-\alpha)$$

Also $P_{H_2}/P_{CO} = (1+\alpha)/(1-\alpha) = 3$

$$\alpha = 0.5$$

$$K_p = (1+0.5)(0.5)/(2-0.5)(1-0.5) = 1$$

$$\log K_p = 0$$

From Figure 2

$$\frac{1}{T} \times 10^4 = 9.5$$

$$T = 1052°K$$

5.1.5 Methanation

The methanation reaction is slow and must be catalyzed to improve the conversion rate. The catalyst is normally nickel based but Pt, Co, Mo, and Fe can be used. The important reaction is:

$$3H_2 + CO \longrightarrow CH_4 + H_2O$$

This reaction is highly exothermic and the temperature must be carefully controlled. To obtain attractive reaction rates, temperatures of 525-675°K are used. If the temperature exceeds 750°K, the catalyst is destroyed. As a result of the requirement for accurate temperature control of the reactor temperature, a number

of reactor configurations are considered.   Several are shown
in Figure 5.9.

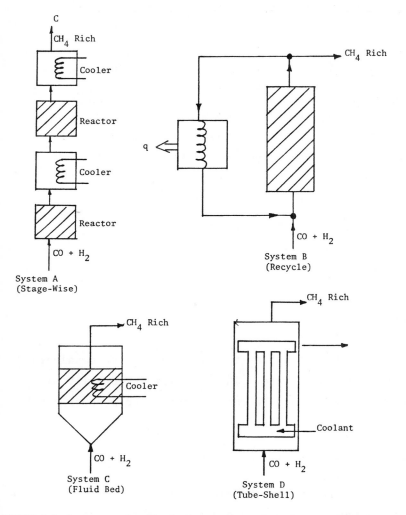

FIGURE 5.9 Systems to Provide Cooling to the Methanation Reaction.

System A shows a series of packed bed stages with inter-cooling. In each stage only a small conversion is achieved. When the gas reaches 700°K, it is passed into a cooling section, cooled and then sent along to another reactor section.

System B recycles most of the gas through a cooler and back into the reactor. The recycle gas serves as an energy absorber and is used to absorb the heat and hold down the temperature.

System C represents a fluid bed where the heat is re-moved directly from the reaction zone.

System D represents a tube bundle where the catalyst is coated on the surface. A cooling stream flows inside the tubes and the reacting gas on the outside.

These systems all accomplish the needed removal of heat and protection of catalyst. The coolant is water that is used to generate steam. This steam can be used in both the gasifier and water-gas-shift reactor.

EXAMPLE 5.5   Consider a reactor stage where the gas entering
               the reactor is 50% converted to CH_4. The gas
               enters at 600°K and it reacts until it reaches
               800°K. This gas then enters a cooler where it
               cools to 600°K and generates steam at 1 atm and
               400°K.

               a) Determine the % conversion leaving the reactor.
               b) Determine the steam produced.

SOLUTION 5.5

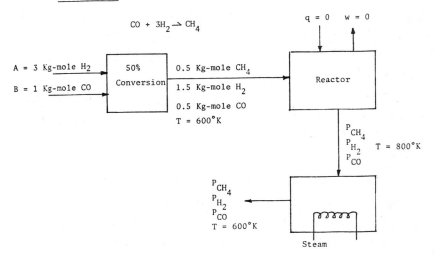

$$CO + 3H_2 \rightarrow CH_4$$

Let α be the CH_4 formed in the reactor.

Component	In	+	Generated	-	Consumed	=	Out
CH_4	0.5	+	α	-	0	=	P_{CH_4}
H_2	1.5	+	0	-	3α	=	P_{H_2}
CO	0.5	+	0	-	α	=	P_{CO}

$$P_{CH_4} = 0.5 + \alpha; \quad P_{H_2} = 1.5 - 3\alpha; \quad P_{CO} = 0.5 - \alpha$$

Stream Identification	Mass Flow	T	P	$\bar{H}_C$	$\bar{H}_T$	$\bar{H}_\lambda$	$\bar{H}_P$	$\bar{H}$	$m\bar{H}$
Units on Terms	[Kg-mole]	[°K]	[bar]			[J/Kg-mole]			[J]
Inlet CH_4	0.5	600		890.8×10^6	13.3×10^6	0	0	904×10^6	452×10^6
H_2	1.5	600		68.3×10^6	8.8×10^6	0	0	77.1×10^6	116×10^6
CO	0.5	600		67.6×10^6	9.0×10^6	0	0	76.6×10^6	38.1×10^6
$\sum a_i \bar{H}_{A_i} = 606 \times 10^6 [J]$									
Exit CH_4	$0.5+\alpha$	800		890.8×10^6	24.9×10^6	0	0	915.7×10^6	915.7×10^6 $\times (0.5+\alpha)$
H_2	$1.5-3\alpha$	800		68.3×10^6	14.7×10^6	0	0	83×10^6	$83 \times 10^6 \times$ $(1.5-3\alpha)$
CO	$0.5-\alpha$	800		67.6×10^6	15.3×10^6	0	0	82.9×10^6	$82.9 \times 10^6 \times$ $(0.5-\alpha)$
$\sum b_i \bar{H}_{B_i} = (623 \times 10^6 - 1248 \times 10^6 \alpha) [J]$									
$\Delta H = \sum b_i \bar{H}_{B_i} - \sum a_i \bar{H}_{A_i} = (623 \times 10^6 - 1248 \times 10^6 \alpha) - 606 \times 10^6 = 0$									

$$\alpha = 0.0136$$

Leaving the reactor:

$$P_{CH_4} = 0.5 + 0.0136 = 0.5136$$
$$P_{CO} = 0.5 - 0.0136 = 0.4864$$
$$P_{H_2} = 1.5 - 3(0.0136) = 1.4592$$

From a balance over the steam generator (since there is no chemical reaction, H_C terms need not be included).

Stream Identifi-cation	Mass Flow	T	P	$\overline{H}_C$	$\overline{H}_T$	$\overline{H}_\lambda$	$\overline{H}_p$	$\overline{H}$	$m\overline{H}$
Units on Terms	[Kg-mole]	[°K]	[bar]		[J/Kg-mole]				[J]
Inlet									
$P(CH_4)$	0.5136	800			24.9×10^6			24.9×10^6	12.8×10^6
$P(CO)$	0.4864	800			15.3×10^6			15.3×10^6	7.4×10^6
$P(H_2)$	1.4592	800			14.7×10^6			14.7×10^6	21.5×10^6
$\sum a_i \overline{H}_{A_i} = 41.7 \times 10^6 [J]$									
Exit									
$P(CH_4)$	0.5136	600			13.3×10^6			13.3×10^6	6.8×10^6
$P(CO)$	0.4860	600			9.0×10^6			9.0×10^6	4.4×10^6
$P(H_2)$	1.4592	600			8.8×10^6			8.8×10^6	12.8×10^6
$\sum b_i \overline{H}_{B_i} = 24 \times 10^6 [J]$									
$\Delta H = \sum b_i \overline{H}_{B_i} - \sum a_i \overline{H}_{A_i} = 24 \times 10^6 [J] - 41.7 \times 10^6 [J] = -17.7 \times 10^6 [J]$									

The heat lost by the gas cooling went to generate steam at 400°K

$$\Delta\overline{H}_{Steam} = \overline{H}_\lambda + \overline{H}_T = 44 \times 10^6 + 3.4 \times 10^6 = 47.4 \times 10^6$$
$$m\Delta\overline{H} = 17.7 \times 10^6$$
$$m = 17.7 \times 10^6 / 47.4 \times 10^6 = 0.37 \text{ Kg-mole Steam}$$

5.1.6 Clean-Up

There are two places in the generalized flow sheet, Figure 5.1, where measures must be taken to clean-up a product stream. These are:

a) Prior to the water-gas-shift reactor. The gas product at this point may contain tar, phenols, and ammonia that must be removed. Tar and other condensable compounds adversely affect heat transfer and may plug equipment. If this gas is to be considered as a fuel source to a gas turbine even sub-micron particles which will damage turbine blades must be removed.

b) Prior to the methanation reactor. The catalyst in this reactor is extremely sensitive to sulfur and small amounts of S will destroy the catalyst effectiveness (called poisoning). For this reason, a clean-up system or guard chamber is essential. It is also necessary to remove CO_2 from the gas if pipeline quality gas is to be produced.

The clean-up system prior to the water-gas-shift usually consists of some combination of water scrubbers, cyclones, oil scrubbers, sand filters, and electrostatic precipitators.

5.2   PROCESSES FOR SUBSTITUTE NATURAL GAS

Provided the discussion above, it is possible to enlarge upon the general flow sheet provided in Figure 5.1 and locate the clean-up steps. This is shown in Figure 5.10. It shows typical temperatures and pressures. It is possible to have both the carbon steam and water-gas-shift reactions take place in the same reactor by providing large excess steam to the gasification reactor.

Some of the various systems receiving serious consideration are discussed below.

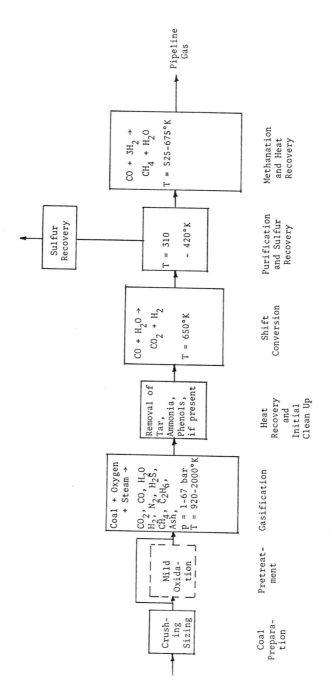

FIGURE 5.10   Schematic Diagram of a Coal-Gasification Process

5.2.1 Synthane Process

A key feature of this process, being developed at
Pittsburgh Energy Research Center, is that pretreatment of
caking coals is integrated with gasification. Another feature is that gas with a high methane content is produced
directly in the gasifier. Coal, crushed to -20 mesh size,
is pretreated by means of high pressure steam and oxygen to
prevent agglomeration of caking coals. The pretreated coal
is then gasified in a fluidized bed gasifier. The incoming
coal free falls through the hot gases rising from the fluid
bed, devolatilizes and increases the methane yield. Unreacted
char (about 30% of available carbon) obtained from the gasifier
as a product can be burned to produce steam. The product gas
is passed through a venturi scrubber and a water scrubber to
remove tar and dust. The ratio of hydrogen to carbon monoxide
is adjusted to 3/1 in a shift-converter. The gas mixture is
then purified by a potassium carbonate process and methanated.
Details of the process are shown in Figure 5.11.

5.2.2 Bigas

The Bigas process uses pulverized coal and is a two stage,
high pressure, oxygen blown system developed by Bituminous
Coal Research. The main advantages of this process are:
1. As the process utilizes an entrained bed gasifier,
    pretreatment of caking coal is not required.
2. The reaction conditions in the upper stage of the
    gasifier are such that no tar and oil are formed
    in the gasification step; and

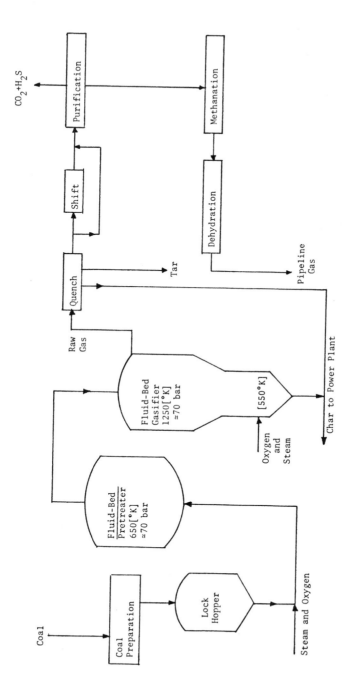

FIGURE 5.11 Sythane Fluid Bed Gasification Process.

3. All the coal charged into the process is consumed;
the principal by-products are slag and sulfur.

A schematic diagram of this process is provided in
Figure 5.12. The coal and steam enter the upper section of
the reactor where it is entrained in the raw gas from the
gasifier. In the region where mixing occurs the temperature
is about 1470°K and the coal is devolatilized and produces
methane and char. The char containing gas leaves at 1200°K
and is passed to a cyclone where the char is separated and
fed back to the lower reactor or gasifier zone. The temperatures reach 1950°K where any ash is converted to a molten
slag.

5.2.3 Hygas

A schematic diagram of Hygas process developed by Institute
of Gas Technology is shown in Figure 5.13. A key feature
of this process is the use of hydrogen and steam for hydrogasification of pretreated coal. Another feature is the use of
aromatic oil to feed coal into the gasifier at high pressures.

The Hygas process uses a series of fluidized beds. The
coal-oil slurry is fed to the top stage that vaporizes the
oil in the coal-oil mixture. The oil is recovered by quench-
ing the product gas. The oil condenses and is returned to the
slurry preparation section. The coal passes to the next
section which represents a low temperature hydrogasification
section. The char that does not react continues downward to
the high temperature gasification section. For the hydrogasification reaction to occur, it is necessary to provide for
hydrogen rich gas. There are several versions of the Hygas
process that differ in the way that this hydrogen is produced.

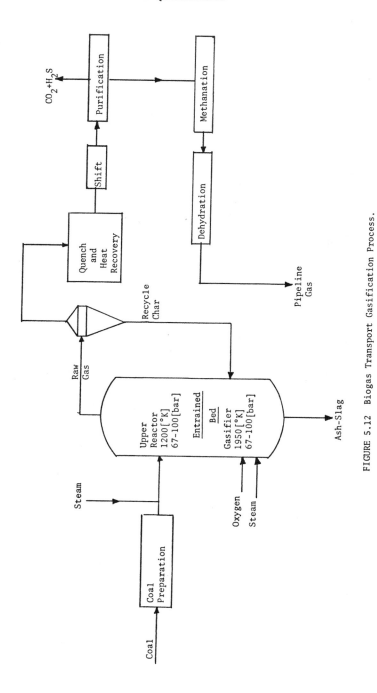

FIGURE 5.12 Biogas Transport Gasification Process.

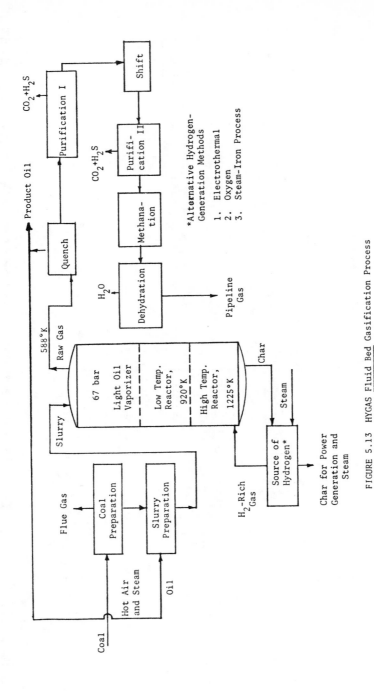

FIGURE 5.13 HYGAS Fluid Bed Gasification Process

Three systems to produce hydrogen referred to as electrothermal, oxygen, or steam-iron are considered. (They are not discussed here.)

5.2.4 CO_2 Acceptor Process

The carbon dioxide acceptor process, under development by the Conoco Coal Development Company, is as shown in Figure 5.14.

Some of the advantages of this process are:

1. Air is used to supply oxygen for combustion in regenerator where acceptor is heated and calcined. Thus, oxygen plant is not required.

2. Product gas clean up requirements are minimized as acceptor reacts with both H_2S and CO_2.

However, pretreatment of caking coal is necessary and control of solid transport may prove difficult in this process.

The earlier discussion provided for the movement of inert solids between a gasification reactor and a combustion reactor. All of the heat was carried by the sensible heat of the carrier. The key difference between these systems and the CO_2-acceptor is the circulating solids undergo a chemical reaction that releases or absorbs heat. This reduces the amount of solids that must be circulated between the reactors.

The carrier in the CO_2 acceptor undergoes the following reactions:

$$\text{Gasifier:} \quad MgO \cdot CaO + CO_2 \longrightarrow MgO \cdot CaCO_3 + \text{Heat}$$

$$\text{Combustor:} \quad MgO \cdot CaCO_3 + \text{Heat} \longrightarrow MgO \cdot CaO + CO_2$$

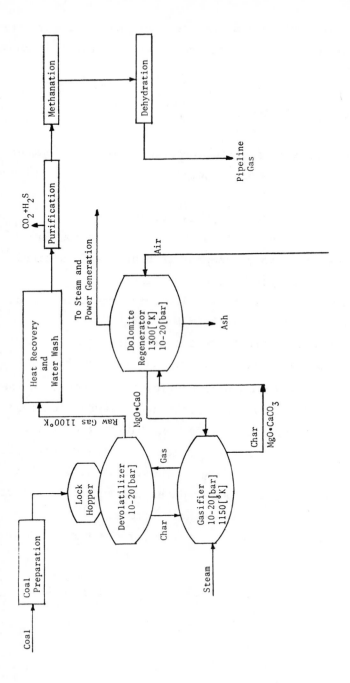

FIGURE 5.14 CO_2 Acceptor Fluid Bed Gasification Process.

5.2.5 Other Processes

Only a few of the very large number of systems for coal conversion being considered are shown above. Which system or systems will prove to become the system of choice for coal gasification is not known. The ones briefly discussed may or may not prove to be one of those that may find wide use if coal gasification will prove a reality. Table 5.2 provides some pertinent data on a few systems. It provides the gas composition out of the primary reactor unit.

TABLE 5.2

High Btu Gasification Processes
Typical Gasifier Raw Product Characteristics

| | Mole Percent | | | | | | | | | Gasifier | |
	CO	CO_2	H_2	H_2O	CH_4	C_2H_6	H_2S	N_2,ETC.	HHV J/M^3	Bar	°K
LURGI	9.2	14.7	20.1	50.2	4.7	0.5	0.6	–	$1.1x10^7$	30	640+
KOPPERS-TOTZEK	50.4	5.6	33.1	9.6	0	–	0.3	1.0	$1.1x10^7$	1.01	1780
WINKLER	25.7	15.8	32.2	23.1	2.4	–	0.25	0.8	$1.0x10^7$	1.01	1100+
HYGAS-OXYGEN	18.0	18.5	22.8	24.4	14.1	0.5	0.9	0.8	$1.4x10^7$	67	590
HYGAS-STM-Fe	7.4	7.1	22.5	32.9	26.2	1.0	1.5	1.4	$2.1x10^7$	67	590
HYGAS-ELEC	21.3	14.4	24.2	17.1	19.9	0.8	1.3	1.0	$1.6x10^7$	67	590
CO_2 ACCEPTOR	14.1	5.5	44.6	17.1	17.3	0.37	0.03	1.0	$1.6x10^7$	10	1100
SYNTHANE	10.5	18.2	17.5	37.1	15.4	0.5	0.3	0.5	$1.5x10^7$	67	1250
BI-GAS	22.9	7.3	12.7	48.0	8.1		0.7	0.3	$1.4x10^7$	67	1200
HYDRANE	3.9	–	22.9	–	73.2	–	–	–	$3.1x10^7$	67	1075
MOLTEN SALT	26.0	10.3	34.8	22.6	5.8	–	0.2	0.3	$1.2x10^7$	81	1200
MOLTEN IRON	69.7	–	9.6	–	20.0	–	–	0.7	$1.7x10^7$	1	1650

5.3   COAL LIQUEFACTION

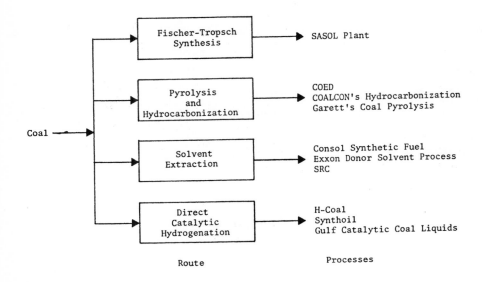

FIGURE 5.15   Classification of Liquefaction Processes

Coal can be used as a raw material to provide a non-polluting liquid fuel suitable for firing power plant boilers. The advantages of coal liquefaction compared to other coal conversion alternatives are:

1)   Thermal efficiency of a liquefaction process (70-75%) is usually higher than that of a gasification process (60-65%). [Energy in fuels generated/Energy in coal]

2)   The product obtained is storable and can also be transported economically as a result of high energy density.

3)   Amount of water required for liquefaction is small compared to a gasification plant of the same size. Thus, the plant site selection is less restrictive.

4) Almost any type of coal can be liquified. Bituminous coals, which are difficult to gasify due to their caking properties, are suitable for liquefaction.

5) Syncrude obtained from a liquefaction process can be used as a raw material for the production of wide variety of products like gasoline, naphtha, diesel, chemicals, etc.

The general reaction for coal liquefaction is

$$C_nH_{.7n} + \frac{.5n}{2} H_2 \rightarrow C_nH_{1.2n}$$

Bituminous + Hydrogen → Synthetic
   Coal   Gas   Crude Oil

The requirement for producing a synthetic crude is to increase the H/C ratio from 0.7/1 found in bituminous coal to a ratio of 1.2/1. This may be done in several ways as depicted in Figure 5.15. Each of these routes differs substantially, and a generalized process flow sheet cannot be drawn as was done for gasification to synthetic natural gas. The processes for liquefaction are not as far along toward commercialization as the processes for gasification.

Two general approaches are used. One is to force hydrogen onto the C in the solid structure and produce a liquid. The second approach is to break-down the coal into hydrogen and carbon monoxide and shift the CO/H_2 mixture and catalytically react this gas to build up a liquid petroleum product.

5.3.1 Fischer-Tropsch Synthesis

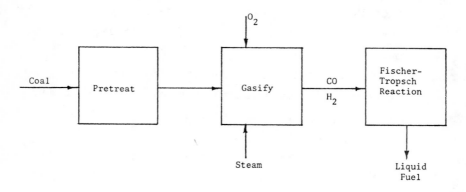

FIGURE 5.16   Fischer-Tropsch Liquefaction

The Fischer-Tropsch synthesis process is shown in Figure 5.16 and differs little from the various systems described for the generation of synthetic natural gas. The coal is pretreated, and gasified to produce CO and H_2. The CO, H_2 mixture is adjusted in a water-gas-shift reactor to provide the desired H_2/CO ratio for the reaction

$$nCO + (2n + 1)H_2 \xrightarrow{\text{Catalyst}} C_n H_{2n + 2} + nH_2O$$

The product is not a single liquid material but a complex mixture of many hydrocarbons. The character of the liquid is controlled by the H_2/CO ratio, the catalyst (typically Fe, Co, Ni, Ru, ZnO or ThO_2) and the reactor temperature.

Any of the reactor systems discussed in the previous chapter that provided a CO-H_2 gas that was not diluted with N_2 could serve to provide the material for the Fischer-Tropsch synthesis reaction. The only commercial plant in operation today to produce synthetic oil from coal is located at Sasolburg, South Africa and uses Fischer Tropsch synthesis to produce

gasoline from coal. In this process coal is gasified with
steam and oxygen in Lurgi moving bed gasifiers at 25 bar.
Raw gas obtained is treated to remove CO_2, H_2S and other impurities. The purified gas is used for both the fixed bed and
the Kellogg synthesis units (fluidized bed) packed with iron
catalyst. The products obtained from this plant include LPG,
gasoline, kerosene, diesel fuel, alcohol, fuel oil, wax, tar,
creosote, phenol and ammonium sulfate depending on the catalyst
and operating conditions.

The Fischer-Tropsch hydrocarbon synthesis process can
also be used to produce methanol by employing catalysts such
as copper-zinc, chrominum and zinc-chromium-oxide.

$$CO + 2H_2 \xrightarrow{cat} CH_3OH$$

Once the coal is gasified and reduced to CO and H_2,
these two compounds may then be reformed to provide a wide
range of hydrocarbons and can be used as the starting materials for a complete petrochemical industry.

5.3.2 Pyrolysis or Carbonization Processes

Pyrolysis or carbonization shown in Figure 5.17 is the
thermal decomposition of coal in the absence of air which
results in the formation of char, liquids and gaseous products. Liquids and gases containing higher H/C ratio than
original coal are thus recovered. This process does not
convert all of the carbon to liquids. A significant portion
is converted to gas and a major portion becomes a char product (the char may be converted to gas by reacting with steam).

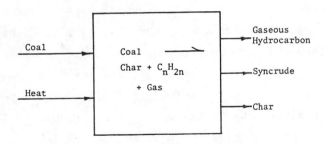

FIGURE 5.17   Pyrolysis Process

The relative yield of liquid and gaseous products depend on the volatile matter of coal, reaction temperature and the residence time of coal in the pyrolyzer. Liquid yields can be significantly increased by flash carbonization which uses very short residence time to prevent the decomposition of liquids into gaseous products. The rapid heating technique also minimizes the repolymerization of the pyrolysis products and usually results in the production of acetylene-rich gas. Three processes that produce liquid in significant amounts are discussed below. They are (1) COED, (2) Garrett's Coal Pyrolysis and (3) Coalcon's Hydrocarbonization. All produce large amounts of char that must be burned. Coalcon produces much less than the first two.

COED PROCESS*: The COED process converts coal to gas, oil, and char through heating in multistage fluidized beds, as shown in Figure 5.18. In this process, the coal is first crushed and dried, then is pyrolyzed in a series of four fluidized-bed reactors with successively higher temperatures. The temperature of each fluidized-bed reactor is just below the maximum temperature to which the coal can be heated without agglomerating.

———————————————

*Descriptive material from here to end of chapter was taken in part from Quarterly Report of ERDA, July-September, 1975.

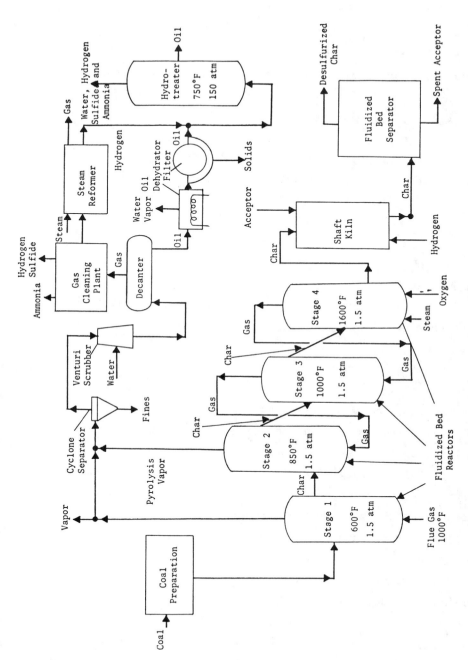

FIGURE 5.18 COED Process Schematic

The number of stages and the operating temperature vary with
the agglomerating properties of the coal. Heat for pyrolysis
is provided primarily by burning a portion of the char with
oxygen in the presence of steam in the fourth stage of pyrolysis.
(Nitrogen is used for start-up to fluidize the first stage
until enough flue gas is available.) Hot gases from the fourth
stage flow countercurrent to the char. These gases, which
provide the fluidizing medium for the third and second stages
of pyrolysis, are then passed to a product recovery system
where the gas and oil are produced.

Gas and oil are recovered from vapors coming off from
the second pyrolysis stage. These vapors pass into a cyclone
which removes the fines. The vapors leaving the cyclone are
then quenched directly with water in a venturi scrubber to
condense the oil, and the gases and oils are separated in a
decanter.

The gas is desulfurized in the gas cleaning plant, then
enriched with steam by a steam reformer. Part of the product
gas is converted to hydrogen and used in the process; the
balance of the product gas can either be scrubbed and sold as
fuel gas or converted to pipeline gas or hydrogen.

The oil from the decanter is dehydrated and filtered in
a rotary pressure precoat filter. The solids-free oil is then
pressurized and mixed with hydrogen in a fixed-bed catalytic
reactor (hydrotreater). The hydrotreater removes nitrogen,
sulfur, and oxygen (which are reacted with hydrogen to form
ammonia, hydrogen sulfide, and water) and produces a heavy
synthetic crude oil with a specific gravity of 0.9.

The char produced by the process is desulfurized in a
shaft kiln. In the kiln, hydrogen is added to the char, which
produces hydrogen sulfide; the hydrogen sulfide is then absorbed

by an acceptor, such as calcined limestone or dolomite. After desulfurization, the char and spent acceptor (which can be regenerated) are separated in a continuous fluidized-bed separator. Product char can be reacted with steam and oxygen in a gasifier to generate low-Btu gas. The rank of coal processed and the marketability of the end product determines the final use of the char and gas.

GARRETT PROCESS: In Garrett's Coal Pyrolysis Process flash carbonization takes place in an entrained bed. The pulverized coal is heated rapidly to 1150°K with hot recycle char. The residence time is less than two seconds. A portion of the char is burned with air to provide heat for pyrolysis. Gas is separated from tar and then purified. This gas can be utilized as a fuel (2.6×10^7 J/m^3) or methanated to pipeline gas. A portion of the gas can also be used for the production of hydrogen required for hydrotreating of tar to produce liquid products.

COALCON'S HYDROCARBONIZATION PROCESS: Pyrolysis of coal in a hydrogen atmosphere or hydrocarbonization can significantly increase the gaseous and liquid hydrocarbon yield. This is shown in Figure 5.19. Presence of hydrogen minimizes the repolymerization of pyrolysis product. The amount of hydrocarbons produced from a given quantity of coal is higher at higher hydrogen pressures.

Pulverized coal is fed to the fluidized bed hydrogenation reactor operated at 37 bar and 830°K. Residence time of gas and solid in the reactor is 25 seconds and 25 minutes, respectively. The gas leaving the reactor is passed through cyclones to remove the entrained solids which are recycled to the reactor. The liquid product is separated into four basic streams: overhead gas, light liquid, heavy liquid and wastewater. The overhead

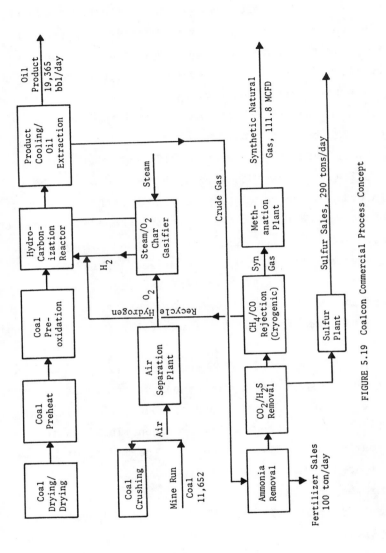

FIGURE 5.19 Coalcon Commercial Process Concept

gas is cooled, purified and further separated into hydrogen, hydrocarbon and methanation synthesis gas in a cryogenic processing unit. Additional hydrogen required is manufactured by gasification of char in Koppers-Totzek gasifiers. The final products obtained include synthetic natural gas, LPG, light oil, and heavy oil.

Table 5.3 provides some data on each process. It can be seen from the product description that there are many products including major amounts of char in two processes (in fact char is the major product). The economics and thermal efficiencies depend upon effective utilization of the non-liquid products.

5.3.3 Solvent Extraction and Catalytic Hydrogenation

In these processes the coal is heated in an organic material in the presence of hydrogen to dissolve the organic matter. The process is complex and no detailed description is attempted. A few important conclusions can be drawn.

Figure 5.20 shows the equilibrium coal conversion versus temperature curve for coal. This figure shows that the reaction goes essentially to completion at temperatures above 650°K.

The reaction rate is represented by the equation

$$\text{Rate} = k\, C_{coal} pp_{H_2}$$

The rate is proportional to coal concentration and partial pressure of hydrogen.

TABLE 5.3 Examples of Coal Pyrolysis Processes

Process	COED	Coalcon Hydrocarbonization	Garrett Carbonization
Developed by	FMC Corporation	Union Carbide	Garrett Research and Development Co., California
Status	pilot plant can process 26 tons/day and hydrotreat 30 barrels of coal derived oil/day.	2790 tons/day demonstration plant is being designed	3.6 tons/day pilot plant
Coal size	16 mesh	60 - 325 Mesh	
Type of Reactor	4 fluidized beds	fluidized bed	Entrained Bed Pyrolyzer
Temperature	590°K, 730°K, 810°K, 1090°K	830°K	1150°K
Pressure	1-1.5 bar	37 bar	1.0 bar
Hydrogen consumed SCF/Ton MAF coal	oil product is hydrotreated in a fixed bed reactor containing Ni-Mo catalyst at 750°K and 2500-3100 psi	12,886	
Products	59.5% char 19.3% oil 15.1% gas (1.86×10^7 J/m^3) 6.1% liquor based on Illinois #6 Coal	SNG: 139 m^3/Ton LPG: 72 m^3/Ton Light Oil: 0.48 Bbl/Ton Heavy oil: 0.95 Bbl/Ton	Tar (35%) Product Char (56.7%) Gas (2.6×10^7 J/m^3) (6.5%) Liquor (1.8%)
Thermal Efficiency	65%	68%	

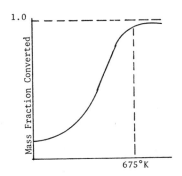

FIGURE 5.20 Approximate Equilibrium
Conversion

SOLVENT REFINED COAL (SRC) PROCESS*: The SRC process
converts high-sulfur, high-ash content coals to low-sulfur,
low-ash liquid fuel. Figure 5.21 shows a schematic of the
process. The coal is first pulverized and mixed with a coal-
derived solvent in a slurry mix tank. The slurry is mixed
with hydrogen, which is produced by other steps in the process,
and is then pumped through a fired preheater and passed into
a dissolver where about 90 percent of the moisture- and ash-
free coal is dissolved. Several other reactions also occur
in the dissolver: the coal is depolymerized and hydrogenated,
which results in an overall decrease in product molecular
weight; the solvent is hydrocracked to form lower-molecular-
weight hydrocarbons that range from light oil to methane; and
the organic sulfur is removed by hydrogenation in the form of
hydrogen sulfide.

From the dissolver, the mixture passes to a separator
where the gases are separated from the slurry of undissolved
solids and coal solution. The raw gas is sent to a hydrogen
recovery and gas desulfurization unit. The hydrogen recovered
is recycled with the slurry coming from the slurry mix tank.

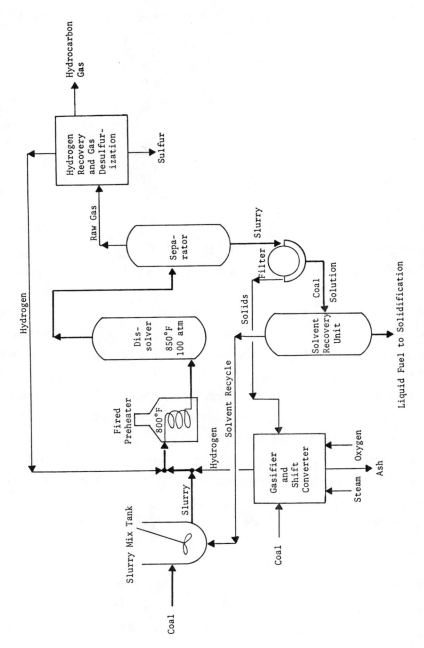

FIGURE 5.21 SRC Process Schematic

Hydrocarbon gases are given off and the hydrogen sulfide is
converted to elemental sulfur.

The slurry of undissolved solids and the coal solution
pass to a filtration unit where the undissolved solids are
separated from the coal solution. In the commercial-scale
process, the solids are sent to a gasifier-converter where
they react with supplemental coal, steam, and oxygen to produce hydrogen for use in the process. The coal solution
passes to the solvent recovery unit and the final liquid product, solvent-refined coal, is produced. The solvent-refined
coal has a solidification point of 350°F to 400°F and a heating value of about 16,000 Btu per pound.

SYNTHOIL PROCESS*: The Synthoil process is a hydrode-
sulfurization process that converts high-sulfur coal to a low-
sulfur, low-ash synthetic fuel oil. A schematic of the process
is shown in Figure 5.22. The coal is first crushed, ground,
and dried; it is then mixed with a portion of the product
oil from the process. The slurry produced is combined with
hydrogen produced during the process and fed into a fired
preheater, which contains ceramic pellets to improve heat
transfer to the slurry. From the preheater, the slurry enters
a fixed-bed catalytic reactor packed with catalyst pellets of
cobalt molybdate on silica-promoted alumina. The mixture is
then cooled and passed to a separator where the liquid and unreacted solids are separated from the gases.

The liquids and unreacted solids leave the bottom of
the separator and pass into a centrifuge, where the solids
are separated and fed into a pyrolyzer. Part of the liquid
leaving the centrifuge is recycled to the mixer to continue
the process; the remainder is the product oil of the process,
a nonpolluting fuel oil. The solids are further pyrolyzed

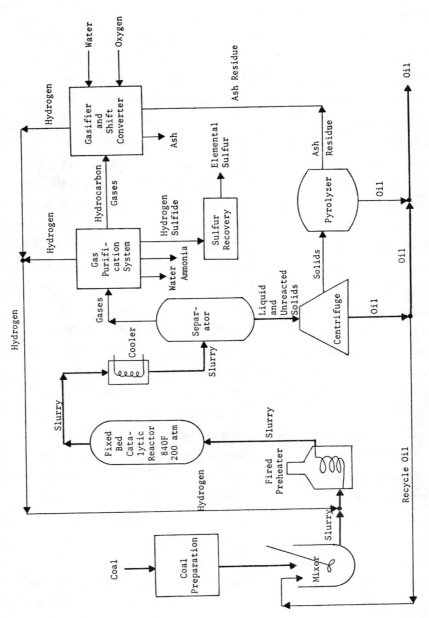

FIGURE 5.22 Synthoil Process Schematic

and yield an additional quantity of nonpolluting product oil,
as well as an ash residue. This ash residue, which contains
some carbonaceous material, is sent to a gasifier and shift
converter.

The gases coming off the top of the separator are sent
through a gas purification system where they are separated
into five product streams. Water and ammonia are withdrawn
separately; hydrogen sulfide is sent to a sulfur recovery
system that yields elemental sulfur as the product; the hydrocarbon gases are then fed to the gasifier and shift converter
where they react with steam and oxygen, together with the ash
residue, to form hydrogen for the process; the final product
stream from the gas purification system is hydrogen, which is
mixed with the hydrogen produced in the gasifier and fed into
the slurry stream entering the preheater.

Hydrogen propels the slurry so violently through the
fixed-bed catalytic reactor that plugging of the bed by the
mineral matter in the coal is prevented as the coal becomes
liquified. The turbulence of the slurry promotes mass and
heat transfer in the slurry, which in turn promotes hydro-
desulfurization and liquefaction.

H-COAL PROCESS*: The H-Coal process is a catalytic
hydroliquefaction process that converts high-sulfur-content
coal to boiler fuels that will not exceed emission standards
and to syncrude. A schematic of the process is provided in
Figure 5.23. Coal is crushed to minus 60 mesh, dried, and
then slurried with recycled oil and pumped to a pressure of
about 200 atm. Compressed hydrogen is added to the slurry,
and the mixture is preheated and charged continuously to the
bottom of the ebullient-bed catalytic reactor. The upward
passage of the internally recycled reaction mixture liquid-

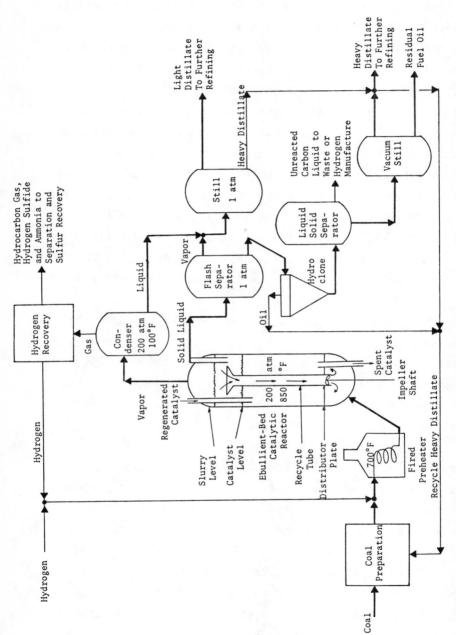

FIGURE 5.23 H-Coal Process Schematic

solid separation unit is not needed. To produce as major products clean fuel gas and low-sulfur residual fuel oil, the temperature and pressure in the ebullient-bed reactor are lowered, and less hydrogen is required.

Features of the H-Coal process are:

° The temperature of the reactor zone is easily maintained at a constant level because the temperature of the incoming slurry can be controlled by adjusting the temperature of the fired preheater. Furthermore, the temperature of the recycled light oils is low enough that additional cooling, using an internal or external cooling system, is not necessary.

° The activity level of the catalyst is essentially constant because the catalyst can be added and withdrawn as necessary.

° Because the operating temperature of the reactor zone and the catalyst activity level can be controlled, the product quality is consistent.

° Both low-sulfur boiler fuel and refinery feed-stock can be produced from the process.

Before the H-Coal process can become commercially and economically competitive, an adequate supply of hydrogen must be generated from the process itself. The H-Coal process requires between 14,000 and 20,000 standard cubic feet of hydrogen for each ton of coal processed, depending on the type of oil produced. Some of the hydrogen required has been removed from the scrubber gas from the ebullient-bed reactor and recycled; additional hydrogen has been purchased. In the commercial operation, hydrogen will be manufactured. Because the H-Coal process converts about 90 percent of the carbon contained in coal to a liquid, the feed to hydrogen manufacture could be a liquid rather than a solid. This suggests that commercial hydrogen manufacturing processes could be adapted. The solid cake char could also be used to produce hydrogen.

Another of the principal unsolved problems in the H-Coal process (and in other liquefaction processes) is the effective separation of the solids from the fuel products. In the H-Coal process, hydroclones have been proven capable of removing about two-thirds of the solids and are useful in partially separating the solids so that the nondistilled liquids can be recycled for slurrying the coal. HRI has also been investigating other methods of separating solids from liquids, such as magnetic separation, filtration, centrifugation, and solvent precipitation. Magnetic separation has shown limited effectiveness. Separation of the solids by filtration has not been conclusive, although filtration rates of 150 pounds per hour per square foot and relatively dry filter cakes have been achieved by using a continuous drum pressure filter.

There are many additional systems not discussed. Some of the information on the systems discussed and other systems are presented in Table 5.4.

TABLE 5.4 Example of Catalytic Hydrogenation Processes

Process	H-coal	Synthoil	Gulf Catalytic Coal Liquids	SRC	CSF	Exxon Donor Solvent Process
Developed by	Hydrocarbon Research, Inc.	U.S. Bureau of Mines	Gulf Research and Development	1) Pittsburg and Midway Coal Co. 2) SSI and EPRI	Consolidated Coal Company	Exxon
Catalyst	CO/MO	Pellets CO-MO on Silica-Alumina				
Coal size	-80 mesh	-100 mesh		-200 mesh	-14 mesh	-30 mesh
Wt% coal in slurry	50%	35%	30%-40%			
Solvent to coal ratio				1.5 - 3		1.2 - 4.0
Type of reactor	Ebullated bed	Fixed bed		Co-current Tublar Dissolver	Extraction vessel	
Temperature(°K)	730	700-720	700	720-725	660	645-755
Pressure (bar)	202	135-170	202	81-101	27	20-170
% coal dissolved	>90	>90	90-91	>90	63	
Hydrogen consumed SCF/Ton dry coal	18,600	10,350-14,306	15,950-22,620	12,600	16,300	
Method of Solid separation	Filtration	Centrifuge	Hydroclones Filtration	Filtration	Hydroclones	Distillation
Products obtained	Synthetic crude oil 3.87 Bbl/Ton	Fuel Oil 2.99-3.27 Bbl/Ton of coal; S: 0.6%; Ash: 1.3%	Light oil 0.46-0.75 Bbl/Ton; Fuel oil 2.78-3.71 Bbl/dry ton coal	SRC, calorific value is about 3.7×10^7 Btu/Kg. About 90% of original S is removed. Ash = 0.1%	Naphtha 0.52 Bbl/Ton; Fuel Oil 1.52 Bbl/Ton	Low sulfur fuel oil, Naphtha blending components

Problems for Chapter 5.0

1. Repeat Example 5.1 using air as the source of oxygen.

2. Determine the amount of C required in the combustion reac-
    tion in Example 5.2. Air is used as the source of oxygen.
    Determine the heating value and composition of gas in
    Example 5.2 and compare it to that found in Example 5.1.
    Compare the efficiency of the 2-bed and 1-bed systems
    (efficiency $\equiv \Delta H$ in synthesis gas/ΔH_C in total C used).
    This comparison does not account for the power required
    to produce O_2 from air. What would this do to the effi-
    ciency in both cases?

3. Assume that all of the $CO + H_2$ coming from Example 5.3
    is shifted and converted to methane. How much CH_4 could
    be produced compared to the amount of CH_4 used at the
    power station?

4. Repeat Example 5.1 using cellulose in place of C as a
    solid fuel. This reduces the amount of steam used.

    $$C_6H_{10}O_5 + H_2O \longrightarrow 6CO + 6H_2$$

5. Assume that methanation is carried out at 650°K in a
    Fluid Bed Reactor. The heat is removed by generating
    steam. (Assume steam to be same as Example 4.6).
        (a) Evaluate lw.
        (b) Evaluate lbs steam/lb CH_4 formed.
        (c) Compare steam generated to steam needed to pro-
             duce the $CO + H_2$ in the gasifier section.

6. In Example 5.1 and 5.2 the reaction

    $$C + O_2 \longrightarrow CO_2 \qquad\qquad (1)$$

    was used to provide the heat needed by the reaction

    $$C + H_2O \longrightarrow CO + H_2 \qquad\qquad (2)$$

    Looking at Table 5.2 the reaction $C + 2H_2 \longrightarrow CH_4$ also
    gives off heat. Assume that this reaction can be encour-
    aged to take place and the H_2 used comes from reaction (2).
    Repeat Problem 5.1 assuming the heat comes from the C +
    $2H_2 \longrightarrow CH_4$ reaction.

7. In the Bigas Generator char is reacted to produce syn-
    thesis gas. Using the temperatures given in Figure 5.12
    and assuming steam conditions to be saturated at 75 bar,
    determine the gas composition of the gas produced and the

oxygen requirements for the case where 50% excess steam
is used. Determine the size of the oxygen plant and
steam plant needed for a plant consuming 6000 tons/day
of a bituminous coal (assume all fixed carbon goes to
char and must be gasified).

6.0 NUCLEAR ENERGY

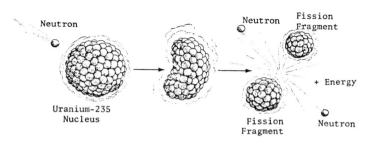

FIGURE 6.1 Fission of U-235

The discovery of the neutron fission of uranium in 1938 followed by the dramatic development of the nuclear reactor and weapons technology during the World War II Manhattan Project, ushered in the prospects of a vast new major energy resource. During the late 1940's and early 1950's work progressed rapidly toward the development of naval nuclear propulsion plants, culminating with the construction and successful testing of the nuclear submarine, <u>Nautilus,</u> in 1955. Military applications of nuclear technology (which include nuclear

fusion as well as fission) continue to take up a significant
portion of nuclear energy research and development programs.
The primary efforts of the next several decades will be directed toward the development and efficient utilization of
nuclear technology as a viable alternative to fossil fuel
based central electrical generating stations.

The first practical demonstration of civilian-commercial
applications of nuclear power was the Shippingport, Pennsylvania, pressurized water reactor (PWR) power plant, which
commenced operation in 1957 at a rated output of 60 Mw(e).
Light water moderated reactor plants have dominated the U.S.
nuclear energy scene, with the technology now advanced to
the point that an 1100 Mw(e) generating plant is typical.
In 1976 the operational nuclear generating capacity in the
U.S. was approximately 40,000 Mw(e) which accounted for 8
percent of the national electrical generating capacity.

Prior to a discussion of the principles behind nuclear
power, the nuclear power plant is compared to a standard
fossil fuel power plant. Figure 6.2 provides a schematic of
a nuclear fueled and fossil fueled electric utility plant.
The basic difference in the two systems is that the boiler of
the fossil fuel power plant is replaced by the nuclear reactor and heat exchanger shown in the containment shell of the
nuclear reactor system (shaded area in Figure 6.2).

Some of the other differences noted from this diagram
include:

1. The fossil fuel system is fed a continuous stream
    of fuel. A 1,000 Mw station would consume about
    240 metric tons of coal (or 53,000 barrels of oil
    or 790,000 ft^3 of natural gas) per hour. This
    amounts to 2×10^6 tons of coal/year. In contrast

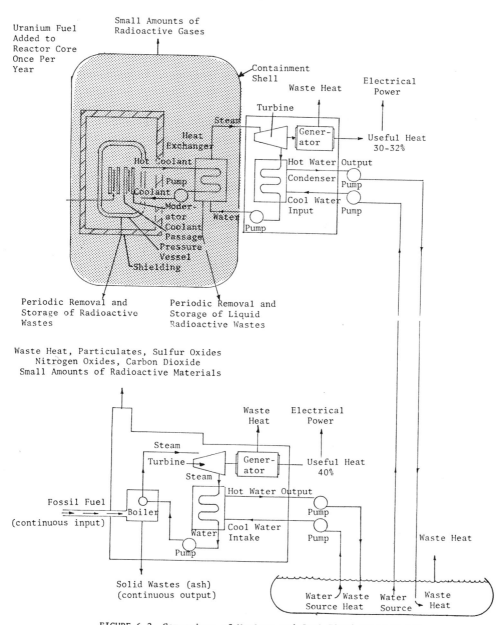

FIGURE 6.2 Comparison of Nuclear and Coal Fired Power Plants

the nuclear plant is loaded about once a year with
100 metric tons/year of uranium oxide pellets.

2. The fossil fuel system, without pollution control
    equipment, will emit on the order of 340,000 metric
    tons of SO_2, NO, and particulate per year along
    with a small amount of radioactive material. In
    contrast the nuclear facility emits a small amount
    of radioactive gas.

3. The fossil fuel plant will dump 60 units of heat
    for every 40 units of electric power. The nuclear
    power plant will dump 70 units for every 30 units
    of electrical power. The nuclear plant must dump
    about one third more heat than the fossil fuel plant.
    Part of the energy in the fossil fuel plant goes to
    the atmosphere (with products of combustion). The
    nuclear plant discharges all of the waste heat into
    rivers or streams.

4. The turbine efficiency is higher in the fossil fuel
    plant. Because of the temperature drop in the heat
    exchangers and because the reactor requires liquid
    water in the reactor, it cannot operate in the super-
    critical (water above its critical temperature and
    pressure) region.

5. The difference of greatest concern is the periodic
    discharge of radioactive wastes in a nuclear plant
    versus solid ash for a fossil fuel plant.

The thermal output of a fossil-fueled power plant is
dependent upon the controlled release of chemical energy
in the combustor. The heat release in a nuclear plant is
based on the controlled release of nuclear energy in the
reactor core.

Although the consequences of a chemical or steam explosion in a fossil-fired plant may be frightening to contemplate, the dangers would be short lived in terms of environmental and human damage; however a steam explosion or core meltdown in a nuclear plant could result in hazards and consequences lasting many, many years.* For these latter reasons, nuclear reactors are required to have numerous multiply-replicated safety and automatic shut down systems. Moreover, a careful scrutiny by industry and regulatory agencies concerning the design and testing of reactor safety systems as required in licensing, permit, and hearing procedures is often responsible for the eight-to-ten year time span between a utility's initial decision to build a nuclear plant and its operation at full design power.

Prior to examining the design and operational aspects of nuclear-based electrical generating stations, some basic concepts of nuclear reactions and radioactivity will be presented. Although the primary focus of the chapter is on the present-day fission reactor technology, some brief discussions will be made concerning the breeder reactor and the prospects of energy generation by controlled nuclear fusion processes.

---

*It should be emphasized that nuclear reactors will not explode like an atomic bomb. Many proponents of nuclear power believe that it is the "mushroom cloud" image of nuclear power that has hindered public acceptance of nuclear technology.

6.1 NUCLEAR ENERGETICS

In the previous chapters only physical and chemical
changes occurred. The chemical atoms always maintained their
identity and were conserved. No change in the nucleus of the
atom occurred. Figure 6.3 depicts a carbon atom.

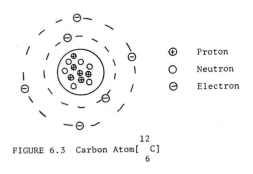

FIGURE 6.3 Carbon Atom[$^{12}_{6}$C]

It is composed of a very dense central core called the
nucleus that contains a number of protons (positive charged
particles) and neutrons (neutral particles). The correct
designation is:

$$^{12}_{6}C$$

The C designates the atom, the subscript 6 is the number of
positive charges or protons and the superscript 12 the number of neutrons plus protons (nucleons). In general the representation of any atom may be written as follows:

$$^{A}_{Z}X \qquad\qquad\qquad (6.1)$$

X = Element

Z = Number of protons

A = Number of nucleons, i.e., protons + neutrons

A-Z = Number of neutrons

The number of electrons in orbits around the nucleus is equal to the number of protons, A, which provide for the atom to be electrically neutral.

The electrons are bound to the nucleus because of the positive charge of the nucleus. When a chemical reaction occurs, these electrons are rearranged. This rearrangement may give off or take up some energy.

EXAMPLE 6.1   Determine the energy given off per molecule of methane for the chemical reaction in MeV (Million electron volts)

$$CH_4 + 2O_2 \longrightarrow CO_2 + 2H_2O \quad (\Delta H = 212.8 \text{ Kcal/g-mole})$$

SOLUTION 6.1

212.8[K-cal/g-mole] x 1/6.023 x 10^{23}[molecules/g-mole] x

2.62 x 10^{22}[eV/K-cal] x 1/10^6[MeV/eV]

= 9.2 x 10^{-6}[MeV/molecule] or 9.2[eV/molecule]

In a chemical reaction the energy released is on the order of several eV (electron volts).

For a nuclear reaction, there is a change in the configuration within the nucleus. Changes in the nucleus usually give rise to values of several MeV (million electron volts). The governing equation for evaluating these changes is

$$E = mc^2 \qquad\qquad (6.2)$$

where m = mass

c = velocity of light (3 x 10^8 m/sec)

E = energy

EXAMPLE 6.2   Determine the energy equivalence of 1 atomic mass unit (amu) in MeV, Kwhr, and Btu. By definition, the $^{12}_{6}C$ isotopic (atomic) mass is 12.000000 amu per atom.

SOLUTION 6.2

By definition

12.00[amu/atom] x 6.023 x 10^{23}[atoms/g-mole] x

1/12[gram-mole/grams] = 6.023 x 10^{23}[amu/g]

or 1.66 x 10^{-24}[g/amu]

From Equation 6.2

$$E = mc^2$$

$$= 1.66 \times 10^{-24}[g] \times (3.0 \times 10^8[m/sec])^2$$

$$= 1.49 \times 10^7[g\text{-}m^2/sec^2]$$

$$= 1.49 \times 10^{-10}[Kg\text{-}m^2/sec^2] = 1.49 \times 10^{-10}[J]$$

$$1[J] = 6.242 \times 10^{12}[MeV]$$

$$E = 1.49 \times 10^{-10}[J] \times 6.242 \times 10^{12}[MeV] = 931[MeV]$$

$$1[MeV] = 4.45 \times 10^{-20}[Kw\text{-}hr]$$

$$1[MeV] = 1.518 \times 10^{-16}[Btu/MeV]$$

$$E = 931[MeV] \times 4.45 \times 10^{-20}[Kw\text{-}hr] = 4.143 \times 10^{-17}[Kw\text{-}hr]$$

or

$$E = 931[MeV] \times 1.518 \times 10^{-16}[Btu/MeV] = 1.413 \times 10^{-13}[Btu]$$

A few constants that are useful for calculations in this chapter are given in Table 6.1.

TABLE 6.1
Useful Nuclear Constants

Atomic Mass Unit (amu)	1.66×10^{-22} Kg
Mass of Proton	1.00759 amu
Mass of Hydrogen ($_1^1H$)	1.00814
Mass of Neutron	1.00898 amu
Energy Equivalent to an (amu)	931 MeV - 4.143×10^{-17} Kw-hr
	- 1.413×10^{-13} Btu

In writing balanced nuclear equations, it is necessary to balance both the number of protons and the number of neutrons. For example,

Reactants   Products

$$_2^4He \quad + \quad _4^9Be \quad \longrightarrow \quad _0^1n \quad + \quad _6^{12}C \qquad (6.3)$$

(Helium)   (Beryllium)   (Neutron)   (Carbon)

From Equation 6.3 it can be noted the number of protons and nucleons balance.

$$A_{in} = 4 + 9 = 13 \qquad A_{out} = 1 + 12 = 13$$
$$Z_{in} = 2 + 4 = 6 \qquad Z_{out} = 0 + 6 = 6$$

Writing this out in general terms:

$$\sum(\text{charge}) \text{ reactants} = \sum(\text{charge}) \text{ products} \qquad (6.4a)$$

$$\sum(\text{total nucleons}) \text{ reactants} = \sum(\text{total}$$

nucleons) products $\qquad (6.4b)$

In a nuclear reaction, if the atomic masses of the reactants and the products are accurately known, the energy associated with the reaction can be calculated by examining the sum of the masses appearing on the left (reactant) and right (product) sides of the reaction equation. Two situations occur:

(a) The sum of the masses on the left exceeds the sum of the masses on the right. A portion of the reactant mass has been converted to energy and the reaction is exothermic.

(b) The sum of the masses on the left is less than the sum of the masses on the right. The reaction is endothermic and will not take place unless energy is supplied. In nuclear reactions the energy is supplied in the form of kinetic energy or radiation.

Table 6.2 provides a bookkeeping system to keep track of the energy-mass relations for a nuclear reaction.

If ΔM is positive, the reaction is exothermic and the energy released. If ΔM is negative, the reaction is endothermic, and the energy must be supplied if the reaction is to take place.

TABLE 6.2
Nuclear Reaction Accounting Sheet

Reaction: $D + B \rightarrow R + S$

	Input			Output		
	D	B	Total	R	S	Total
A (nucleons)						
Z (protons)						
A-Z (neutrons)						
Mass (amu)						

ΔM = Mass In - Mass Out

$\Delta E = 931 \Delta M$

EXAMPLE 6.3   Determine the energy change for the reaction:

$$^{4}_{2}He + {}^{9}_{4}Be \longrightarrow {}^{1}_{0}n + {}^{12}_{6}C$$

$$^{4}_{2}He = 4.00260 \,[amu]$$

$$^{12}_{6}C = 12.000 \,[amu]$$

$$^{9}_{4}Be = 9.01218 \,[amu]$$

SOLUTION 6.3

	Input			Output		
	D	B	Total	R	S	Total
A	4	9	13	1	12	13
Z	2	4	6	0	6	6
A-Z	2	5	7	1	6	7
Mass	4.00260	9.01218	13.01478	1.00898	12.000	13.0098

$\Delta M = (13.01478 - 13.0098) \,[amu]$

$\Delta E = 0.005 \times 931 = 4.70 \,[MeV]$

Reaction is exothermic.

An important concept in nuclear energetics is that termed "mass defect." The nucleus is made up of protons and neutrons but the mass of the nucleus is less than the mass of the protons and neutrons taken separately. This is the "mass defect" and may be written:

$$\Delta M = Z(1.00814) + (A - Z)(1.00898) - M \qquad (6.4)$$

where M is the mass of the nucleus

EXAMPLE 6.4   Determine the mass difference between the carbon nucleus and the protons and neutrons making up the nucleus.

$$6 \, _{0}^{1}n \; + \; 6 \, _{1}^{1}p \longrightarrow \, _{6}^{12}C$$

The $_{6}^{12}C$ weight = 12.0000 [amu]

SOLUTION 6.4

	Input			Output		
	D	B	Total	R	S	Total
A	6	6	12	12	0	12
Z	0	6	6	6	0	6
A-Z	6	0	6	6	0	6
Mass	6 x 1.00898	6 x 1.00814	12.09565	12.0000	0	12.0000

$\Delta M = 0.10272$

$E = 95.6$ [MeV]

E/Nucleon = $95.6 / 12 = 8.0$ [MeV]

The value of ΔM from equation may be divided by the total number of nucleons to give the mass defect per nucleon

Mass Defect per Nucleon = $\Delta M/A$

This is often plotted against the number of nucleons (A). An

alternate presentation is to multiply the mass defect per
nucleon by 931[MeV/nucleon] to obtain the "binding energy" per
nucleon. This is shown plotted against the number of nucleons,
A, in Figure 6.4.

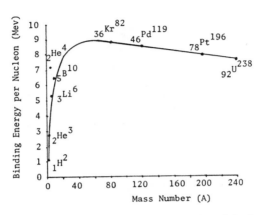

FIGURE 6.4 Binding Energy per Particle in Nuclei

Using Figure 6.4, the difference between fission and
fusion may be described. Fission as the name implies is the
result of breaking up a large nucleus into smaller nuclei.
If a large molecule Z = 200 is broken into smaller ones, say
two nuclei of 100, the binding energy moves from that of 8.0
MeV/nucleon to one of 8.8 MeV/nucleon. There is an increase
in the binding energy per nucleon of about 0.8 MeV/nucleon.
Since there were 200 nucleons, the energy change in the fission
reaction was:

$$(8.8 - 8.0)[MeV/nucleon] \times 200[nucleons] = 160[MeV]$$

This is the amount of energy given up _if_ a nucleus of
200 broke into two nuclei of 100 nucleons each. More useful
information is provided in the next example that looks at U^{235}

which is an atom that does break-down into smaller nucleus.
The break-down is called fission.

More accurate values may be obtained by using data regarding the fission event. This is shown in the next example.

EXAMPLE 6.5   A particular fission of $^{235}_{92}U$ by a thermal (0.025eV)
neutron yields the fission fragments $^{137}_{56}Ba$ (barium -137)
and $^{97}_{36}Kr$ (krypton -97). Calculate the energy released
by this reaction.

The reaction is

$$^{1}_{0}n + ^{235}_{92}U \longrightarrow (^{236}_{92}U) \longrightarrow ^{137}_{56}Ba + ^{97}_{36}Kr + 2^{1}_{0}n$$

Atomic mass data: $^{235}_{92}U$ = 235.0439 [amu], $^{137}_{56}Ba$ = 136.9061 [amu]

$^{97}_{36}Kr$ = 96.9212 [amu]

SOLUTION 6.5

		Input			Output		
	D	B	Total	R	S	T	Total
A	1	235	236	137	97	2x1	236
Z	0	92	92	56	36	0	92
A-Z	1	143	144	81	61	2	144
Mass	1.00898	235.0439	236.05288	136.9061	96.9212	2 x 1.00898	235.8454

ΔM = 0.2075

E = (0.2075) (931) = 193 [MeV]

About 200 MeV of energy is released in a typical
fission event.

The fusion reaction, as the name implies, refers to the
case where two small nuclei combine to form a large nucleus.
The same logic that allowed the amount of energy given off in
a fission reaction can be used for the fusion reaction. The

product of fusion has a higher binding energy per nucleon than the reactants.

EXAMPLE 6.6　Estimate the energy from the fusion of $_1^2H$ + $_1^2H \longrightarrow _2^4He$
$_2^4He$ weighs 4.00387 [amu]; $_1^2H$ weighs 2.01473 [amu].

SOLUTION 6.6

	Input			Output		
	D	B	Total	R	S	Total
A	2	2	4	4		4
Z	1	1	2	2		2
A-Z	1	1	2	2		2
Mass	2.01473	2.01473	4.02946	4.00387		4.00387

$$\Delta M = 4.02946 - 4.00387 = 0.0256$$
$$\Delta E = 0.0256 \times 931 = 23.8 [MeV]$$

The discussion above was shown that energy is released during fission or fusion. The question not answered is "in what form is this energy?" The table given below breaks down the energy from a typical fission reaction.

Kinetic energy of fission fragments	-166 MeV
Prompt γ-rays at fission	- 7 MeV
Kinetic energy of fission neutrons	- 5 MeV
β- particles from decay of fission fragments	- 5 MeV
γ-rays from decay of fission fragments	- 7 MeV
Neutrinos	- 10 MeV
TOTAL	-200 MeV

The major portion of the energy leaves with the fission
fragments and neutrons <u>as kinetic energy</u> (171 MeV). Another
14 MeV leave as γ-rays. How γ-rays carry energy will be discussed below. Beta particles (electrons) that are given off
when a nucleus seeks its stable condition, amount to 5 MeV.
The neutrinos are extremely light particles (little mass),
have no charge, and are extremely non-reactive. The energy
they contain is lost.

<u>EXAMPLE 6.7</u>   Calculate the fission rate (no. of fissions
per second) required to produce 1 watt of
thermal power.

<u>SOLUTION 6.7</u>

1 watt = 1[J/S]

1[J/Sec] x 6.242 x 10^{12}[MeV/J] x 1/190[Fission/MeV]

= 3.285 x 10^{10}[Fissions/Sec]

Gamma (γ) rays are <u>quanta</u> (bundles) of electromagnetic
energy which have no charge or mass. As such, they are physically identical to other forms of electromagnetic radiation
such as radio waves, visible light, ultra-violet light, and
x-rays. Such forms of energy are characterized by their wavelength and frequency, and all satisfy the expression:

$$\lambda \nu = c \qquad\qquad\qquad (6.5)$$

where λ = wavelength in meters, ν = frequency in
sec^{-1}, and c = speed of light, 2.998 x 10^{8}m/sec

γ-rays have extremely high frequencies (hence, short wavelengths).
The relationship between their energy and wavelength can be
determined by:

$$E = hc/\lambda \qquad\qquad\qquad (6.6)$$

where h = Planck's constant (4.136×10^{-21} MeV-sec)

γ-ray energies of interest in nuclear energy technology
typically range from a few tenths MeV to several MeV and
arise from the following sources: (a) they are released
instantaneously in the fission process (prompt γ's), (b) they
accompany other radioactive decay processes. They constitute
an important source of energy in the overall energetics of
the fission process.

The gamma rays are most often emitted with a discrete
energy level rather than an energy spectrum. This may be
seen in Figure 6.5. The atom or nucleus may temporarily
rest on any of the energy steps shown but not in between.

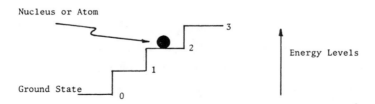

FIGURE 6.5 Excited States for Nucleus or Atom

When the atom drops from one level to another (one step to
another), a fixed ΔE occurs. The energy leaves as a gamma ray
with this energy value.

When two particles collide in an "elastic collision" both
the kinetic energy and momentum are conserved. Consider a
head-on collision between a neutron and a nucleus of mass M.
It will be assumed that the nucleus is at rest and the neutron
is moving with a velocity v_o. The collision is shown in Figure 6.6.

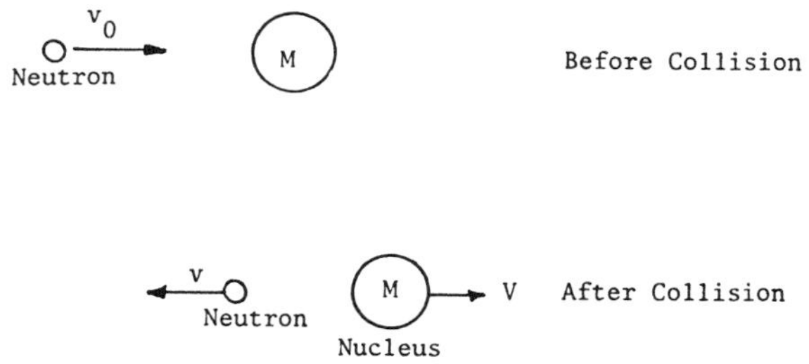

FIGURE 6.6 Result of Headon Collision of Neutron
and Nucleus of Mass, M

From a balance on kinetic energy:

$$1/2mv_o^2 + 0 = 1/2mv^2 + 1/2MV^2 \qquad (6.7)$$

From a momentum balance

$$mv_o = mv + MV \qquad (6.8)$$

The mass of the neutron is about one amu and the mass of the
nucleus is given by its mass number, A. Letting m = 1, M = A,
and solving for the ratio of the kinetic energy of the neutron
before and after the collision gives:

$$\frac{E \ (after)}{E \ (before} = \left[\frac{A - 1}{A + 1}\right]^2 \qquad (6.9)$$

6.2 RADIOACTIVITY

When an atom or nucleus is not in a stable ground state,
it seeks to return to its stable condition. Figure 6.4 shows
an atom in an energy state above the ground level. It seeks
to return to its ground condition. Figure 6.5 depicts the
energy levels as a set of stairs. The atom may reside on

any level, but will eventually roll down the stairs until
it reaches the ground state. It will remain at the ground
state. An atom or nucleus in the unstable condition must seek
to rearrange itself in a more stable state (ground state).
The tendency to change to a more stable condition results in
"radioactivity."

Radioactive decay can be characterized in three ways:
(a) by the type of radiation emitted, (b) by the rate at
which the radiation is emitted, and (c) by the energy of the
radiation emitted. The rate (dN/dt) at which all radioactive
atoms (N) emit radiation is governed by the relation:

$$\frac{dN(t)}{dt} \; \alpha \; N(t) \qquad\qquad (6.10)$$

Simply stated, the instantaneous rate of decay is proportional
to the number of radioactive atoms present at time (t). Equation 6.10 can be written as

$$\frac{dN(t)}{dt} = -\lambda N \qquad\qquad (6.11)$$

where λ = decay constant (proportionality constant), and is
a constant characteristic of a given radioactive specie.
The negative sign in Equation 6.11 denotes the fact that
N(t) decreases as time increases. Separating variables and
integrating

$$\int_{N_o}^{N} dN/N = -\lambda \int_{0}^{t} dt$$

$$\ln N/N_o = -\lambda t$$

or

$$N = N_o \, e^{-\lambda t} \qquad\qquad (6.12)$$

Perhaps a more familiar term to the reader in characterizing the rate of radioactive decay is the half-life, $T_{1/2}$. By definition, this is the time required for a given number of radioactive nuclei to decay to one-half their original amount. Using Equation 6.12, when $t = T_{1/2}$, $N = 1/2(N_o)$ and

$$1/2(N_o) = N_o e^{-\lambda T_{1/2}}$$

Solving this relationship for $T_{1/2}$ gives the relationship between $T_{1/2}$ and λ:

$$T_{1/2} = 0.693/\lambda \tag{6.13}$$

There are four types of radiation which are important in nuclear power generation processes: alpha (α) particles, negative beta (β^-) particles, gamma (γ) rays, and neutrons ($_0^1 n$). Each is discussed below.

Alpha (α) particles are helium ($_2^4 He$) nuclei stripped of the orbital electrons and have a +2 positive charge. They are emitted most often by the atoms having atomic number and mass greater than lead, which included the naturally occurring elements up to uranium and the transuranium (artifically produced) elements above uranium in the atomic table. An example of α-decay is

$$_{88}^{226} Ra \xrightarrow{\hspace{3cm}} _{86}^{222} Rn + _2^4 He \tag{6.14}$$
$$T_{1/2} = 1600 \text{ years}$$

the disintegration of radium -226 to radon.

EXAMPLE 6.8   Compute the energy in MeV of the $_2^4$He nucleus (alpha
                particle) emitted in the radioactive decay of radium -226
                to radon -222 (Eq. 6.14). Assume that since the radon -222
                is heavy and relatively immobile, that all the energy
                released will be vested in the kinetic energy of the
                α-particle. Calculate the speed in mph of the helium
                nucleus at the instant of decay.

        The atomic masses are: $_{88}^{226}$Ra = 226.0245 [amu]

                                    $_{86}^{222}$Rn = 222.0175 [amu]

                                    $_2^4$He = 4.0026 [amu]

SOLUTION 6.8

        $_{88}^{226}$Ra $\longrightarrow$ $_{86}^{222}$Rn + $_2^4$He

        ΔM = 226.0245 [amu] - (222.0175 + 4.0026) [amu]

        ΔM = +0.0044 [amu] (positive-an exothermic reaction)

        Kinetic energy of the α-particle

        $_2^4$He = 0.0044 [amu] x 931 [MeV/amu] = 4.1 [MeV] or

        6.57 x 10^{-13} [J]

        For the alpha particle, m = (4.0026 amu) (1.66
        x 10^{-27} kg/amu) = 6.64 x 10^{-27} kg.

        v = $\sqrt{2K.E./M}$

        v = $\left[\dfrac{(2)(6.57 \times 10^{-13} J)}{6.64 \times 10^{-27} kg} \right]^{1/2}$ = 1.41 x 10^7 m/sec.

        $\dfrac{(1.41 \times 10^7 \text{ m/sec})(3600 \text{ s/h})}{(1609 \text{ m/mile})}$ = 3.15 x 10^7 miles/hr (mph)

        This calculation neglects relativistic effects, which
        would cause a very slight increase in the mass of the
        α-particle as it approaches the speed of light.

Beta particles ($_{-1}^0$e) are high energy electrons that are
emitted from the nucleus by the reaction

$$_0^1n \longrightarrow _1^1p + _{-1}^0e \qquad\qquad (6.15)$$

The number of neutrons in a nucleus decreases by one and the
number of protons increases by one and a beta particle is
ejected from the nucleus.

They are typically emitted by nuclei which have a "too high" neutron to proton ratio to be stable. This is almost always the case with fission fragment nuclei, and there are a few cases of β^- emission among naturally occurring isotopes, e.g., radioactive carbon -14 which decays via the reaction

$$^{14}_{6}C \longrightarrow {}^{14}_{7}N(\text{stable}) {}^{+}_{-}{}^{0}_{1}e\,(0.156 \text{ MeV}) \qquad (6.16)$$

Note the application of conservation of charge and nucleon mass in Equation 6.16, where the β^- particle is symbolically written as the electron of -1 charge and essentially zero mass (the actual electron mass is approximately 1/1800 amu). β^- particle energies can range from a few hundredths of an MeV to several MeV depending upon the energetics of the decay reaction.

6.3   INTERACTION WITH MATTER

There are several types of reactions that nuclear particles participate in. Several of these are discussed in this section.

When a charged particle attempts to pass through matter they interact because of the charge. When heavy particles with high velocity attempt to pass through matter, they react with electrons in the matter, strip the electrons from the matter and ionize the atoms. This takes up energy and slows the particle down rapidly. Several types of charged particles reactions are encountered.

The fission fragments are highly positive charged, nuclei. Because of the charge and size, they are stopped within fractions

of a millimeter of the site of the fission. They are, thus,
contained in nuclear fuel material, and their entire kinetic
energy is transformed to "nuclear heat" generation in the
fuel.

Alpha particles $_2^4$He are not very penetrating. As in
the case of fission fragments their charge and relatively
heavy mass cause them to quickly give up their energy to
surrounding atoms via interactions with atomic electrons.
In fact, most α particles can be stopped by a sheet of paper,
and a 5 MeV α particle can only penetrate 3-4 cm of air at 1
atmosphere and room temperature before losing its energy.

The physical mechanism for <u>attenuating</u> (stopping) β^-
particles is Coulomb force interactions with atomic electrons.
Even the most energetic β^- particles can be attenuated by a
millimeter thick aluminum sheet. As an item of comparison,
3 MeV α particles are stopped by 1.7 cm thick layer of air,
whereas approximately 13 m of air is required to attenuate
3 MeV β^- particles.

γ-rays are highly penetrating forms of radiation, since
they possess no mass and their zero charge makes them immune
to interactions with the atomic electrons of the medium
through which they are passing unless they directly collide
with the electron. Coulomb forces do not exist. The mechanisms for attenuating gamma rays is complex. Basically, they
lose their energy by making "scattering" collisions with atomic
electrons and gradually transferring and losing their energy
in the process.

In general, the ability of a material to attenuate γ-rays
increases with the density of the material. Thus, it takes
a smaller thickness of lead than iron to stop a given quantity
of γ-rays, a smaller thickness of iron than water, etc.

The capability of a given material to attenuate γ-rays can be measured by knowing its mass attenuation coefficient, usually written as (μ/ρ). It is a function of the electronic structure of the material's atoms and is dependent upon the energy of the gamma particles themselves. Values of (μ/ρ) have been tabulated for many materials and gamma energies. The units of (μ/ρ) is cm^2/g, where ρ = density of the material in g/cm^3. γ-attenuation in a material can be computed (to a first approximation) by use of the equation

$$I(x) = I_0 e^{-(\mu/\rho)\rho x} \qquad\qquad (6.17)$$

Here I_0 is the intensity of the γ-radiation striking the surface of a material, e.g., the number of γ's hitting a 1 cm^2 area of the surface each second, $I(x)$ is the intensity at some distance x cm into the medium, ρ is the medium density in g/cm^3, and (μ/ρ) is the mass attenuation coefficient.

EXAMPLE 6.9   Calculate the thickness of air, water, iron, and lead required to reduce the intensity of a beam of 3 MeV γ-rays to 99.99% of its original intensity. Also determine the wavelength of 3 MeV γ-rays and compare this to that of visible light which is in the range of 4,000 Å - 7,000 Å (1°A = 1 angstrom unit = 1 x 10^{-10}[m]. Then compute the energy range associated with the visible light spectrum of radiation. Use the following data for 3 MeV gamma rays.

Material	Density (g/cm^3)	$(\mu/\rho)(cm^2/g)$
Air	1.29 x 10^{-3}	0.0357
Water	1.00	0.0396
Iron	7.87	0.0361
Lead	11.34	0.0421

SOLUTION 6.9   Reduction of intensity by 99.99% implies that at x, the desired material thickness, $I(x) = 0.0001\ I_0$. Then, from (6.17):

$0.0001\ I_0 = I_0 e^{-(\mu/\rho)\cdot\rho x}$ - 9.21 = $-(\mu/\rho)\rho x$

$x = 9.21/[(\mu/\rho)\cdot\rho]$

For air: $x = 9.21/(0.0357[\text{cm}^2/\text{g}] \cdot 1.29 \times 10^{-3}]) = 200,000 \text{ cm}(1.24 \text{miles})$
  Water: $x = 9.21/(0.0396[\text{cm}^2/\text{g}] \cdot 1[\text{g/cm}^3]) = 233 \text{ cm} (7.64 \text{ ft})$
  Iron: $x = 9.21/(0.036[\text{cm}^2/\text{g}] \cdot 7.87[\text{g/cm}^3]) = 32.4 \text{ cm} (1.06 \text{ ft})$
  Lead: $x = 9.21/(0.0421[\text{cm}^2/\text{g}] \cdot 11.34[\text{g/cm}^3]) = 19.3 \text{ cm} (7.60 \text{ in.})$

Using (6.6)

$$\lambda = hc/E = 4.136 \times 10^{-21}[\text{MeV-sec}] \cdot 2.998 \times 10^{8}[\text{m/sec}]/3.0[\text{MeV}] = 4.13 \times 10^{-13}[\text{m}]$$

$$\lambda = 4.13 \times 10^{-13}[\text{m}] \times 1/10^{-10}[\text{A/m}] = 4.13 \times 10^{-3}[\text{A}]$$

The 3 MeV γ-ray has a wavelength a factor of 10^{-6}
shorter than visible light.

Energy of visible light, using (6.16) again is

$$E_{4000\text{A}} = hc/\lambda = 4.136 \times 10^{-21}[\text{MeV-sec}] \cdot 10^{6}[\text{eV/MeV}] \cdot 2.998 \times 10^{8}[\text{m/sec}]/$$
$$4 \times 10^{-7}[\text{m}] = 3.10[\text{eV}]$$

Similary, $E_{7000\text{A}} = 1.77[\text{eV}]$

It is evident that visible light is much less energetic than γ-radiation, and thus, possess little penetrating power (as is well-known by the existence of opaque materials). γ-radiation, on the other hand, is capable of penetrating and depositing energy in all matter. It should be noted that when γ-radiation is attentuated, it surrenders its energy to the atoms and molecules of the medium through which it is passing. This may appear in the form of heat and/or induced chemical-molecular structural changes. It is this latter effect that makes γ-rays, and indeed all forms of nuclear radiation, dangerous to biological and human tissues.

The discussion of the reaction of neutrons with matter is the basis to nuclear fission and is discussed in detail in the next section.

6.4 NUCLEAR REACTIONS

Three important fission reactions are:

$$\,_{0}^{1}n + \,_{92}^{235}U \xrightarrow{\text{fission}} \,_{Z_1}^{A_1}X + \,_{Z_2}^{A_2}X + \text{neutrons}$$

$$\,_{0}^{1}n + \,_{92}^{233}U \xrightarrow{\text{fission}} \,_{Z_1}^{A_1}X + \,_{Z_2}^{A_2}X + \text{neutrons}$$

$$\,_{0}^{1}n + \,_{94}^{239}Pu \xrightarrow{\text{fission}} \,_{Z_1}^{A_1}X + \,_{Z_2}^{A_2}X + \text{neutrons}$$

In the fission process, the heavy nucleus reacts with a neutron, and an intermediate nucleus is formed which exists for the order of 10^{-15} seconds before it splits into two nuclei (fission fragments) of lower mass with the simultaneous release of additional neutrons. The net release of energy is approximately 200 MeV per fission. It is worth noting that these reactions are initiated with neutrons having average kinetic energy of 0.025 eV. This is the energy possessed by air, or any gas molecules, in thermal equilibrium at room temperature (20°C). Thus, for a 1/40 eV (thermal) neutron energy investment and the presence of fissionable isotopes 200 MeV/fission energy output can be achieved.

Of the three fissionable nuclei, only uranium -235 is a naturally occurring isotope. Uranium -233 and plutonium -239 are isotopes which can be produced from thorium -232 and uranium -238 respectively. The number of neutrons yielded in a fission process will vary, depending upon the identification of the fission fragments $\,_{Z_1}^{A_1}X$ and $\,_{Z_2}^{A_2}Y$. Multiple combinations of fission fragments are possible in a given fission reaction, with most likely appearing nuclei to be those having nuclear masses of 90 and 140 amu. The average number of neutrons

emitted per fission are: 2.54 for uranium -233, 2.46 for
uranium -235, and 2.88 for plutonium -239.

Neutrons, although primarily viewed as a reactant particle
in the fission process, are also considered a form of nuclear
radiation.

Neutrons interact with the nucleus of atom via absorption,
scattering, and fission. When absorbed, they frequently produce radioactive isotopes (usually β^- and γ emitters); and
when scattered, the neutrons are reduced in energy, having
transferred some of their energy to the scattering nucleus.

Fission is the primary source of neutrons; however,
there is a small class of "neutron rich" fission fragments
which seek a more stable neutron-proton balance by radioactive decay of neutrons. These neutrons are known as
delayed neutrons, and play an important role in the control
and safety of a nuclear power reactor system. The nuclei
from which they are emitted are called precursors, and have
half-lives ranging from 0.23 sec. to 55.72 sec.

Neutron attenuation is governed by a complex relationship
of nuclei density in a material, neutron energy, and the probability that a neutron of given energy will be absorbed or
scattered by a specific isotope nucleus. This last item is
represented by a parameter known as the microscope cross
section. There have been literally thousands of neutron cross
section measurements made for most known isotopes at neutron
energies ranging from several ten-thousandths of an eV to
greater than 10 MeV. It is beyond the scope of this section
to discuss cross-section concepts in detail; nevertheless,
as a broad rule of thumb, the probability of neutron absorption by a given nuclear species decreases with increasing
neutron energy, while the scattering probability remains

fairly constant and independent of neutron energy. It is
through accurate knowledge of neutron cross sections that
reactor designers are able to specify the type, density,
and distribution of materials in a nuclear reactor so that a
controlled chain reaction can be initiated and maintained.
Such information is also required to properly design neutron
shields for reactor systems so that neutron radiation to the
environment is minimized.

6.5 NUCLEAR REACTORS

A nuclear reactor is basically a device wherein a controlled fission chain reaction can be maintained to produce a
requisite amount of thermal energy; this energy being extracted via a coolant medium, which is ultimately converted
to electrical energy (in the case of an electrical generating
plant). In a controlled fission chain reaction, a constant
population of neutrons must be maintained in the reactor with
fission.

Figure 6.7 shows the basic elements of a thermal nuclear
reactor. The shaded portions represent fuel elements. The
area surrounding these elements contains moderator. The fission neutrons are generated at a high energy (about 2 MeV).
At these high energies, they do not react with U^{235} to cause
further fission. They must be slowed down (lose kinetic
energy). This is the purpose of the moderator. The sustained
fission process consists of

    a)   Fission of uranium -235 in the fuel generating fast
           neutrons (high energy) [2 Mev].

    b)   Fast neutrons leave the fuel element and collide
           with the moderator and slow down (low energy) [0.025 eV].

c) Slow neutrons find their way to the fuel element
    and cause uranium -235 to fission.

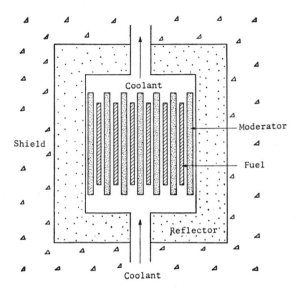

FIGURE 6.7 Schematic Reactor

This cycle is complicated by the fact that
a) Some of the neutrons find their way out of the
    reactor zone and are lost.
b) Some of the neutrons are absorbed in other mate-
    rials other than uranium -235 and cannot cause the
    U -235 to fission.

Figure 6.8 shows the history of 100 neutrons resulting
from the fission of uranium -235. For the system shown in
Figure 6.8, the number of fast neutrons leaving the bottom
block in the flow sheet is

$$100 \times 1.04 \times 0.952 \times 0.899 \times 0.944 \times 0.904 \times 1.33 = 101$$

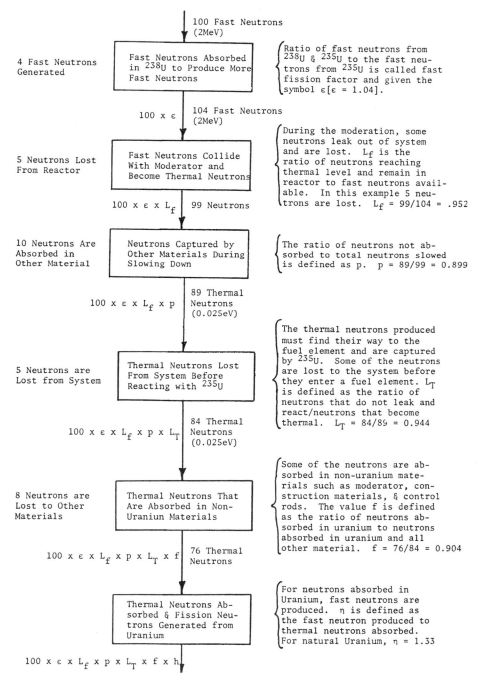

FIGURE 6.8 Fission Cycle

For 100 fast neutrons entering the process, 101 exit. These
101 become the input to the next cycle. After this cycle

$$(101)(1.01) = 102$$

are present. For each cycle, more neutrons are present. The
energy production is increasing and will continue to increase.

The increase in neutron population with each cycle, or
generation, is given by the multiplication factor

$$k_{eff} = \varepsilon \, \eta \, p \, f \, L_{th} \, L_f \tag{6.18}$$

Each of these terms was defined in Figure 6.8. For the system shown in Figure 6.8 the value of $k_{eff} = 1.01$. For a
steady-state system, it will be necessary to change one of
the factors in Equation 6.18 to provide for a k_{eff} of 1.0.
The usual way would be to insert some material that will capture the thermal neutrons. This will decrease f, the thermal
utilization factor. This is done by inserting   rods
made of this material into the bed.

If the reactor were <u>infinite in size</u> L_f and L_{th} = 1.0 (no
chance to leak out)

$$k_{\infty} = \varepsilon p \, f \, \eta \tag{6.19}$$

and be effective multiplication factor may be written

$$k_{eff} = k_{\infty} \, L_f \, L_{th} \tag{6.20}$$

If $k_{\infty} > 1.0$, it means that for an infinite size reactor
for each neutron cycle there would be more neutrons. The
neutron population will grow. The value of L_{th} and L_f (that
depend on reactor size) may be chosen to make $k_{eff} = 1.0$.
The values of L_{th} and L_f may be calculated from the equations.

$$L_f = e^{-B^2 \tau} \tag{6.21}$$

$$L_{th} = 1/(1 + L^2 B^2) \tag{6.22}$$

where B^2 is a function of geometry
L^2 and τ depend on moderator.

TABLE 6.3
Evaluation of B^2

Shape of Reactor	
Parallelepiped	$B^2 = (\Pi/a)^2 + (\Pi/b)^2 + (\Pi/c)^2$
	where, a, b, c, are side length
Sphere	$B^2 = (\Pi/R)^2$
	where R is the radius
Cylinder	$B^2 = (2.405/R)^2 + (\Pi/H)^2$
	where H is height and R the radius

EXAMPLE 6.10   For a light moderated highly enriched uranium reactor
$k_\infty = 1.5$. Determine the size of a spherical reactor
to give $k_{eff} = 1.0$. For water $\tau = 33cm^2$, $L^2 = 2.49$.

SOLUTION 6.10

$$k_{eff} = k_\infty L_{th} L_f$$

$$1.0 = 1.5 \ (e^{-33B^2})/(1+2.49B^2)$$

$$B^2 = 0.014 = (\Pi/R)^2$$

$$R = 29.5cm$$

The term σ_a is a measure of the relative ease that an
atom will absorb a neutron. σ_a has units of area and may be
thought of as a "target." The larger the target area, the
greater the probability of a projectile to hit it. If N is
the number of atoms or molecules/unit volume, the product of
$N\sigma_a$, called microscopic cross section, has units cm^{-1}.

In order to evaluate the value of f a few terms must be identified.

$$\Sigma_a^C = \sigma_a^C N_C \text{ control rod cross section}$$

$$\Sigma_a^M = \sigma_a^M N_M \text{ thermal absorption cross section of moderator.}$$

$$\Sigma_a^U = \sigma_a^U N_U = \text{ thermal absorption cross section of fuel}$$

$$\Sigma_a^S = \sigma_a^S No = \text{ thermal absorption cross section}$$
$$\text{of structural material}$$

EXAMPLE 6.11   Determine the macroscopic cross-section Σ for U^{235}. The density of U is 18.7/cm^3 and σ_a = 0.048 barns (1 barn = 10^{-24}cm^2).

SOLUTION 6.11

$$\Sigma^U = N\sigma_a = 18.7[g/cm^3] \times 1/235[g\text{-atoms}/g] \times$$
$$6.023 \times 10^{23}[\text{atoms}/g\text{-atom}] \times 0.048[\text{barns}/\text{atom}]$$
$$\times 10^{-24} [\text{barn}/cm^2] = 0.023 \text{ cm}^{-1}$$

The term Σ when appearing as ratios, may be interpreted to be a probability that the event will take place. The value of f, may be obtained by relative probability neutron will be absorbed in fuel.

$$f = \frac{\Sigma_a^U}{\Sigma_a^U + \Sigma_a^M + \Sigma_a^S + \Sigma_a^C} \tag{6.23}$$

The nuclear reactor may be controlled by inserting a control rod. The control rod is made of a material that absorbs thermal neutrons readily. This increases Σ_a^C which in turn reduces f, which in turn lowers k_∞ and k_{eff}. By lowering control rods into a reactor the value of Σ_a^C and f

can be minipulated to control the reactor. The delayed neutrons allow time for control rods to be inserted and withdrawn
to control the nuclear reactor.

Natural uranium contains only 0.72% $^{235}_{92}$U. The rest is
$^{238}_{92}$U. The value of η is 2.08 for pure $^{235}_{92}$U and 1.33 for
natural uranium. The value of ε = 1.0 for highly enriched
reactors.

6.6 SAFETY

The debate over nuclear safety is likely to rage for many
years. No answer acceptable to a large majority is likely to
be found. In a later chapter on cost-benefit analysis some
of the difficulties of a decision that may benefit today's
generation, but risk future generations, are discussed.
There is no effort made in this chapter to answer the question
on nuclear safety--it is far too complex a problem.

The probability of a catastrophic nuclear accident is
small. On the other hand, the damage resulting from an
accident is extremely large. To determine the yearly cost
of nuclear accidents the following equation may be used:

$$\text{Probable yearly cost of nuclear accidents} = \text{Yearly probability of failure per reactor} \times \text{Number of reactors} \times \text{Cost of a failure}$$

$$(6.24)$$

As the probability of failure approaches zero and the cost of
an accident approaches infinity, this equation becomes

$$\text{Probable yearly cost of nuclear accidents} = 0 \times \infty = \text{indeterminate}$$

By the laws of math, the cost becomes indeterminate. There
is no clear answer.

Nuclear reactions are a source of large quantities of radioactive isotopes. The unit of radioactivity is the curie (named after Madame Curie) and is 3.7×10^{10} dis/sec. What biological damage can be done by radioactive decay depends on the type and energy of radiation. Except for some limited (but significant) medical application, it is safe to state "Radioactivity is dangerous to your health," and should be avoided.

Some idea of the problem can be obtained by considering the waste problem from a single 1,000,000 Kw reactor. At 30 percent efficiency, this comes to 300 Mw electrical output. This results in 3.125×10^{19} fissions/sec or 9.86×10^{26} fission events/year. This reactor fissions two metric tons in a five-year period. If 1 Kg of U^{235} is fissioned, 999 grams of radioactive materials are generated. Some have a short half-life and can be held a short period of time and are rendered harmless.

Others have long half-life and are around for years. Table 6.4 shows some of the particularly hazardous isotopes.

TABLE 6.4
Some Isotopes Produced in Fission Reactors
That Are Particularly Hazardous to Man

Isotope	Half-life*	Critical organ
Iodine-129	17,000,000 years	Thyroid
Plutonium-239	24,000 years	Total body but especially lungs
Strontium-90	28 years	Bone
Cesium-137	27 years	Total body
Hydrogen-3 (tritium)	12 years	Total body
Krypton-85	11 years	Lungs, skin
Iodine-131	8 days	Thyroid

*Isotopes must be stored for periods of 10 to 20 times their half-life before they decay to safe levels.

Three general techniques are used to dispose of radioactive
material: (a) dilute and disperse, (b) delay and decay, and
(c) concentrate and contain.

Only the last can be used for the high level of activity
from a used nuclear reactor core. Storage for extremely long
periods is necessary. This persistent problem must be resolved for nuclear energy to be viable.

Figure 6.9 shows a nuclear fuel cycle for a light
water moderated reactor (LWR).

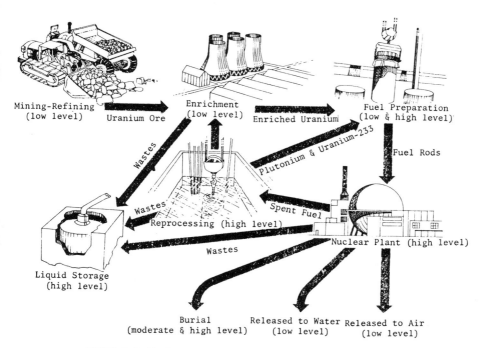

FIGURE 6.9 Nuclear Fuel Cycle (From Living in the
Environment: Concepts, Problems & Alternatives
by G. Tyler Mills, Wadsworth Publishing Co.)

Some of the major risks are given below:

1. Mining and milling of uranium entail hazards that are
    intensifications of natural hazards. With reasonable

precautions, they need not be of major concern.

2. With enrichments used in LWR's (Light Water Moderated Reactors) careful procedures can keep the probability of an accident negligibly small.

3. Radioactive releases from an LWR under normal conditions can be, and are being, kept quite low and so are not a problem.

4. In spite of all precautions, accidents at reactors are always a possibility. Results of a major accident could be serious.

5. The radioactive releases from a fuel reprocessing plant, under normal conditions, are high. For the long haul, both ^{85}Kr and tritium will have to be contained.

6. Long-term storage and high-level radioactive wastes is an unsolved problem and a very serious one.

7. With increasing volumes of high-level radioactive materials moving about the country, a transportation accident involving the release of radioactivity to the environment appears likely to occur sooner or later.

In spite of all of these risks, the nuclear industry has been exceedingly safe. Radioactivity is contained in the reactor by: (a) the uranium fuel is incased in cladding that will not pass the fission products; (b) the uranium fuel is in pellets that contain most of the fission products; (c) the primary coolant loop does not leave the containment vessel (see Figure 6.2) and will not reach the turbine-condenser loop.

In the design of reactor systems all manner of creditable

accidents are considered. The design incorporates considerable
redundancy in safety features--there are parallel safety systems, auxiliary power sources, etc. Several independent
failures at the same time are considered.

In addition to the many safety features, the reactor is
typically protected by an outer shell of two to three feet
thick concrete and approximately 100 feet in diameter. This
vessel can withstand earthquakes, tornadoes, and internal
failures.

Because it is known that one is dealing with a dangerous
system in a nuclear reactor, extreme caution is taken and
nothing is taken for granted. Typically the industries
known to have extreme dangers take precautions and often have
the best safety records.

6.7 BREEDER REACTORS

In evaluating the k_∞ and k_{eff} (Eq.6.19 & 6.20), it was
noted that $^{238}_{92}U$ captured neutrons. When this occurs, the
Urnaium is unstable and decays according to the following
reaction:

$$^{1}_{0}n + {}^{238}_{92}U \longrightarrow {}^{239}_{92}U \xrightarrow{\beta^-} {}^{239}_{93}Np \xrightarrow{\beta^-} {}^{239}_{94}Pu$$

Two beta particles are emitted and plutonium-239 is produced.
This is a nuclear fuel. It is far easier to separate than to
remove $^{235}_{92}U$ from natural uranium. Uranium-238 is said to be
fertile as it can produce a fissionable material. Natural
occuring thorium is another fertile material and produces
Uranium-233, a fissionable material, by the reaction.

$$n + {}^{232}_{90}Th \longrightarrow {}^{233}_{90}Th \xrightarrow{\beta^-} {}^{233}_{91}Pu \xrightarrow{\beta^-} {}^{233}_{92}U$$

Breeding requires two or more neutrons per fission. One
neutron is needed to keep the reaction going and the second
neutron to replace the fuel just fissioned with a "new" fuel
from the fertile material. If more than 1 atom of "new fuel"
is produced then there is a net increase in fuel in the reactor.

The value of η (the number of neutrons produced per
neutron absorbed in fuel) is given in Table 6.5 for uranium
thorium and plutonium for both fast and slow neutrons.

The fissionable material that can be produced is given by

$$C = \eta - 1 - L \tag{6.25}$$

where L is neutrons lost

1 is neutron needed to sustain reaction.

For an infinite size reactor L = 0 and

$$C_{max} = \eta - 1 \tag{6.26}$$

TABLE 6.5
Comparison of Fast and Slow Neutrons on η, C

	Thermal Neutron Fission			Fast Neutron Fission		
	233_U	235_U	239_{Pu}	233_U	235_U	239_{Pu}
η	2.3	2.1	2.1	2.4	2.2	2.6
C_{max}	1.3	1.1	1.1	1.4	1.2	1.6

It is clear that ^{239}Pu using fast neutrons represent the best
possible conversion. Because σ's are smaller, fast breeder
reactors will be large and will contain no light elements
that will serve to moderate the neutrons.

Figure 6.10 is a schematic diagram of a liquid metal
breeder reactor. An intermediate nonradioactive sodium heat

transfer loop is used to isolate the radioactive sodium of
the reactor core from the steam turbine portion of the plant.

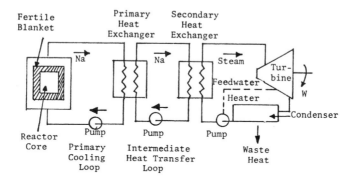

FIGURE 6.10 Schematic Diagram of a Liquid Metal Breeder
            Reactor Power Plant

Reactor coolant temperatures of 600-650°C (1100-1200°F) are
considerably higher than the temperatures possible for water-
cooled reactors, but they are still much less than the boiling
temperature of sodium. These reactor coolant temperatures
result in steam temperatures on the order of 540°C (1000°F).
Despite the unavoidable temperature drop of the intermediate
heat transfer loop, thermal efficiencies comparable to fossil-
fuel plants may be achieved. Surrounding the reactor core is
a blanket of fertile material, which, for breeding plutonium-
239, is natural uranium. Fast neutron fission of uranium-238
and plutonium-239 of the blanket result in a small contribution to the total power of the reactor.

6.8   ENERGY CONVERSION

The power cycle for a nuclear system is essentially the
same as a conventional system. The only significant difference
is in the discharge of the waste heat. In the nuclear reactor
system there is no stack emissions and there is no loss of
energy as heat products of combustion. As a result, the nuclear
system will discharge more energy to rivers and/or lakes.

For water-moderated reactor systems, it is necessary to
maintain liquid in the reactor to serve as a moderator.
Therefore, the temperatures are limited and provide for lower
Carnot Efficiencies.

6.9   FUSION

The discussion has thus far been limited to the fission
reaction. In contrast to breaking up large nuclei, energy can
be given off if two small nuclei are fused. Figure 6.11
shows two such reactions. In the first, a deuterium nucleus
(similar to hydrogen but contains a neutron in the nucleus)
reacts with a second deuterium atom and forms a helium nucleus
plus a neutron. Several MeV(26) are released in this reaction.
In the second reaction the deuterium nucleus combines with a
tritium nucleus and forms helium. In this reaction 17.6 MeV
are released.

There are several immediate advantages from fusion reactions. The radioactive release is much less serious than
for fission. Only tritium is radioactive. Also, there is a
large abundance if isotopes of hydrogen in water.

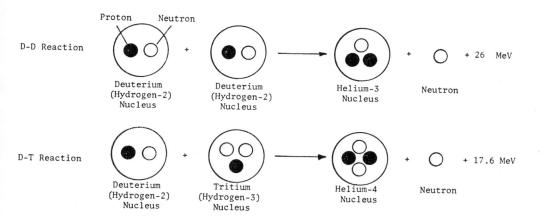

FIGURE 6.11 Fusion

EXAMPLE 6.12   Determine the equivalent gallons of gasoline to provide the
same energy as 1 gallon of seawater if all the deuterium
was reacted (Ratio $_1H^1/_1H^2$ is 6,000/1)

SOLUTION 6.12

1[gallon seawater] x 8.33[lb/gal] x 1/2.2[Kg/lb]

x 1/18[Kg mole H_2O/Kg]x6.023x10^{27}[molecules/Kg-mole] = 1.27x10^{27}[molecules]

1.27x10^{22}[molecules] x 2[atoms H /molecule H_2O] x 1/6001[$_1^2H/_1^1H$]

= 4.2 x 10^{23} [atoms of deuterium]

4.2x10^{23}[$_1^1H$] x 26[MeV/$_2^1H$] x 1.9 x 10^{-16}[J/MeV] x 1/1.318x10^8 [gallon

gasoline/J] = 15.8 gallon gasoline

The problem with the fusion reaction is that temperatures
on the order of 50 x 10^6°C must be achieved for the fusion
reaction to occur. No material can withstand these temperatures and some technique must be developed to "contain" the
reaction. The techniques for achieving this are not discussed.

Figure 6.12 is a conceptual configuration of a fusion
reactor. Of interest is the use of a conventional power cycle
and the ensuing loss of energy. Development of the fusion

reactor would not provide <u>infinite energy resources</u> because
of the resulting thermal pollution (the ultimate pollutant).

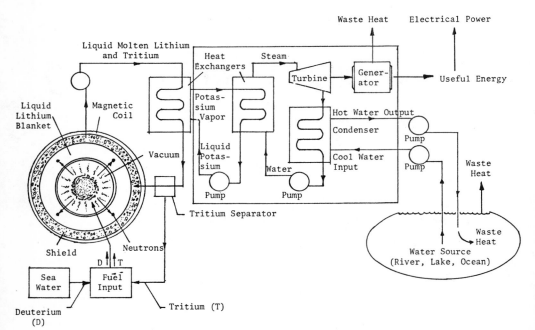

FIGURE 6.12   A Nuclear Fusion Power Plant Using Deuterium
and Tritium (D-T) As Fuel

Problems for Chapter 6.0

1. Calculate the mass of $^{235}_{92}U$ fissioned and consumed during the one day operation of a uranium-235 fueled reactor running at 1 Mw(th) power.

2. A uranium-235 fueled light water reactor power plant produces a daily average of 900 Mw(e) power at an over-all thermal efficiency of 30%. How much ^{235}U would be consumed by the plant annually? Compare this with the tonnage of coal which would be required for a similar plant if the coal has a heating value of 14,000 Btu/lb.

3. The energy of formation of water is 54,500 cal/g-mole. What is this in eV/molecule?

4. Compute the megawatt-days of energy produced by the fission of all ^{235}U contained in one ton of natural uranium where the initial concentration of ^{235}U is 0.7%.

5. The half-lives of two radioisotopes are one and two days, respectively. Find the relative concentrations after 2 hours, 1 day, 1 month, and 1 year for mixtures with an initial concentration of 1:1, 1:10, and 10:1.

6. What is the mass equivalent to a 1 MeV photon? What is the wave length?

7. What is the minimum amount of kinetic energy an α-particle must have to cause the reaction

$$^{14}_{7}N + {}^{4}_{2}He \longrightarrow {}^{17}_{8}O + {}^{1}_{1}H \ ?$$

$^{14}_{7}N = 14.003074$ AMU $\qquad\qquad$ $^{17}_{8}O = 16.999133$ AMU

$^{4}_{2}He = 4.002603$ AMU $\qquad\qquad$ $^{1}_{1}H = 1.007825$ AMU

8. The mass of $^{56}_{26}Fe$ is 55.93494 amu. Determine the mass defect and binding energy per nucleon and compare to Figure 6.4.

9. Consider the energy given off by the following fission event:

$$^{1}_{0}n + {}^{235}_{92}U \longrightarrow {}^{140}_{55}Cs + {}^{93}_{37}Rb + 3\ {}^{1}_{0}n + Q$$

    (a) Does the charge on number of nucleons balance?
    (b) Estimate the energy released by this fission.

10. Determine the total number of radioactive atoms in a sample decaying at 10^{10} atoms/min with a half-life of 60 days.

11. Determine the size of a ^{235}U light water moderated reactor. The number of water molecules to uranium molecules is 500/1. The cross-section for thermal neutrons is 678 barns for uranium and 0.66 for water. The value of η may be calculated from the following information.
   (a) Number of neutrons produced from fission of uranium 235 is 2.5.
   (b) Total cross section of U-235 for thermal neutrons is 678 barns. This includes neutrons that cause fission and neutrons that are absorbed and do not cause fission.
   (c) Cross-section of U-235 for thermal neutrons that cause fission is 580 barns.

   η = 2.5(580/678) = 2.12

   For highly enriched reactors ε = 1.0 and p = 1.0. Assume that the reactor is spherical. (For water τ = 33 cm^2, L^2 = 8.3(1-f))

12. The size of the reactor from Problem Number 4 is what is termed a bare reactor. In practice the reactor core (the portion containing the fissionable material) is surrounded by a reflector. The purpose of this reflector is to reflect neutrons that escape the core back into the core. This is shown in Figure 5.1.

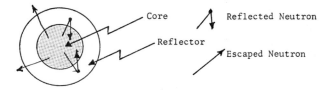

Assume that the addition of a reflector reduces the leakage of fast neutrons by 20% and thermal neutrons by 40%. What would the size of the core be in this case?

13. For Problems 11 and 12 calculate the critical mass. The core may be assumed to have the density of water.

14. The critical size of a spherical bare reactor (water moderated) is found to have a radius of 40 cm^2.
    (a) What is the ratio of atoms of uranium to atoms of water?
    (b) What is the critical mass?
    The system uses highly enriched uranium. See Problems 11 and 12 for additional information and assumptions.

15. The energy equivalent of matter is given by the well-known equation E = mc^2.
    (a) Calculate the quantity of matter that must be converted to energy to supply the 1972 U.S. electrical energy requirement (thermal) for one day. (P = 6 x 10^{11}W).
    (b) If ^{235}U is used for a fission process, 200 MeV of energy is obtained for each atom. Determine the mass converted to energy by one ^{235}U atom. What is the fraction of the atom's initial mass that is converted to energy?
    (c) Determine the number of metric tons of ^{235}U needed for the energy of part (a).
    (d) Uranium ore consists of an oxide of uranium (U$_3$O$_8$). Only .711% of the uranium atoms are fissionable isotopes, ^{235}U. The remainder are primarily ^{238}U isotopes. Rich uranium ore contains by mass, approximately .25% uranium oxide. Determine the number of metric tons of ore necessary for part (a).
    (e) What is the number of metric tons of coal necessary for the same energy requirement?

16. Uranium-238 is known to decay with a half life of 4.51 x 10^9 years.
    (a) Calculate the number of disintegrations per second for one kilogram of uranium-238.
    (b) In the chain of decay, one of the intermediate products is radon gas. Assuming radon gas (mass number, 222) is provided at the same rate at which ^{238}U decays, calculate the volume rate of production from one kilogram of ^{238}U for a pressure of one atmosphere and a temperature of 20°C.

17. A reactor consists of a homogeneous mixture of heavy
     water and pure ^{235}U. The value of $B^2 = 10^{-3}$ cm^{-2}.
     Find the ratio of number of moderator molecules to
     fuel atoms. Use the equation

$$k_{eff} = k_\infty \frac{e^{-B^2 r}}{1 + B^2 L^2}$$

$$k_\infty = \eta f$$

$$L^2 = L_{D_2O}^2 (1-f)$$

Given:

U^{235} $\sigma_a = 694b$ $\eta = 2.07$

D_2O $\sigma_a = 2.6 \times 10^{-3} b$ $L = 115$ cm $\tau = 125$ cm^2

18. The half-life of Co^{60} is 5.2 years. A sample of 400 mg
     of pure Co^{60} is purchased for medical purposes. How
     many mg will remain after 20 years?

7.0 SOLAR ENERGY

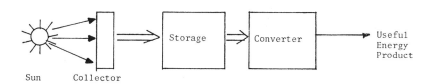

FIGURE 7.1 Elements of a Solar System to Produce Useful Energy.

The amount of solar energy incident upon the earth is somewhere on the order of $200,000 \times 10^{12}$ watts (J/sec) which is over 30,000 times man's current needs for power. Solar energy represents a major energy source and the merits of solar must be considered as an energy resource.

The overwhelming benefit provided by solar energy is that it represents a continuous energy source, if used it does not contribute to the energy released in our biosphere, it does not "belong" to any country or group of countries and it is free.

The effective utilization of solar energy has many serious problems that must be resolved if it is ever to become a major energy resource.

The major problem inherent in the utilization of solar energy is that the energy is dispersed rather than concentrated in small areas where it may be easily collected and used to produce work. As in previous chapters, emphasis was placed on producing work as the essential energy form most needed by an industrial society. The low energy flux makes large scale power generation difficult.

The incidence of solar radiation over any area is not constant. It varies with time of day, weather conditions, and time of year. The demand for energy does not correspond in time with the availability of solar energy. Solar energy cannot be called for "on demand" and consumers will not accept energy on a "when available" basis. The mismatch of supply-demand curves is a serious problem.

The regions where solar energy is most abundant and where large land areas necessary for economical power generators are not the regions where power is needed. Most of these are in desert regions where little industry is located. Often the reason for the scarcity of industry is due to a scarcity of water. This lack of water needed as a heat sink for power cycles seriously affects the capability to produce power in the desert region.

In order to use solar energy on a large scale, serious problems must be overcome; but, successful resolutions would reap major benefits.

7.1 EARTH ENERGETICS

The earth is in a unique relationship to the sun. In this section the thermal energy balance between the earth and

the sun and the universe is explored. To study the thermal
balance, it is necessary to apply the First Law of Thermodynamics over the biosphere. The biosphere is defined here
as the area between the earth surface and a distance of about
10 miles out from the surface. This was chosen as the system
as it represents the volume where man lives and must maintain
in a condition that will support life.

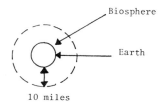

FIGURE 7.2 Earth's Biosphere

For the thermodynamic analysis, Figure 7.2 may be redrawn
along with the energy input and output terms.

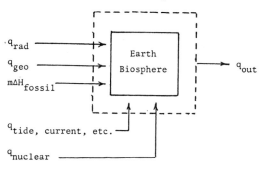

FIGURE 7.3 Thermodynamic System for Analysis

The energy balance for the system shown in Figure 7.3 is:

$$q_{rad} + q_{geo} + m\ \Delta H_{fossil\ fuel} + q_{tide,\ current} +$$
$$q_{nuclear} = q_{out} \qquad\qquad (7.1)$$

In this equation:

q_{rad} — the energy received by the biosphere as a result of radiation from the sun

q_{geo} — the energy reaching the biosphere by conduction through earth's surface

$q_{tide, current}$ — the energy reaching the biosphere by natural phenomana (tide, wind, volcano, ocean currents, etc.)

$m\Delta\overline{H}_{fossil\ fuel}$ — the energy brought into the earth's biosphere and released by combustion of fossil fuels

$q_{nuclear}$ — the energy brought into the earth's biosphere and released by nuclear reaction

q_{out} — the energy given off by the biosphere

Energy moves about within the system (biosphere), but these are the primary ways that they can enter and leave the system defined.

The evaluation of $m\Delta\overline{H}_{fossil\ fuel}$ is obtained from an estimate of the product of all of the fossil fuels removed from the earth, m, and the heating value of these fuels, $\Delta\overline{H}_{fossil\ fuel}$.

The value of $q_{tide, current, etc.}$ is small (in comparison to other terms and is neglected.

The term q_{geo} is a result of conduction from the earth's interior to the earth's surface. There are other energy inputs that come from the interior of the earth. These are due to volcanic action and gysers. These are not major contributors to the energy entering the biosphere and are not likely to change significantly unless man begins to bring "heat to the surface" in large amounts.

The value $q_{nuclear}$ refers to the heat released by man resulting from nuclear activities.

The value q_{rad} represents the amount of energy the biosphere receives from the sun. This value is obtained from the equation

$$q_{rad} = (A)\ (S)\ (a) \tag{7.2}$$

    A = the area of the incoming solar energy inter-
        cepted by the earth. This is πR^2 (See Figure 7.4).

    S = the energy flux of the sun. This is termed "the
        solar constant" $(0.136\ w/cm^2)$.

    a = the fraction of the energy that is intercepted
        that is absorbed in the biosphere.

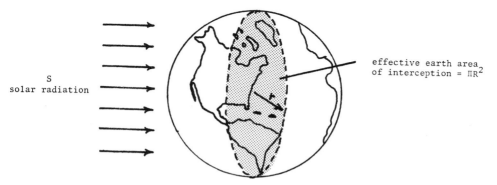

S
solar radiation

effective earth area
of interception = πR^2

FIGURE 7.4 Effective Cross Section of the Earth, πR^2
Where R is the Earth's Radius

Table 7.1 shows the order of magnitude of each of the
various energy input terms.

TABLE 7.1

Estimates of Energy Input Values (1970)

Input	Value [Watts]
q_{rad}	$174{,}000 \times 10^{12}$
$q_{geo} + q_{tidal,\ current,\ misc}$	35×10^{12}
$q_{nuclear} + m\Delta\bar{H}_{fossil\ fuel}$	6×10^{12}
TOTAL	$174{,}041 \times 10^{12}$

It is evident that all input terms in Equation 7.1 are small compared to q_{rad}, the energy received from the sun. Man's activity contributes 0.0035% to the input terms.

A more useful way to write Equation 7.1 would be to collect terms according to whether they are man-made or naturally occurring energy inputs. This is done in Equation 7.3:

$$q_{man} + q_{natural} = q_{out} \qquad (7.3)$$

$$\text{where } q_{man} = m\Delta\overline{H}_{fossil\ fuel} + q_{nuclear}$$
$$q_{natural} = q_{rad} + q_{geo} + q_{tide,current}$$

In this equation, geothermal energy brought out of the earth's interior <u>by man</u> is not a natural energy source, but a man-made energy source and is part of q_{man}

Until the value of q_{man} becomes much larger, Equation 7.3 may be written:

$$q_{man} + q_{natural} = q_{rad} = q_{out} \qquad (7.4)$$

The energy flux to the system (biosphere) from the sun is 0.136 watts/cm^2. This is the value of S in Equation 7.2. The area (A) of the earth intercepting this energy flux is ΠR^2. A so-called "black body" absorbs all of the radiation reaching it. For a black body, a = 1.0. The earth, however, is not a black body. Portions are covered by water, green matter and snow (polar ice cap). The fraction of the radiation reaching a surface that is reflected is termed an "albedo." The value of "a" (in Eq. 7.2) is (1-albedo).

Figure 7.5 shows the fraction of solar energy reaching the biosphere. Of 100 units of energy reaching the biosphere, 50% is intercepted by the cloud cover (satellite pictures

show about 50% of the earth to be cloud covered) and of this
50%, 46% is reflected (the albedo for clouds is about 0.46).
Therefore, 23 units are reflected back out of the earth's
biosphere. The air molecules and dust particles reflect an
additional six units of energy. Seventeen additional units
are adsorbed in the biosphere atmosphere.

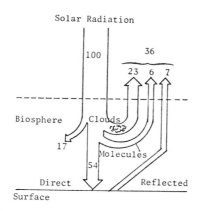

FIGURE 7.5 Distribution of Energy Coming from Sun

From this figure it can be seen that the average albedo for
the earth's surface is about 0.36. Some values for albedo's
are given in Table 7.2. The values used in Figure 7.5 are
consistent with these values in Table 7.2.

Substituting these values into Equation 7.2 the energy
received in the biosphere from the sun is:

$$q_{rad} = \Pi R^2 \ (0.136 \ watt/cm^2)(1-0.36) \qquad (7.5)$$

Having considered how the biosphere received heat from
the sun it is time to look at how it loses heat in order to
maintain a thermodynamic balance. Energy leaves as it arrives

in the form of electromagnetic radiation. The amount of
energy leaving the earth's surface can be obtained by using
Stefan's Law.

$$q_{out} = (A)(\sigma \epsilon T^4) \qquad (7.6)$$

A = the total surface area of the earth $(4 \Pi R^2)$

σ = Stefan-Boltzman Constant $(0.567 \times 10^{-11}$ watts/$cm^2 T^4)$

ϵ = the earth emissivity

T = absolute T in $^\circ$K

Table 7.2

Albedo for Various Objects

Object	Albedo(%)
Clouds	
Thin serous	20
Altostratus-Altocumulus	50
Cumulus	70
Open Water	3-8
Open Water, polar regions	25
Snow	30-70
Arable land	10-15
Deciduous forest	18
Desert	30

The emissivity is the fraction of the energy that is given off
from the earth relative to the amount that would be given off
by a "black body." It is equal to the absorptivity of the
earth (0.87). Therefore,

$$q_{out} = 4 \Pi R^2 (0.567 \times 10^{-11})(0.87)(T^4) \qquad (7.7)$$

Substituting Equation 7.5 and 7.7 into 7.4 and solving for
temperature gives:

$$T^4 = \frac{\Pi R^2 Sa}{4\Pi R^2 \sigma \varepsilon} = \frac{S}{4\sigma} = \frac{a}{\varepsilon} = \frac{0.136[\text{watts/cm}^2]}{4 \times 0.567 \times 10^{-11}[\text{watts/cm}^2 T^4]} \times \frac{0.64}{0.87}$$

$$T = 258°K \ (-15°C)$$

This temperature value is significantly lower than the aver-age temperature of the earth. The reason for this is now explored.

The radiation from the earth is not free to pass through the atmosphere to space. Large fractions are absorbed. Since energy cannot accumulate in space, the energy absorbed is re-radiated. This energy re-radiated is not all directed away from the earth. Much of it is returned to the earth's sur-face. One way to approach this problem is to consider the earth's atmosphere (as far as the earth radiation is con-cerned) to have a shell surrounding it. This shell will ab-sorb all of the energy heading outward from the earth. It will then re-radiate the energy. Fifty percent will head outward to space and 50 percent will be sent back to earth. This is shown in Figure 7.6.

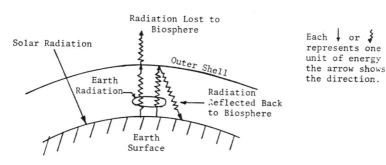

FIGURE 7.6 Re-Radiation of Earth Radiation

From Figure 7.6 it can be seen that the earth surface receives two units of energy: one from the sun and one reflected from

the outer shell. The earth, in turn, radiates two units to
retain thermal balance. The outer shell intercepts two
energy units from the earth and then re-radiates one to space
and one back toward the earth. Using this model, the average
earth temperature may be re-evaluated.

$$T = 258°K(2)^{1/4} = 307°K \qquad (34°C)$$

The increase in temperature from 258 to 307°K was due to this
hypothetical outer shell.

The shell considered above stopped all radiation from
the earth. In the actual case the shell may be considered
to have holes in it and allow some of the earth's radiation
to pass. Increasing the concentration of CO_2 in the atmosphere tends to decrease the fraction of radiation that escapes
through these holes (i.e., closes the holes). This is termed
the "greenhouse effect."

EXAMPLE 7.1   Consider the case where the shell is 50% holes.
               Estimate the average earth temperature.

SOLUTION 7.1   The earth must emit 1.5 times the amount without a
               shell. 0.5 units pass through the shell. The other
               unit is absorbed. 0.5 units are re-emitted to space
               and 0.5 units back to earth.

$$T = 258°K \ (1.5)^{1/4} = 285.5°K \ (12.5°C)$$

The presentation of the idea of the earth being surrounded
by shell (with or without holes) that absorb radiation leaving
the earth may be viewed as a mode of scientific soothsaying.
It allows for an explanation of what has been observed and
prediction of what shall happen if certain changes take place.
The development of a "picture" that can help explain what we
observe is termed modeling. The model for the earth energetics
provided above is far too simple than that needed to completely

explain a very complicated system. The simple model presented is useful and can provide some insight into what might
happen if certain future events occur.

The temperature value obtained earlier is a reasonable
approximation to the earth's average temperature. Land surface
alterations could have an effect on the earth's temperature.

EXAMPLE 7.2   Massive desert land converted to arable land would alter the
earth reflectivity. From Table 7.2 the albedo for
desert is 30 and arable land 10-15. Assume that the
conversion from desert decreases the earth albedo
by 1%. Estimate the effect on the earth's temperature.

SOLUTION 7.2   From Figure 7.5 the 54 units of energy reaching the
earth (0.13) (54) = 7.0 is reflected. If the earth
albedo increased from 0.13 to 0.14, the amount reflected would be (0.14) (54) = 7.6. The overall
albedo would become (23 + 6 + 7.6)/100 = 0.366
[1 - 0.366 = 0.634]

$$T^4 = [(0.136)(0.634)/(4)(0.567)(0.87)] \times 10^{11}$$

$$T = 257.1°K$$

Accounting for the greenhouse effect

$$T = (257.1)(1.5)^{1/4} = (284.5)(11.5°C)$$

The average overall earth temperature decreased
by about 1.0°C. This is in itself significant.
A local effect may be far more severe.

The same type of analysis may be used to determine the effect
of vapor trails caused by jet aircraft. This increases cloud
cover and affects the overall albedo.

EXAMPLE 7.3   Assuming that man increased his energy consumption at a
rate that doubled every 11.5 years. Determine how long
it will be to see the earth temperature rise 1°C.

SOLUTION 7.3   The energy that comes into the atmosphere will be coming
from the sun and that released by man's activities

$$q_{in} = q_{rad} + q_{man} = q_{out}$$

From equation 7.7:

$$q_{out} = 4\pi R^2 (0.567 \times 10^{-11}) (.87) (T^4)$$

Writing this equation once for solar energy only and
once for solar energy and man's activity and taking the
ratio:

$$\frac{q_{rad} + q_{man}}{q_{man}} = \frac{4\pi R^2 (0.56 \times 10^{11})(.87)(T_1^4)}{4\pi R^2 (0.56 \times 10^{11})(.87)(T_2^4)}$$

$$= \frac{(T_1^4)}{(T_2^4)} = \frac{(259°K)^4}{(258°K)^4}$$

$$\frac{q_{rad} + q_{man}}{q_{rad}} = 1.016$$

$$q_{man} = (.016)q_{rad}$$

$$q_{rad} = 174 \times 10^{15} \text{watts} \qquad \text{(From Table 7.1)}$$

$$q_{man} = 2.8 \times 10^{15} \qquad \text{(From Table 7.1)}$$

In 1970 $q_{man} = 0.006 \times 10^{15}$

q_{man} in Year T $= 2.8 \times 10^{15}$

q_{man} in Year 1970 $= 0.006 \times 10^{15}$

Increase in heat release is $2.8/.006 = 467$

Number of half-lives (n)

$2^n = 467 \qquad n = 8.9$

Years $= 8.9 \times 11.5 = 102$ years

Will occur in $1970 + 102 = 2072$

This section has provided a model based upon the First Law of Thermodynamics. The simple model provides insights into what factors will contribute toward a change in the earth's equilibrium temperature. The contributing factors to this model are:

1. Solar energy constant
2. Cloud cover
3. Earth albedo
4. Greenhouse effect
5. Man's energy contribution

The model takes no cognizance of how energy flows take place within the biosphere. Because the problem of earth energetics is far more complex than shown by the model, any values obtained should be considered to be "suggestive" of what may happen rather than a "proof" of what may happen.

The development of the average temperature used global average conditions. These are important for the discussion of global problems. It is also important to appreciate the local conditions in order to consider the effective use of solar energy. The solar constant was given as 0.136 watts/cm^2. The area this was incident upon was only 1/4 of the earth surface area (see Figure 7.4). The average value of radiation received by the earth is 0.136/4 = 0.034 watts/cm^2. This accounts for daylight and nighttime hours.

This value of 0.034 watts/cm^2 is not evenly distributed over the earth surface. Not all of the earth surface receives the same amount of radiation. It depends upon the inclination of the earth to the sun's rays. This is shown in the following diagram:

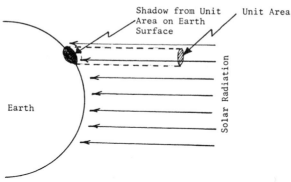

The radiation passing through the unit area may be distributed over a much larger area of the earth surface area. An easy way to picture this is to consider the unit area to block all of the radiation and the shadow cast upon the earth would be the area that would have received this radiation.

Another factor is the cloud cover that intercepts the solar energy headed toward the earth surface. This reflects much of

the radiation away from the earth. Figure 7.7 shows the
yearly average flux reaching the continental United States.

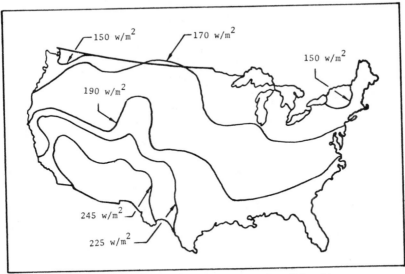

FIGURE 7.7 Yearly Average of Solar Flux Densities in
w/m^2 (24-hour Average)

7.2 SOLAR COLLECTORS

The energy from the sun is not useful as an energy source
and must be converted to another energy form before it can be
used to produce power.

The energy input from the sun is in the form of electromagnetic radiation (like γ-rays discussed in the chapter on Nuclear
Energy). The limited discussion used here considers elotromagnetic
radiation "photons" to be small concentrated "packets of energy,"
called quanta. The amount of energy carried is given by Equation
6.6.

$$E = h\nu = h/\lambda$$

where h is Plank's Constant
ν is frequency
λ is wave length

The greater the frequency the greater the energy carried by a
photon.

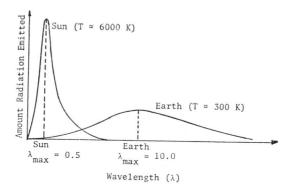

FIGURE 7.8 Energy Spectrum from the Earth and the Sun

Photons are emitted by any hot object. The wave length
of the photon that is most frequently emitted is provided by
Wein's Law

$$\lambda_{max}\, T = 2.9 \times 10^{-3}\,[m°K] \tag{7.8}$$

For hotter bodies λ is smaller and the amount of energy carried
by each photon is greater. Not all of the photons come off
with a single value but follow a spectrum. Figure 7.8 shows
a spectrum for T = 6,000°K and 300°K. These temperatures were
chosen because they represent the characteristic temperature
of the sun and earth respectively. The earth receives energy
from the sun at short wave lengths (and high quanta energy)
and emits energy at long wave lengths (and low quanta energy).
The value λ at the maximum of these curves are at about 0.5
microns for the sun and 10 microns for the radiation from
the earth.

The energy reaching the earth is in the form of photons
of various wavelength (various energy) distributed about a

most frequent value of 0.5 microns (10 eV) and ranging from about 0.2 microns to over 3 microns.

It is this energy that must be captured in any solar collector system. The type of collector systems considered will be:

1. Photosynthetic collectors
2. Flat plate collectors
3. Focusing collectors
4. Satellite collectors

The solar-cell will be considered separately under conversion systems. This is a direct conversion system that does not require the conversion of photons to alternative energy form.

Table 7.3 provides the ranking of these collectors in terms of:

1. stage of development
2. efficiency of collection to produce work
3. complexity of collector construction

Table 7.3

Ranking of Various Collectors
(Rank 1 is preferred)

System/	Stage of Development	Efficiency	Complexity
Photosynthetic collector	1	4	1
Flatplate collector	2	2	2
Focusing collector	3	3	3
Satellite collector	4	1	4

The photosynthetic collector in green plant life has been the source of the fossil fuels used today. Nature has taken care of the construction of these collectors. The

efficiencies of these collectors is quite low. The flat plate
can be built any time economics are attractive. The satellite
collector has much higher efficiencies because they may be
located in a position that always receives the sun's radiation,
but has many technical problems that have not been proven.

The next few sections describe each of these collector
systems.

7.2.1 Flat Plate Collectors

The task of the collector is to convert solar radiation
into heat at a high temperature and at a high efficiency.
This is done by maximizing the amount of solar radiation
received and minimizing the energy losses. Consider the ideal-
ized system shown in Figure 7.9

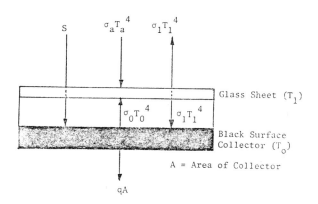

FIGURE 7.9   Ideal Single Sheet Flat Plate Collector

The black surface absorbs all radiation reaching it. Heat is
removed from this black collector. A sheet (or film) covers
the collector surface. In Figure 7.9

S = represents the solar flux

T_a = ambient temperature

T_1 = sheet temperature

q   = energy removed per unit of collector area.

T_0 = collector temperature

σ = Stafan Boltzman Constant

For the ideal case it is assumed that the sheet covered the collector is completely clear (passes all radiation) to radiation that comes directly from the sun (λ small) and is completely black (stops all radiation) toward all radiation coming from the atmosphere or the collector (λ large). Referring to Figure 7.9, the collector receives energy directly from the sun, and from radiation of the cover sheet. It loses energy as heat and by radiation to the cover sheet. The cover sheet receives no energy directly from the sun but does intercept all of the radiation from both the atmosphere and the collector. It gives off radiation to the surroundings and returns radiation to the collector. Equal amounts are radiated from the cover sheet to the collector and space.

The relationship between the heat product, ambient temperatures and efficiencies can be obtained from the First Law of Thermodynamics written over the total system and the collector.

<u>Energy balance over the collector and cover sheet:</u>

Energy in = Energy out

Solar energy in + Energy radiated from atmosphere =

Heat out + Energy radiated by cover sheet

$$SA + A\sigma T_a^{\,4} = Aq + A\sigma T_1^{\,4} \qquad (7.9)$$

<u>Energy balance over the collector:</u>

$$SA + A\sigma_1 T_1^{\,4} = Aq + \sigma_a T_0^{\,4} \qquad (7.10)$$

Adding Equations 7.9 and 7.10 (setting $\sigma = \sigma_1 = \sigma_2$)

$$s(SA - q) = \sigma A(T_0^4 - T_a^4) \tag{7.11}$$

If more than a single sheet of cover material is used, this equation becomes

$$(n + 1)(SA + q) = \sigma A(T_0^4 - T_a^4) \tag{7.12}$$

The efficiency of the collector system may be defined as the ratio of the heat recovered/to the direct solar energy received. If $\eta \equiv Aq/AS$, Equation 7.12 becomes

$$T_0^4 = T_a^4 + (n + 1)(S/\sigma)(1 - \eta) \tag{7.13}$$

As the value of η increases, the temperature of the collector decreases. Thus, as more energy is drawn from the collector, the temperature decreases. From previous discussions the lower temperatures provide for lower Carnot conversion efficiencies to produce work. Thus, a compromise is made between quality and quantity of energy removed.

EXAMPLE 7.4   Determine the maximum temperature attainable from
               a single and double sheet flat plate collector when
               the ambient temperature is 0°C. Determine the tem-
               perature when 50 percent of the energy is collected.
               For this problem take S = 190 watts/m².

SOLUTION 7.4   Maximum temperature is determined when $\eta = 0$. From
               equation 7.14.

$$T_0^4 = T_a^4 + (n + 1) S/\sigma$$
$$T_a = 273\,°K$$
$$\sigma = 5.67 \times 10^{-8} \ [W/m^2 °K^4]$$

For n = 1:
$$T_0^4 = (273)^4 + 2[190/0.567 \times 10^{-7}] = 12.2 \times 10^9$$
$$T_0 = 332\,°K(59°C)$$

For n = 2:
$$T_0^4 = (273)^4 + 3[190/0.567 \times 10^{-7}] = 353(80°C)$$

At 50% capacity, η = 0.5 and there values become:

For n = 1:

$$T_o^4 = (273)^4 + (1)(190/.567 \times 10^{-7}) = 308°K(35°C)$$

For n = 2:

$$T_o^4 = (273)^4 + (1.5)(190/.567 \times 10^{-7}) = 320°K(47°C)$$

The previous example showed that only modest temperature increases were obtained for flat plate collectors. As the temperature goes up, the wave length becomes less and the cover sheets no longer trap all of the radiation and some passes through the cover sheet (it develops energy leaks). Therefore, high temperatures are not possible. Under favorable conditions, temperatures of 100°C can be achieved by flat plate collectors. The temperatures that flat plate collectors can achieve are modest and fall rapidly when heat is removed from the collector.

7.2.2 Focusing Collectors

The purpose of a focusing collector is to concentrate the solar energy into a small area to increase the temperature. The principle of focusing the sun's energy to provide high temperatures is well known. Archimedes attempted to set a fleet of sailing ships afire by using a reflecting surface (the venture was not successful). Two general types of reflectors are used, cylindrical or spherical. These are shown in Figure 7.10.

The temperature that may be reached by these concentrating collectors is due primarily to the fact that the solar energy received over a large area is concentrated into a small area.

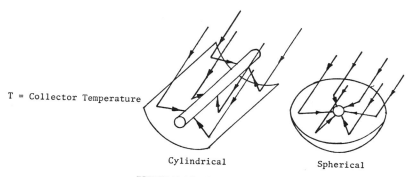

T = Collector Temperature

Cylindrical   Spherical

FIGURE 7.10 Cylindrical and Spherical Collectors

An energy balance over the system shown above gives:

$$AS = A_c T^4 \sigma + q \qquad (7.14)$$

or

$$T^4 = AS/A_c \sigma - q/A_c \sigma \qquad (7.15)$$

where A_c is collector area.

Defining the ratio of q/AS as the efficiency, η.

$$T^4 = AS[1 - \eta]/A_c \sigma \qquad (7.16)$$

Equation 7.16 differs from the Equation 7.13 for flat plate in two most important aspects. There is no term accounting for the energy the collector receives from the atmospheric radiation. When this radiation strikes the reflector surface, it is not oriented in the direction necessary to focus it on the central collector. The small amount of atmospheric radiation that reaches the central collector is negligible. The focusing collector, therefore will not operate when it is not positioned so that the photons strike the mirror in a given direction. For this reason the focusing collector must track the sun accurately. On a cloudy day the sun's radiation is scattered and the focusing collector becomes virtually useless.

EXAMPLE 7.5   Repeat Example 7.4, but assume that the ratio of areas is 100/1.

SOLUTION 7.5

$\eta = 1.0$

$T^4 = AS/A_C\sigma = 100(190/5.67 \times 10^{-8}) = 3.35 \times 10^{11}$

$T = 760°K(488°C)$

$\eta = 0.5$

$T^4 = (A^S/A_C\sigma)(1 - \eta) = 3.35 \times 10^{11}(0.5) = 16.75$

$T = 640°K(367°C)$

The focusing collectors can achieve much higher temperatures, but require direct radiation and a complex mechanical tracking system. Higher Carnot Efficiencies for conversion to work are obtained.

7.2.3 Satellite Collectors

A major problem with solar collectors lies in the inconsistent rate that the solar collectors receive radiation. A drastic variation between day and night, summer and winter, clear and cloudy days occurs. All of these can be eliminated by placing the collecting surface in orbit. If a satellite were placed in orbit on a period of one day (which matches the earth's rotation) it would appear stationary above a point on the earth's surface. The satellite would receive continuous radiation 24 hours a day and there will be no interference from the clouds and other atmospheric phenomena. The energy may be concentrated and the energy "beamed" down to earth. Many technical problems must be resolved before serious consideration can be given to this advanced concept.

7.2.4 Photosynthetic Conversion

An alternative collecting system to one composed of mirrors and reflectors is that provided us by nature. All of the fossil fuel used and available today was a result of the reaction:

$$H_2O + CO_2 + light \longrightarrow CH_2O + O_2 \qquad (7.17)$$

This reaction converts energy from the sun to organic matter, removes CO_2 from the atmosphere and replenishes the atmosphere with O_2. When the organic matter (carbohydrate) burns it releases the energy stored.

EXAMPLE 7.6 When carbohydrate burns it releases 112,000 calories/g-mole. Estimate the energy required to form the carbohydrate molecule.

SOLUTION 7.6

112,000[calories/g-mole] x 1/6.023 x 10^{23}[g-mole/molecules] x 4.186[J/calories] x 1/1.602 x 10^{-19} [eV/J] = 4.86[eV/molecule]

The mechanism of formation of a carbohydrate molecule is complex, but it is known that 8-10 photons are required to form one molecule. Only light in the visible range is used by the plant. The radiation in the red light range is most effective. Using this information it is easy to estimate the efficiency of photosynthetic conversion to organic matter. Assuming 9 photons of red light (1.9 eV/photon) are required to provide one carbohydrate unit, the total energy required is:

9[photons] x 1.9[eV/photon] = 17.1[eV]

The efficiency is thus: η = 4.86[eV]/17.1[eV] = 0.28 (28%)

EXAMPLE 7.7   Compare the efficiency of photosynthetic conversion
to that of a Carnot engine running on radiant energy
used by plant life.

SOLUTION 7.7   According to Wein's Law (Equation 7.8), the temperature corresponding to 1.9[eV]photons can be calculated.

$\lambda T = 2.9 \times 10^{-3} [m^\circ K]$

Using equation

$E = hC/\lambda$

$\lambda = hC/E$

$\qquad = 6.625 \times 10^{-34} [J/Sec] \times 1/1.6 \times 10^{-19} [eV/Joule]$

$\qquad \times 3 \times 10^{8} [m/Sec]/1.9[eV] = 6.5 \times 10^{-7} [m]$

$T = 2.9 \times 10^{-3} [m^\circ K]/6.5 \times 10^{-7} [m] = 4461^\circ K$

Carnot Efficiency = $1 - (298/4461) = 0.93$

Efficiency Ratio = $(0.93/0.28) = 3.32$

Not all of the energy from the sun is captured by plants.
Only about 45% of the solar spectrum is in the range that is
collected by plants. Of the fraction that is collectable,
some is reflected. From Table 7.2 it can be estimated that
about 15% of the radiation is reflected by plants. The overall conversion efficiency of solar energy to plant life is
then

$\qquad (0.28)(0.45)(0.85) = .107$ or 10.7%

This value represents the rate of conversion during a growing
period when the foliage covers the earth. If it is assumed
that there is a 6-month growing season, and "on the average"
only one-half of the earth is covered by green plants, the
efficiency is further reduced:

$\qquad 0.107(.5)(.5) = 0.026$ or 2.6%

This value may be considered a higher value for conversion of sunlight into vegetable matter. Table 7.4 shows values of tons/vegetable matter per acre for several crops.

The values above are for field grown species. If the plants can be grown in controlled environments, the temperature, nutrients, and CO_2 concentration can be set to provide maximum yield. This has been done with bluegreen algae with yields up to 40 tons/acre being achieved. The term biomass has been coined to describe the total vegetable matter.

7.3 STORAGE

The demand for energy varies with time of day, season, weather conditions, and the region of the country. The availability of energy does not follow the same pattern. In order to provide the energy user with energy when needed, it is necessary to arrange for energy storage. For example, the automobile stores extra energy which has been generated during its operation for use at times when the driver needs it. The most critical time at which extra capacity is needed is during startup. In a solar heated house, the greatest demand exists during cold, long, winter nights which correspond to the times when there is no energy input from the sun. Storage becomes essential.

In the following discussion, emphasis is on storage of energy by the electrical utility. This may appear somewhat out of place in this chapter on Solar Energy. It is included here for several reasons.

1. Energy storage is commonly associated with solar energy, and one would look in this chapter for any discussion on **energy storage.**

2. The basic reason for storage of solar energy is essentially the same as that for storage by the electrical utility.

TABLE 7.4

Aboveground, Dry Biomass Yields
of Selected Plant Species or Complexes

Species	Location	Yield (tons/acre-year)
Annuals		
Sunflower X Jerusalem artichoke	Russia	13.5
Sunflower hybrids (seeds only)	California	1.5
Exotic forage sorghum	Puerto Rico	30.6
Forage sorghum (irrigated)	New Mexico	7-10
Forage sorghum (irrigated	Kansas	12
Sweet sorghum	Mississippi	7.5-9
Exotic corn (137-day season)	North Caroline	7.5
Silage corn	Georgia	6-7
Hybrid corn	Mississippi	6
Kenaf	Florida	20
Kenaf	Georgia	8
Perennials		
Water hyacinth	Florida	16
Sugarcane	Mississippi	20
Sugarcane (state average)	Florida	17.5
Sugarcane (best case)	Texas (south)	50
Sugarcane (10-year average)	Hawaii	26
Sugarcane (5-year average)	Louisiana	12.5
Sugarcane (5-year average)	Puerto Rico	15.3
Sugarcane (6-year average)	Philippines	12.1
Sugarcane (experimental)	California	32
Sugarcane (experimental)	California	30.5
Sudangrass	California	15-16
Alfalfa (surface irrigated)	New Mexico	6.5
Alfalfa	New Mexico	8
Bamboo	South East Asia	5
Bamboo (4-year stand)	Alabama	7
Abies sacharinensis (dominant species) and other species	Japan	6
Cinnamomum camphora (dominant species) and other species	Japan	6.8
Fagus sylvatica	Switzerland	4.3
Larix decidua	Switzerland	2.2
Picea abies (dominant species) and other species	Japan	5.5
Picea omorika (dominant species) and other species		6.4
Picea densiflora (dominant species) and other species	Japan	6.1
Castanopsis japonica (dominant species) and other species	Japan	8.3
Betula maximowicziniana (dominant species) and other species	Japan	3
Populus davidiana (dominant species) and other species	Japan	5.5
Hybrid poplar (short-rotation)		
Seedling crop (1 year old)	Pennsylvania	4
Stubble crop (1 year old)	Pennsylvania	8
Stubble crop (2 years old)	Pennsylvania	8
Stubble crop (3 years old)	Pennsylvania	8.7
American sycamore (short rotation)		
Seedlings (2 years old)	Georgia	2.2
Seedlings (2 years old)	Georgia	4.1
Coppice crop (2 years old)	Georgia	3.7
Black cottonwood (2 years old)	Washington	4.5
Red alder (1-14 years old)	Washington	10
Eastern cottonwood (8 years old)		3
Eucalyptus sp.	California	13.4
Eucalyptus sp.	California	24.1
Eucalyptus sp.	Spain	8.9
Eucalyptus sp.	India	17.4
Eucalyptus sp.	Ethiopia	21.4
Eucalyptus sp.	Kenya	8.7
Eucalyptus sp.	South Africa	12.5
Eucalyptus sp.	Portugal	17.9
Miscellaneous		
Algae (fresh-water pond culture)	California	8-39
Tropical rainforest complex (average)		18.3
Subtropical deciduous forest complex (average)		10.9
Puckerbrush complexes (average)	North Carolina	2.2
Puckerbrush complexes (average)	Maine	4.4
World's oceans (primary productivity)		6

Reference: "Effective Utilization of Solar Energy to Produce Clean Fuel," Stanford Research Institute, June, 1974.

3. The text emphasizes the need for energy in the form of
work as the essential ingredient of the industrial society. The use of energy in the form of heat receives
little attention.

Figure 7.11 shows a load demand curve (power use curve) for
a typical utility for a one week period. There are several terms
on this graph which are discussed below.

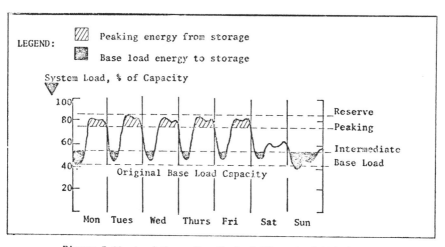

Figure 7.11 Load Curve For Typical Electrical Utility

1. Base Load - this represents generating capacity from
continuously operating sources and provides for the
greatest portion of the electric demand. The generat-
ting units used to provide this capacity are normally
the newest, largest, most efficient units owned by the
utility.

2. Intermediate Load - Additional capacity must be available to provide for demand above the base load. The capacity is needed regularly but not continuously and is
provided by additional generating units.

3. Peak Load - There are peak demand periods which require
additional capacity on occasion. This extra capacity
is usually the most expensive and least efficient power
produced. The most important characteristic of these
units is that they must start and stop quickly.

4. Reserve Capacity - The utility must have some additional
    capacity above the peak load. If there is a breakdown
    in any of the generating units the reserve capacity may
    be needed. With increased demand for electricity, the
    reserve must provide this capacity while new units are
    being built. A new generating station takes many years
    to build and start.

As a result of varying demand the total installed generating capacity must be twice the average rate that is needed. In
order to make maximum use of the more efficient base line units,
it will be necessary to control energy use or to provide energy
storage or both. Control of energy use may be accomplished to
a degree by offering incentives which make it more attractive
to operate during off-peak hours.

Figure 7.11 is useful in showing how energy storage can increase the use of base line units. The solid area represents
the time that the energy load is less than the base load capacity. If storage were available, this extra energy could be
stored and used during periods of high demand. The use of
stored energy is shown by the shaded areas. The peaking units,
which are less efficient and more expensive, would not be needed. If the storage units were placed near the user rather than
the power plant, large transmission lines would not have to handle the peak load.

There are many ways to arrange for energy storage. Water
may be pumped to an elevated reservoir during low demand periods.
For peak demands the water could be used in a hydroelectric
plant to produce electricity. This is referred to as hydrostor-
age, or pumped storage. It is necessary to have the proper
terrain for the needed reservoir capacity to be at sufficient
height. Underground pumped storage may also be used. The elevated reservoir is at ground level and the hydrogenerator placed
in a subterranean cavern.

Compressed air storage is an alternate use for underground caverns. Figure 7.12 shows one example of a pumped air system.

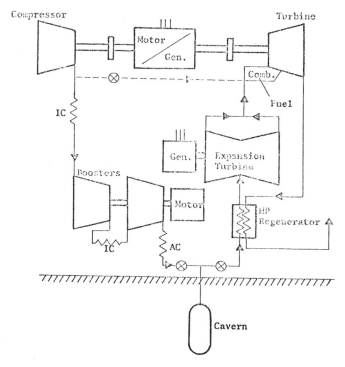

Figure 7.12 Pumped Air Storage System

Air is withdrawn from a cavern where it is stored at about 1200 psi. It is dropped to about 600 psi and goes to a heat recuperator where it is heated by the hot gas exhausting from the gas turbine. It then passes to a combustion unit where the high pressure gas is heated. This high temperature, high pressure gas is expanded in a gas turbine which drives a generator to produce electricity.

During the storage portion of the cycle, a motor receives electricity from an electrical utility. It runs a compressor system which raises the pressure to 1250 psi. The compressed gas is sent to storage. Because the compressed air has a small amount of energy per unit volume, extremely large storage capacity is required. Storage tanks would be uneconomical.

The available reservoir systems of sufficient size are divided into three types:

1. Aquifers - the top of the reservoir is "cap rock" which is impermeable to the air being stored. They are commonly used to store methane. From the same formation, the withdrawal rate would be much greater than for methane. When used for storage the system must have high permeability.

2. Solution Mined Salt - For many years salt has been mined by passing a water stream into a salt deposit. The salt dissolves, and a salt brine is brought to the surface. This leaves large voids in the salt formation which are excellent for gas storage.

3. Conventional Mining - Not favorable to gas containment.

The pumped storage systems store the energy as potential energy. In contrast to these systems is thermal storage where the energy stored can be used in a heat engine. The storage system must be at high temperature in order to use the energy to generate electricity. The Carnot Cycle efficiency will limit the performance of an engine that must produce work from a heat reservoir.

Lower temperature storage has a place in heating where work is not produced or in tasks such as heating feed water to a boiler. Thermal storage systems may store energy solely as sensible heat or may store energy as latent heat (heat associated with phase change).

Materials to store energy as sensible heat should have the following properties:

1. High specific heat

2. High density

3. High chemical stability

4. Inexpensive

5. Non-hazardous

6. Environmentally acceptable

The primary candidate for sensible heat storage is water. It
has high specific heat, high density, is inexpensive, non-hazardous, and environmentally acceptable. Unfortunately, it has a
high vapor pressure and extremely high pressures are required to
maintain the liquid form at high temperatures.

Thermal systems using latent heat are attractive as they
have an energy storage density of 2 - 5 times that of water.
They also have the advantage that the change between solid-liquid
occurs at constant temperature. The characteristics for materials
to be used for latent heat thermal storage are:

1. High heat of fusion

2. High specific heat

3. High density

4. Melting point temperature of 500 - 1000°C

There are several problems to overcome in removing heat from
storage. When energy is removed the material solidifies around
cooling tubes. The thermal conductivity is low, and thermal
resistance becomes large. The heat exchanger also undergoes
severe forces resulting from differences between liquid and
solid.

The use of the battery offers many advantages. They are
easy to install, responsive to demand and may be distributed
in the system to minimize transmission loads. The typical lead-
acid battery can be built at low cost but cannot go through

more than a limited number of severe cycles. They are meant to
be charged and held there most of their lives. This is not the
condition for electrical energy storage systems being discussed.
Systems are being developed that may resolve this problem.

The storage of energy as hydrogen is basically a "chemical
energy storage system". The hydrogen is produced by using the
electricity to split water into hydrogen and oxygen.

$$2H_2O \rightarrow 2H_2 + O_2$$

EXAMPLE 7.8   Determine the overall efficiency for a process that
uses electricity to produce hydrogen during low
electrical demand periods. The hydrogen is then used
during peak periods as a fuel in a gas fired furnace.
Assume that 60% of the electrical energy is converted
to hydrogen and the electrical generating stations
are 40% efficient.

SOLUTION 7.8   Basis: 100 units of methane burned.

100 units of CH4
$\downarrow$ x 0.4

40 units of electricity
$\downarrow$ x 0.6

24 units of hydrogen
$\downarrow$ x 0.4

9.6 units of electricity

Only 9.6 units of energy are distributed to user for
every 100 units of CH4 consumed.

Storage of hydrogen once produced can be accomplished in
several ways:

1. Compress and store as gas - similar to natural
   gas storage.

2. Liquify and store at low temperature (cryogenic
   process).

3. Bind chemically as a metal hydride.

In contrast to storage of potential energy, the flywheel
can store energy as kinetic energy. This system has been used
for centuries. The size of flywheels are limited by the

strength-to-density ratio of available materials of construction.
They would have to be extremely large to handle the storage for
an electric utility.   Table 7.5 gives some characteristic values
for many types of energy storage systems.

TABLE 7.5

Relative Amounts of Energy Released from Storage

	J/Kg
(1) From wound clocksprings	0.39×10^2
(2) Potential energy in 1,000ft (305 m)	3.0×10^3
(3) Gas turbine from 100 psia (6.9bar)	6.3×10^3
(4) Flywheel just below bursting speed	11.4×10^3
(5) Twisted rubber bands	58.0×10^3
(6) Fully-charged battery	213.5×10^3
(7) Rocks $\Delta T = 150°F$ (83°C)	60.7×10^3
(8) Water $\Delta T = 150°F$ (83°C)	242.6×10^3
(9) Chemical hydration	209×10^3
(10) Freezing water	334×10^3
(11) Steam condensation	$2,340 \times 10^3$
(12) Exploding gunpowder	$2,900 \times 10^3$
(13) Exploding dynamite	$5,340 \times 10^3$
(14) Burning firewood	$15,780 \times 10^3$
(15) Burning alcohol	$27,850 \times 10^3$
(16) Burning fuel gas	$18,560 \times 10^3$
(17) Burning crude petroleum	$37,130 \times 10^3$
(18) Burning rubber bands	$45,200 \times 10^3$
(19) Burning gasoline	$46,400 \times 10^3$
(20) Burning hydrogen	$119,000 \times 10^3$

Reference: Energy Reference Handbook, Editor, N. C. McNerey and T. F. P.
Sullivan, Eq Government Institutes, Inc.

EXAMPLE 7.9   Determine both the mass and the volume to store
enough solar produced energy to run a 1,000 MW
power plant for 24 hours using the following
storage system:

(a) pumped storage, 1,000 ft.
(b) compressed air storage (100 psia)
(c) flywheel storage
(d) biomass (30% conversion biomass to electricity)
(e) hydrogen (35% conversion biomass to electricity)

SOLUTION 7.9

1,000 [Mw] x 1,000 [Kw/Mw] x 3,600,000 [J/Kw-Hr]

x 24 [Hr] = 3.6×10^{12} [J]

Using Table 7.5

(a) 3.6×10^{12}[J] x 1/3,000 [Kg/J] x 1,000 [g/Kg]

x 1.0 [cm^3/g] x $1/10^6$ [m^3/cm^3] = 1.2×10^6 [m^3]

(b) 3.6×10^{12} [J] x 1/6300 [Kg/J] = 5.7×10^8 [Kg]

@ STP ρAIR = 1.29 [g/l]

1.29 [g/l] x (100/14.7)[Psi/Psi] = 8.78 [g/l]

5.7×10^8 [Kg] x 1,000 [g/Kg] x 1/8.78 [1/g] x

1/1,000 [m^3/l] = 64.9×10^6[m^3]

(c) 3.6×10^{12} [J] x 1/11,400 [Kg/J] x 1,000 [g/Kg]

x 1/8 [cm^3/g] x $1/10^6$ [m^3/cm^3] = 3.94×10^4 [m^3]

(d) 3.6×10^{12} [J] x 1/0.3 [$J_{biomass}/J_{elec.}$] x

1/15,780,000 [Kg/J] x 1,000 [g/Kg] x 1/0.6[cm^3/g]

x $1/10^6$ [m^3/cm^3] = 1,267 [m^3]

(e) 3.6×10^{12} [J] x 1/0.35 [$J_{H_2}/J_{elec.}$] x

1/119,000,000 [Kg/J] = 86,435 [Kg]

@ STP ρH_2 = 0.088 [g/l]

0.088 [g/l] x 100/14.7 [Psi/Psi] = 0.6 [g/l]

86,435 [Kg] x 1,000 [g/Kg] x 1/0.6 [1/Kg] x

1/1,000 [m^3/l] = 1405 [m^3]

SUMMARY TABLE

	Mass (Kg)	Volume (m^3)
Pumped Storage	1.2×10^9	1.2×10^6
Compressed Air	5.7×10^8	64.9×10^6
Flywheel	3.2×10^8	0.04×10^6
Biomass	8.7×10^4	1.27×10^3
Hydrogen	8.6×10^4	1.41×10^3

7.4 CONVERSION

Once the solar radiation is converted to thermal energy
any of the systems discussed in Chapter 2 can be used to provide
the "work energy" needed by the industrial society. These systems are limited by the Carnot Efficiency. The energy from the
sun can be used to produce forms of energy other than thermal
energy. Some of these are discussed below.

7.4.1 Solar Cell

The solar cell, or more precisely, the photo-voltaic cell;
converts solar energy directly into electricity (see Figure
7.11). When two semiconductors, one of the n-type where negative electrons are free to move about and conduct electricity
and one of the p-type where electron holes or positive charges
are free to move about and conduct electricity, are placed in

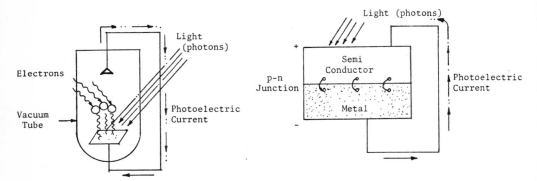

FIGURE 7.13   Functional Diagrams of Two Types
of Photovoltaic Cells

contact with each other, they form a p-n junction. This sets up
a potential difference across the junction. When a photon
of light strikes an atom in the cell, energy is imparted to the
most loosely bound valance electrons. If the energy exceeds
the minimum level necessary to dislodge the electron from the
atom, a free electron of negative charge and an electron hole
of positive charge are produced. If the cell is connected
to an external circuit electric current will flow. The
basic equation may be written

$$hc/\lambda = w + 1/2mv^2$$

where   h = Plank's Constant
        c = velocity of light
        λ = wavelength
        m = mass of electron
        v = electron velocity
        w = photo-electric work function

Materials do not respond to the entire solar spectrum.

The amount of energy that can be obtained depends upon the
work function w and the fraction of the solar spectrum
that is effective in providing free electrons.

The advantage to the solar cell is that it is a direct
conversion system and is not limited by the thermal Carnot
Efficiency.

7.4.2 Conversion of Biomass

There are several alternative processes for the conversion of biomass that were not considered in the schemes for
coal conversion.

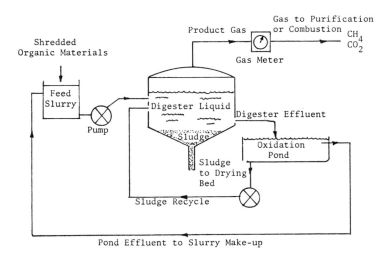

FIGURE 7.14 Schematic Diagram of a Fermentation Plant to Produce
Combustible Gas (mainly methane) from Organic Materials

Figure 7.14 shows a system that converts the organic biomass
to a mixture of methane CH_4, and carbon dioxide, CO_2. The
composition of the gas is about 60 to 65% methane.

The generation of methane from decomposition of organic material in an aqueous media has long been observed as swamp gas or marsh fires. The control of reactor temperature to 30-37°C and control of pH to assure that the system does not become acid will provide a high level of conversion of some biomass materials to methane. The conversion is the result of action of living organisms and has been used in sewage treatment facilities for many years. When the CO_2 is removed, pipeline quality gas is obtained. Such a system may be particularly attractive in the conversion of water plants such as algae and water hyacinths that contain a large fraction of water.

The gasification of biomass to fuel gas follows the same steps described in the section on coal gasification. The only real difference is in the first step, the pyrolysis step. In coal gasification, this played a minor role (depending upon volatile matter) because there was a small hydrogen content. For biomass the pyrolysis step plays the dominant role. As a first approximation, biomass may be assumed to be primarily cellulose.

Biomass - $(C_6H_{10}O_5)n$

Cellulose is a long chain polymeric material and the C_6 grouping given above may be repeated n times. If the biomass contains a small amount of water vapor the biomass may be treated as an equal molal mixture of C and H_2O.

$$(C_6H_{10}O_5)n + nH_2O \longrightarrow 6nC + 6nH_2O \longrightarrow 6nCO + 6nH_2$$

The reaction between the C and the H_2O give an equal molal
mixture of CO and H_2. In a crude sense it may be thought
that the cellulose molecule brings the water for gasification along with it and does not have to be added as it did
in the coal gasification.

The reactions are far more complex than the one shown
and some CH_4, C_2H_6, C_2H_4, CO_2 are produced along with the CO
and H_2. As a fuel for producing gas, cellulose (i.e., biomass)
it is easier to convert than coal. As shown in Table 4.1,
wood (a biomass material) is considered as a young fuel and
has not reached the stability of coal. It is easier to
break down and react.

Cellulose is a long chain polymeric material. It is a very
large molecule, and is held together by chemical bonds. When it
is exposed to a high temperature tnese bonds are broken and the
molecule fractures. Figure 7.15 shows two systems.

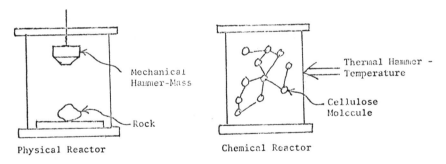

Physical Reactor   Chemical Reactor

Figure 7.15   Analogous Mechanical System and
              Cellulose Pyrolysis System

a) A physical system where a large particle (like a rock)
    is struck by an impact hammer. This imparts energy to
    the rock.

b) A chemical system where a large molecule (like cellu-
    lose) is struck by a thermal hammer. This imparts ener-
    gy to the cellulose.

The rock receiving the energy may relieve the stress placed upon it by breaking into smaller pieces. The cellulose receiving the energy may relieve the stress by breaking into smaller pieces. If the cellulose molecule is broken into small pieces, the products are gaseous: CO, H_2, CH_4, C_2H_4, CO_2, etc. If the molecule breaks into moderate sized pieces containing 6 - 12 carbons, they form a liquid.

From an analogy to the physical system it is possible to suggest the type of chemical system to produce the desired fuel product. To break the rock into small pieces, it is necessary to hit it hard a sufficient number of times to make it a powder. By the analogy, to break up the molecule it is necessary to provide high temperature rapid heating (hit hard with the thermal hammer) and sufficient time in the reactor.

If the cellulose is to be converted to a liquid, different conditions are needed. To break the rock into a few smaller sized pieces (but no powder) care is taken to place the rock correctly and strike it crisply only once or twice. Cutting a diamond is a vivid example. By the analogy, to break up the molecule it is necessary to strike it crisply once or twice. Unfortunately, it is not possible to obtain any desired molecular orientation, and the size of products produced varies widely.

If the rock is struck lightly many times, it does not shatter, but the edges are chipped away and any loose material is broken off. The product is a more stable rock. When the cellulose molecule is struck lightly many times with a thermal hammer, the product is a stable solid char.

By choice of reactor type, temperature and residence time, the products obtained can be varied.

Any of the reactors discussed in Chapter 5 may be used for pyrolysis of biomass. Almost all are being tried. From the

discussion above it appears that the fluid bed would offer advantages in producing a gas. It provides the very high heating
rates to high temperatures (a hard thermal hammer).   To provide for high liquid yields, an entrained system operated at
moderate temperatures can assure that the molecule stays in the
reacting zone a short time (a moderate thermal hammer).

One unique system is shown in Figure 7.16.

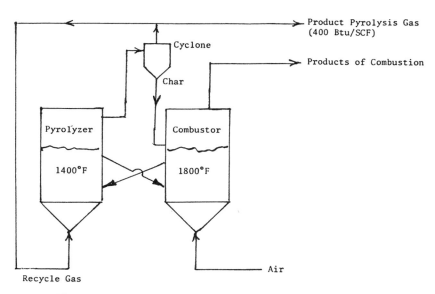

Figure 7.16 Two Fluidized Bed System For Producing a
Fuel Gas From Pyrolysis of Cellulose

The left-hand reactor is a fluid bed where biomass is fed.
It is held at 1400°F. The biomass quickly pyrolyzes and gives
a high gas yield. Some char is made. This is removed from the
gas and transported to the fluid bed on the right where combustion takes place at 1800°F. Transporting the fluidized bed material (which is inert sand) between the two systems provides
the energy needed for the pyrolysis reaction.

The gas is of sufficient quality that it can be burned in

a gas fired boiler with no need to retrofit the boiler or reduce its capacity. It is not diluted with nitrogen and can be used as a raw material for methane, methanol or chemical feed stock.

Biomass may be converted to alcohol (either methanol or ethanol). The conversion to a liquid that may be used as a fuel provides many advantages. Foremost is that it may be transported easily. It may be used to fire boilers, power plants, and as a fuel in internal combustion engines. It is combusted with little pollution. A mixture of ethanol-gasoline is used in Brazil as a fuel for automobiles. It increases octane and reduces pollution.

Methanol is produced from CO/H_2 mixtures in a similar manner to the production of synthetic natural gas. The reaction for synthetic natural gas is

$$3H_2 + CO \longrightarrow CH_4 + H_2O$$

The reaction for methanol is

$$2H_2 + CO \longrightarrow CH_3OH$$

Where the H_2/CO ratio is adjusted to 3/1 for synthetic gas production, the ratio is adjusted to 2/1 for methanol production. Both reactions use a catalyst and take place at elevated temperature and pressure.

Ethanol may be produced by the hydrolysis of cellulose or starch to produce sugar which may be <u>followed by</u> fermentation to produce ethanol.

Hydrolysis reaction:

$$C_6H_{10}O_5 + H_2O \longrightarrow C_6H_{12}O_6$$

    Cellulose   Sugar

Fermentation:

$$C_6H_{10}O_5 \longrightarrow 2CH_3\ CH_2OH + 2CO_2$$

Sugar   Alcohol

The hydrolysis of cellulose, a major portion of most biomass, takes place in an aqueous solution and requires an acid catalyst (usually H_2SO_4). The reaction is not well behaved. The sugar continues to react and form an undesirable product. Several other side reactions other than production of sugar occur. The conversion of biomass cellulose result in severe environmental problems. Once sugar is produced, the conversion of sugar to alcohol is a well known and often practiced conversion.

A more attractive starting material is either sugar (no hydrolysis needed) or starch (enzymatic hydrolysis used). These do not require a harsh chemical treatment and are not as damaging to the environment. The disadvantage is the sugar-starch portion of a plant is only a portion of the total biomass produced in any plant.

Problems for Chapter 7.0

1. The energy plantation concept has been described by
    Szego, C and C. Kemp, Chem Tech, May 1973, p. 276.
    Based on some of the values provided in this paper it
    is possible to estimate the land area required to pro-
    vide a fuel for a 1000 Mw power station.
    (i)   0.4% of solar energy is converted to wood
            biomass 8,500 Btu/lb.
    (ii)   Conversion efficiency of biomass to elec-
            trical energy is 35%.
    (iii)   Load factor (% of the design capacity that
            plant averages over a one-year period).
    (a)   Calculate the tons of biomass/day required for
            power plant.
    (b)   Land area required.

2. Demonstrated advanced agricultural practices have
    exhibited efficiencies of 2% (P.W. Gram, Mech. and
    Engineering, May 1975). Translate this information
    into tons of biomass per acre per year. How does this
    compare to Napier Grass that under ideal conditions
    produces 40 tons/acre of dry biomass?

3. "If through advanced management practices including
    exploitation of modern developments in plant genetics,
    plantations could be operated to produce continuous
    crops at a greater than 3% solar conversion . . ."
    NSF/NASA Solar Energy Panel, "Solar Energy as a Natural
    Resource," December 1972. What percentage of the land
    area would be required to provide for the total elec-
    trical energy needs of the U.S.?

4. What minimum land area would be required in the south-
    west part of the U.S. to provide for 1000 Mw? Repeat
    for a two sheet flat plate collector. Assume that the
    energy received from a focusing collector (ratio of
    areas = 100/1) is used to run a Carnot Engine to pro-
    duce electricity.

5. How many quanta of energy emitted from the earth's
    surface are needed to give as much energy as one quanta
    of energy from the sun (assume that the most probable
    wave length is representative of each spectrum).

6. Determine the maximum efficiencies for conversion of
    solar energy to work for n = 0.5 for the number of
    sheets equal to 1, 2, 3, 4, and 5.

7. In Example 7.4 only the temperature of the collector
    was evaluated. Determine the temperature of each sheet
    for all cases given in Example 7.4.

8. Consider a single film covering a flat plate collector.
    The film does not pass all of the solar energy but
    absorbs 10% of the solar energy. Compare the maximum
    temperature that this system can reach and compare
    it to the ideal system in Example 7.4.

9. Given the following definitions:
    C = constant
    I = solar insolation (Btu/ft^3)
    E = solar energy capture efficiency
    F = fuel value
    Derive an equation for Y[yield in tons/acre-year]. Then
    evaluate C. (This equation is derived by Klass, D. L.,
    Chem Tech, March 1974.)

10. A flat plate solar collector is to be considered as a
    means to drive a pump to irrigate crops. This will be
    used in place of natural gas pumps presently in use for
    irrigation. Consider using a single plate collector
    system that can drive a perfect pump. Make a plot of
    the efficiency (work delivered to pump water/solar energy
    striking collector) as a function of the fraction of
    energy removed from the collector. Repeat for a two
    and three plate collector. If we can achieve 10% of the
    theoretical work, what is the size of collector required
    to pump 1000 gal/min from a well 1500 ft?

11.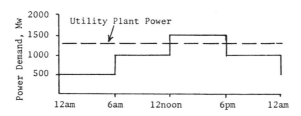

    The above sketch shows an idealized power demand cycle
    for a community. It is proposed to have a pumped water
    storage facility for meeting peak demand. The overall
    efficiency of the storage facility (i.e., electrical
    energy delivered during high demand to electrical energy
    drawn during low demand) is 72%. The power requirement

to be met above plant capacity is the storage plant
capacity:
  (a) Determine the required installed capacity of
       the steam power plant (in Mw).
  (b) Determine the peak power output of the
       storage plant (Mw).
  (c) Calculate the volume of water reservoir
       required for a 100-foot height above the
       turbines neglecting any piping or valve losses.
  (d) What are the major advantages and disadvantages
       of such a pumped water storage facility?

12. Science Magazine, October 14, 1977, reported on "Ocean
     Energy: The Biggest Gamble in Solar Power." They noted
     that the maximum temperatures that are available between
     surface and deep water is about 44°F.
       (a) Evaluate the maximum possible efficiency.
     Actual efficiencies of less than 3% can be achieved.
       (b) Determine the water flow required for a 100
            Mw plant.
     The article points out that 30% of the gross energy
     output is required to move the water.

13. Suppose that sodium is used for a thermal storage sys-
     tem of an electric power plant with an average output
     of 500 Mw (24-hour average). Assume the thermal to
     electric conversion efficiency is 30%. The specific
     heat of sodium is 0.3 cal/°K-g and a temperature change
     of 50°C is acceptable. Determine the mass of sodium
     that would be necessary to provide for 48 hours of oper-
     ation of the power plant. Determine the required diam-
     eter of a spherical vessel which could be used to hold
     the sodium (the density of sodium is 93 g/cm^3).

14. Consider the storage of heat by the following reaction:
     $$Na_2SO_4 \cdot 10H_2O \rightleftharpoons Na_2SO_4 + 10H_2O$$
     This transition occurs at 32.3°C and absorbs or releases
     5 x 10^4 cal/Kg.
       (a) Can this be considered as a system to store
            energy for generating electricity?
       (b) Devise a system using this as energy storage
            for a solar home heat system. In the winter
            the energy needs of a modest sized home are
            90 Kw-hr. Assume one day storage. Recommend
            collector size. Show calculations.
       (b) Compare to a water system used for storage.

15. Consider the use of an underground aquifer to store
    (a) Methane at 1300 psia
    (b) Air at 1300 psia
    This material will be used to produce electricity. The methane will go to a furnace and 42% of the energy will be converted to electricity. The air will go to an expansion turbine that is 90% efficient.
    (a) For a given resevoir which of the systems will store more energy?
    (b) For a fixed electricial energy demand what is the relative flow rate from storage?
    Assume that the storage aquifer pressure is held at 1300 psia by a large reactor column.

16. Consider the possibility of storing hot sand. It has been proposed that a fluidized bed furnace is operated at 1800°F. During low demand hot sand is generated and stored at 1800°F. During peak periods the hot sand is used to preheat the combustion air. The system schematic is shown below.

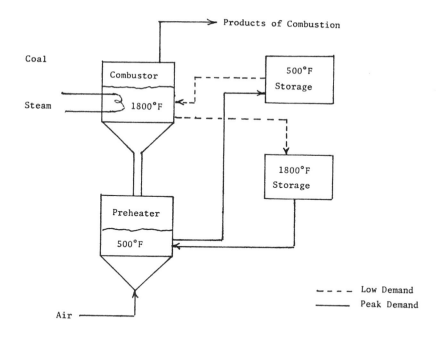

How large a storage system is necessary to retain energy equivalent to 1 ton of coal?

17. It has been proposed that
    thiosulfate pentahydrate is used
    to store energy needed by a solar
    heated home. The solar collec-
    tor can reach 130°F and the temp-
    erature needed to heat the home
    is 110°F. The storage system is
    to provide for 1 day energy de-
    mand of 60,000 Btu/hr.
        (a) How much thiosulfate
            pentahydrate is needed?
        (b) How much water is
            needed?

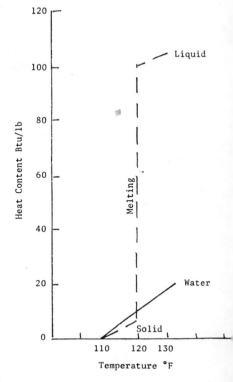

Heat content of sodium thiosulfate
pentahydrate compared with water
(110 F to 130 F)

8.0   ENVIRONMENTAL CONSIDERATIONS

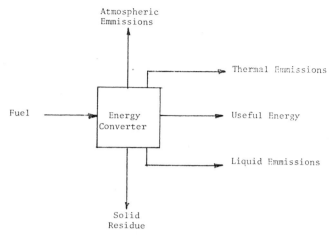

FIGURE 8.1   Products of Energy Conversion

In the process of converting basic fuel resources into useful energy forms that fuel the industrial society, several outputs other than useful energy occur.   These are classified as:

The material in Section 8.3 used "Air Pollution - Its Origin and Control" by K. Wark and C. F. Warner, Pub. IEP a Don-Donnelly (1976) and "Air Pollution" by H. C. Derkins, Pub. McGraw-Hill (1974) for much of the material.

Richard C. Bailie, Energy Conversion Engineering, ISBN 0-201-00840-8

1. Atmospheric Emission--products that leave the converter site by entering the air.

2. Liquid Emission--products that leave the converter site and enter the earth's waterways.

3. Solid Residue--products that leave the converter site in the solid form.

4. Thermal Emissions--energy that leaves the converter site other than as a useful energy product. Thermal energy may leave with liquid, gas or solid stream.

This classification defines the pollutant by the stream which carries it away from the energy conversion site. Thus, a solid pollutant leaving with a gas stream and entering the atmosphere is classified as an atmospheric emission. The same solid may be removed from the gas and leaves with a liquid. It is then classified as a liquid emission. The same solid may be removed from the gas in an electrostatic precipitator as a solid, in which case it leaves as a solid residue.

An alternate way to look at this classification of pollutants is to consider the problem of containment or control of the pollutant. Materials which are gaseous, or are entrapped along with gaseous emissions, cannot be easily contained or directed. They have a large volume per unit of mass. Once emitted to the atmosphere, they mix with the atmosphere and are carried on the air currents. Their movement is not controllable. For this reason, atmospheric pollution recognizes no political, social or economic boundaries. An atomic bomb test in China causes nuclear debris, along with potentially harmful radioactive fallout to pass over parts of the United States. The citizenry can often see atmospheric pollution; they can smell atmospheric pollution, and no exotic instrumentation is required to recognize that it exists in many places.

Liquid emissions do not move so aimlessly, but are directed toward rivers and streams. The volume of liquid that be-

comes polluted is far less than in the case of air pollution,
and may be contained and treated. The United States has always
had a relatively abundant supply of high quality water. There
have been few instances, in recent years, of contaminated water
which have come to public awareness. Liquid, unlike air, flows
in a predictable direction; it can be controlled by dams or in
ponds. It allows discharges from several different points to
be carried to a central treatment plant for clean-up. The sani-
tary sewer system is an excellent example. Water pollution has
not had the high exposure that air pollution has had and is
observed by a smaller proportion of citizens. It is a less ob-
servable source of pollution.

We have seen that atmospheric pollution is carried on the
winds and cannot be contained and that liquid pollution flows
in a more predictable pattern; however, solid waste can be
piled up or buried. It can be hidden by a fence, as it may
offend the aesthetics of many citizens. When it either enters
the atmosphere or leaches into the water supply, it becomes a
serious matter.

The material in this chapter emphasizes the problems
associated with the conversion of fossil fuels. The pollution resulting from extracting the fuels from the earth is
not covered. This is not to suggest that they are not significant. They are often a dominating factor and must be considered in any comprehensive assessment of a potential fuel
candidate.

The arguments provided in this section may be described
as soft when compared to previous discussions based upon the
Laws of Thermodynamics. The Laws of Thermodynamics are invariant, and, given the same set of circumstances, will provide
a single answer. Environmental consequences of any activity
are not so easily evaluated. New environmental standards

are being written by man--not nature. What is considered environmentally acceptable yesterday differs from what is acceptable today and shall differ from what is acceptable tomorrow.
What may be acceptable in the USSR may not be acceptable in the
U.S. What is acceptable in the state of Ohio may not be acceptable in West Virginia.

It is clear that determination of standards of environmental acceptability may differ along political subdivisions. Some
agreement on environmental consequences might be achieved by an
evaluation of health hazards. Those operations shown to be dangerous to health should be eliminated. This has been used in
many cases to curtail certain activities. But there are almost
no operations which are absolutely safe, and there is little evidence that such a criterion is acceptable to most people. Certainly the elimination of tobacco would help save lives, improve
health, and help the environment, but such drastic action is not
foreseen. Similar arguments can be made to terminate automobiles,
coal mines, etc.; all these activities are dangerous to health.
It seems clear that in almost all our activities each of us is
willing to accept some danger to obtain certain goods and servi-
ces we consider desirable. Each of us, consciously or not, has
made a balance between risks and benefits.

8.1   COST-BENEFIT ANALYSIS

It seems clear that any decision which must be made that
makes both favorable and unfavorable contributions to our "quality of life" should weigh the good points against the bad. One
approach, the cost-benefit analysis, allows for comparison of
several alternatives. One simply lists and sums all the costs
(in some consistent units) and all the benefits (in identical
units) of an alternative action. If one then subtracts the

costs from the benefits, a net benefit is obtained. The alternative with the largest net benefit is chosen. There are other
analyses, such as the net benefit ratio, which divides net benefit by total costs and yields a benefit return on an investment.

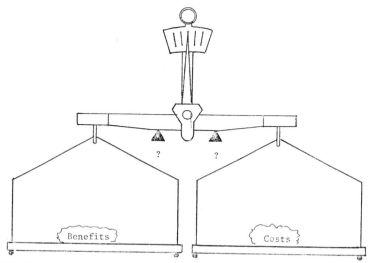

FIGURE 8.2 Balancing Environmental Risk Against Energy Benfits

The difficulty in determining net benefit is the need to
reduce all contributing factors to a common unit so that the
effects may be added and subtracted. Many economists insist
that the only common unit is money, and all contributions must
be in terms of these units. This may or may not be, but the
dollar is used in this discussion to point out the difficulties
of trying to reduce many contributions to a single set of units.
    Consider the alternative of mining coal, burning it in a
utility station near the mine mouth and shipping the electricity
to the east. The benefit is that the electric power needs of
the east are supplied, while the environmental damages (risks)
are primarily focused in the Appalachian region. One group of

people receive the benefits, while another takes the risks.
What is the justification for requiring one group to take all
the risk for the second? The same dilemma exists for nuclear
power. It is not a simple consideration.

The problem becomes more complex when benefits realized
today create risks for tomorrow. Acid mine drainage, a serious
pollution problem in Appalachia, is a result of operations that
have long been abandoned. The nuclear wastes produced today
will remain a risk to succeeding generations. The benefits realized from domestic oil use today must be weighed against the
denial of oil to future generations. Additional ramifications,
such as aesthetic and social effects, must be considered and
quantified. A comparison of human lives with dollars, with
beauty, cannot be made based solely upon scientific principles.

Assessment of cost benefits is not a simple task and may
appear impossible. At best, any analysis would be imperfect.
The alternative is to ignore the wide variety of impacts resulting from a decision to institute any process system. This
would be far worse than an imperfect solution.

It is clear that the utilization of our energy resources
involves environmental risks. It is also clear that energy is
necessary to today's society and provides benefits which we all
enjoy. Risk-benefit analyses are being made at individual and
at governmental levels. Each of us will weight factors differ-
ently in the analysis and come to different answers/conclusions.
We all have our own set of values. There is no single correct
solution. The search for an acceptable solution will be with
us now and for ever more.

The following discussion on pollution does not attempt to
weigh the costs and benefits of energy alternatives. This is
recognized as a cop-out and may ignore the real issue associated with environmental problems. The discussion here is limited

to determining the amount, type and cause of pollution. This
would provide input data for a comprehensive evaluation of environmental impact.

8.2 POLLUTANT REDUCTION

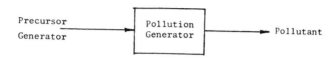

FIGURE 8.3 Generalized Flow Sheet for Generation of Pollution

Figure 8.3 represents the generation of pollution from the
conversion of fossil fuels. The reaction that results in production of pollution may be written:

$$A + B \rightarrow P$$

where A and B are the precursors and P is the pollutant. The
system where A and B are converted to P is termed the pollution
generator. For example, in the combustion of coal accompanied
by the formation of SO_2,

    (a)   Sulfur is a precursor, A

    (b) SO_2 is a pollutant, P

    (c)   Furnace is the pollution generator

There are several general approaches to reducing the amount
of pollution that enters the atmosphere:

1. Remove or reduce the precursor

2. Alter the pollution generator to discourage formation
    of the pollutant

3. Remove the pollutant from the product stream

The Second Law of Thermodynamics provides a theoretical

argument against producing pollution and then attempting to re-
cover it. Once a pollutant is dispersed, it will take a quantity of work to return it to an ordered state. For example,
once sulfur in a fuel such as coal is burned to form SO_2 (gas)
which becomes mixed with air, it takes work to return it to a
solid sulfur material.

The most attractive approach to reducing the SO_2 problem
would be to remove the sulfur from the fuel prior to conversion.
If this cannot be done, then consideration should be given to
designing the conversion system to discourage formation of the
pollutant. If this is not possible, then the only choice remaining is to remove the pollutant from the product stream.

The problem in cleaning up after the pollutant is once
formed is seen in Figure 8.4. This represents a liquid scrubbing system which removes the pollutant from a gas. The pollutant has not been eliminated, but has been transferred to the
liquid. The pollutant is still with us; it has only been transferred to another medium.

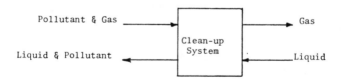

FIGURE 8.4 Liquid Scrubber System

The removal of pollutants <u>after</u> being generated often represents the path most often used. Several examples may be
cited:

(a) Catalytic converters for automobiles

(b) Stack gas clean-up systems for power plants

(c) Charcoal filters on domestic water supplies

The probable reason for this approach in spite of the less desirable potential is that it causes less changes in already existing technology and existing equipment.

To illustrate the problems associated with changing existing technologies to reduce pollution, consider alternatives to reduce the pollution from the internal combustion engine. The list of alternatives to reduce pollution has been provided below. This list is not complete and some of the alternatives are selectively chosen to illustrate the problems of choosing between the "best" theoretical solution, the "best" practical solution and the "best" political solution.

The following alternatives all reduce the level of pollution from the internal combustion engine:

(a) Change the fuel--replace all gasoline fuel with methyl-fuel. This burns clean and the pollutants in the exhaust from an internal combustion engine will be reduced.

(b) Change the engine--replace the spark-ignition internal combustion engine with an engine that produces less pollution.

(c) Clean-up exhaust--add a tail-end clean-up system.

(d) Change engine operation--provide for better operation of the spark-ignition internal combustion engine.

(e) Outlaw automobiles.

Alternative (a) is an example of eliminating the precursor; (b) and (d) are examples of modifying the pollutant generator, and (c) the clean-up of the discharge.

Alternative a: The consequences of changing the fuel are not insignificant. This would reduce pollution from the automobile but it would:

(a) Require new manufacturing plants to produce the methyl-fuel.

(b) Eliminate the need for many existing gasoline manufacturing facilities.

(c) Require some modification to all automobiles.

It is clear that this would cause a large capital expense, large dislocation of labor, etc., and, as discussed in the introduction to this section, the costs may far outweigh the benefits when other less disruptive alternatives are available.

Alternative b: The elimination of the spark-ignition engine may seem desirable but the following factors must be evaluated.

(a) New engine power plant would require a massive re-education program among mechanics and repairmen. The repair and testing equipment would be obsolete.

(b) Massive costs to convert over manufacturing facilities would be involved.

(c) For a reasonable period of time the repair facilities for two general type systems would be required.

Alternative d: The change in conditions of operation of the spark-ignition internal combustion engine has been demonstrated to reduce the pollution. By a more controlled combustion in a lean gasoline-air mixture the level of pollution generation has been reduced. This would have a much smaller effect on the general economy than the first two alternatives.

Alternative c: The tail-end clean-up system (catalytic converter) is the system that has been widely adopted to combat pollution from automobiles. It has added to the cost, created a new industry, required minor modifications of gasoline manufactures and has not caused the massive capital expense and economic dislocation of the first two alternatives.

This discussion has shown that there are several paths toward the solution of any problem. It also showed that the solution that is fundamentally sound may not be the preferred solution when all factors are evaluated. The solution to a pollution problem often is divided into two parts:

1. What to do with existing facilities that pollute.

2. What to do with new facilities that are being designed.

For existing facilities, the only solution may be to provide a tail end clean-up system. For a new system, this should be considered more as a solution of last resort. The preferred solution <u>should always be one of eliminating the generation of the pollutant</u>, rather than producing it and then trying to eliminate it.

With the concern for the environment, many coal systems which emitted SO_2 were converted to natural gas. This eliminated the precursor S from the fuel and thereby reduced the SO_2 problem. With natural gas in short supply, this alternative is being removed as a viable option.

As an alternative to the direct use of coal and production of SO_2, one could carry out the combustion in two steps. This is shown in Figure 8.5.

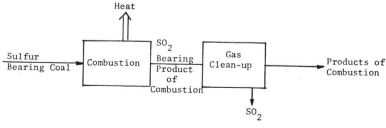

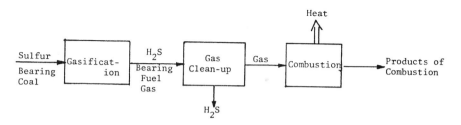

FIGURE 8.5 Alternate systems for burning coal to produce heat and clean products of combustion.

In the first system the conversion of coal to heat is one

step. Sulfur is then removed as SO_2 by a "tail-end" scrubber.

The second process, the conversion, takes place in two

steps. The sulfur leaves the first step as H_2S and is re-

moved. This is preferred because:

1. H_2S is easier to remove than SO_2.

2. The volume of the H_2S bearing stream is much less
    than the SO_2 stream (stream not diluted with N_2).
    Larger volume requires larger equipment.

3. The concentration of the pollutant in the gas stream
    to be scrubbed (H_2S or SO_2) is greater in Alternative
    b. The size of a clean-up system is proportional to
    concentration.

All of these factors favor the two-step combustion process.

This two-step process was discussed in a previous chapter.

Some ability to perform work is lost in the gasification

process. Work is lost but the environment may be less abused

by the second process when all the risks and benefits are con-

sidered.

EXAMPLE 8.1   Consider a fuel composed of 98% C and 2% sulfur. Determine:
               (a) The volume of gas (STP) when burned in 50% excess air.
               (b) The concentration of pollutant for the two-step and one-step
                   process.

SOLUTION 8.1

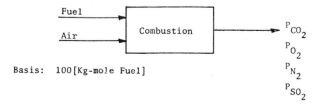

Basis: 100[Kg-mole Fuel]

Theoretical O_2 = 8.23[Kg-mole]

Stream	In	+	Generated	-	Consumed	=	Out
C	8.17	+	0	-	α	=	0
S	0.0625	+	0	-	β	=	0
O_2	8.23x1.5=12.34	+	0	-	$(\beta+\alpha)$	=	P_{O_2}
N_2	(12.34)(79/21)=46.4	+	0	-	0	=	P_{N_2}
CO_2	0	+	α	-	0	=	P_{CO_2}
SO_2	0	+	β	-	0	=	P_{SO_2}

$$\alpha = 8.17$$
$$\beta = 0.0625$$

		% in product gas
P_{O_2} = 4.11		7.0
P_{N_2} = 46.4		79.0
P_{CO_2} = 8.17		13.9
P_{SO_2} = 0.0625		0.11
P_T = 58.74		100.01

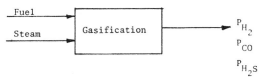

Reactions

(1) $C + H_2O \longrightarrow CO + H_2$ $\alpha = CO$
(2) $H_2 + S \longrightarrow H_2S$ $\beta = H_2S$

Stream	In	+	Generated	-	Consumed	=	Out
S	0.0625	+	0	-	β	=	0
C	8.17	+	0	-	α	=	0
H_2O	8.17	+	0	-	α	=	0
CO	0	+	α			=	P_{CO}
H_2	0	+	α	-	β	=	P_{H_2}
H_2S	0	+	β	-	0	=	P_{H_2S}

α = 8.17, β = 0.0625   % in product gas

		% in product gas
P_{CO} = 8.17		50
P_{H_2S} = 0.0625		0.4
P_{H_2} = 8.17 - 0.0625 = 8.0075		49.6
P_T = 16.34		100.0

Concentration of Pollutant

Combustion $[SO_2]$ = 0.11%
Gasification $[H_2S]$ = 0.40%

Volume 1

Combustion 58.74[Kg-mole] x 22,400[L/Kg-mole] = 1,316,000[L]
Gasification 16.34[Kg-mole] x 22,400[L/Kg-moles] = 366,000 [L]

This is an example of modification of the pollution generator.

A second example of the modification of the pollution generator is the use of a fluid bed combustor. One of the fluidized bed's characteristics is low temperature isothermal operations. No hot spots are developed. In order to form NO_x, a pollutant, it is necessary to achieve high temperatures. Consideration of chemical equilibrium and chemical kinetics show that temperatures in excess of 1525°K are necessary. The fluid bed operates below 1280°K, and NO_x formation is low. The lean burn internal combustion engine achieves significant reduction in pollution levels because of better mixing and lower temperatures.

It is evident that the generation of pollution from the combustion of fossil fuels can be reduced by proper design of the energy converter to discourage the formation of pollutants.

8.3 AIR POLLUTION (See note on page 369)

Air pollution is an integral part of an industrial society. The major source is combustion, combustion essential to the industrial society. This section focuses on the pollutants that result from these combustion sources and what can be done to reduce these emissions.

The regulation of air pollution from combustion sources is not new. A proclamation by King Edward I (1300 A.D.).

> "Be it known by all within the sound of my
> voice, whosoever shall be found guilty of
> burning coal shall suffer the loss of his
> head."

Air quality regulations are not covered. There are, and shall continue to be, pressures to both strengthen and relax regulations. The regulations will respond to many forces that are political and economic, as well as technical.

The rational control of air pollution rests on four basic assumptions.

1. <u>Air is in the public domain.</u> Such an assumption is necessary if air pollution is to be treated as a public problem, of concern not only to those who discharge the pollution but also to those who may suffer as a result.

2. <u>Air pollution is an inevitable concomitant of modern life.</u> There is a conflict between man's economic and biologic concerns; in the past, this conflict was recognized only after air pollution disasters. We need a systematic development of policies and programs to conserve the atmosphere for its most essential biological function.

3. <u>Scientific knowledge can be applied to the shaping of public policy.</u> Information about the sources and effects of air pollution is far from complete, and a great deal of work must be done to develop control devices and methods. Nevertheless, sufficient information is now available to make possible substantial reductions in air pollution levels. Man does not have to abandon either his technology or his life, but he must use his knowledge.

4. <u>Methods of reducing air pollution must not increase pollution in other sectors of the environment.</u> Some industries reduce wastes in the air by dissolving them in water and by pouring the polluted water into streams. For example, one proposal to reduce the sulfur dioxide emitted by coal-burning electrical power plants results in the formation of large quantities of either solid or liquid wastes. Such methods are not true solutions to air pollution problems.

The term <u>air pollution</u> shall include the following:

Aerosol: A dispersion of solid or liquid particles of microscopic size in gaseous media, such as smoke, fog or mist.

Dust: A term loosely applied to solid particles predominantly larger than colloidal, and capable of temporary suspension in air or other gases. Dusts do not tend to flocculate except under electrostatic forces; they do not diffuse, but settle under the influence of gravity, Derivation from larger masses through the application of physical force is usually implied.

Droplet: A small liquid particle of such size and density
as to fall under still conditions, but which may remain
suspended under turbulent conditions.

Fly Ash: The finely divided particles of ash entrained in
flue gases arising from the combustion of fuel. The particles of ash may contain incompletely burned fuel. The
term has been applied principally to the gas-borne ash from
boilers with spreader stoker, underfeed stoker, and pulverized fuel (coal) firing.

Fog: A loose term applied to visible aerosols in which
the dispersed phase is liquid. Formation by condensation
is usually implied. In meteorology, a dispersion of water
or ice.

Fume: Properly, the solid particles generated by condensation from the gaseous state, generally after volatiliza-
tion from melted substances, and often accompanied by a
chemical reaction such as oxidation. Fumes flocculate and
sometimes coalesce. The term is popularly used in reference to any or all types of contaminants, and in many laws
or regulations, with the qualification that the contami-
nant have some unwanted action.

Gas: One of the three states of aggregation of matter,
having neither independent shape nor volume, and tending
to expand indefinitely.

Mist: A term loosely applied to low-concentration disper-
sions of liquid particles of large size. In meteorology,
a light dispersion of water droplets of sufficient size
to be falling.

Particle: A small, discrete mass of solid or liquid matter.

Particulate: A general term meaning "existing in the form
of minute separate particles, either solid or liquid".
Particulate is used interchangeably with aerosol.

In the earlier sections on thermodynamics it was possible
to determine the highest efficiency for any energy conversion.
For air pollution problems, such precise laws are not available.
It is essential to establish what is clean or "normal" air before it is possible to determine what is polluted air. Table

8.1 provides the composition of dry air in areas where the contribution from man's activity is minimal.

TABLE 8.1

Composition of clean, dry air near sea level

Component	Content	
	% by volume	ppm
Nitrogen	78.09	780,900
Oxygen	20.94	209,400
Argon	.93	9,300
Carbon Dioxide	.0318	318
Neon	.0018	18
Helium	.00052	5.2
Krypton	.0001	1
Xenon	.000008	0.08
Nitrous Oxide	.000025	0.25
Hydrogen	.00005	0.5
Methane	.00015	1.5
Nitrogen Dioxide	.0000001	0.001
Ozone	.000002	0.02
Sulfur Dioxide	.00000002	.0002
Carbon Monoxide	.00001	0.1
Ammonia	.000001	.01

Note: The concentrations of some of these gases may differ
with time and place, and the data for some are open to question.
Single values for concentrations, instead of ranges of concentrations, are given above to indicate order of magnitude, not
specific and universally accepted concentrations.
Source: Reprinted by permission from "Cleaning Our Environment - The Chemical Basis for Action", American Chemical
Society, 1969.

Atmospheric air also contains from 1 to 3 percent by volume water vapor and traces of sulfur dioxide, formaldehyde,

iodine, sodium chloride, ammonia, carbon monoxide, methane, and
some dust and pollen.

At the present time neither carbon dioxide nor uncombined
water vapor is considered to be a pollutant. This condition
could change in the future, since the discharge of either substance into the atmosphere in increased quantities may result
in a significant change in the global atmospheric temperature.

It is common practice to express the quantity of a gaseous
pollutant present in the air as parts per million (ppm). Thus

$$\frac{1 \text{ volume of gaseous pollutant}}{10^6 \text{ volumes (pollutant + air)}} = 1 \text{ ppm}$$

$$0.0001 \text{ percent by volume} = 1 \text{ ppm}$$

The weight of a pollutant is expressed as micrograms of pollutant per cubic meter of air. Symbolically,

$$\frac{\text{micrograms}}{\text{cubic meter}} = \mu g/m^3$$

At 25°C and 760 mm Hg (1.01 bar) pressure the relationship between parts per million and micrograms per cubic meter is

$$\mu g/m^3 = \frac{\text{ppm x molecular weight}}{24.5} (10^3) \qquad (8.1)$$

EXAMPLE 8.2   Derive the relationship represented by equation 8.1

SOLUTION 8.2   Volume of lg-mole of gas at 25° C:

22.4 (1/g-mole) @ 0°C or 273°K x 298/273 (°K/°K)

= 24.45 (1/g-mole) @ 25°C or 298°K

$\mu g/m^3$:

By definition - ppm = $\dfrac{\text{volume of pollutant}}{10^6 \text{ (volume of air + pollutant)}}$

or $\dfrac{m^3 \text{ pollutant}}{10^6 \ m^3 \text{ (air + pollutant)}}$

$\mu g/m^3$ = ppm $[\dfrac{m^3 \text{ pollutant}}{m^3 \text{ pollutant + air}}]$ x

1,000 $[\dfrac{1 \text{ pollutant}}{m^3 \text{ pollutant}}]$ x $\dfrac{1}{24.45}$ $[\dfrac{\text{g-mole pollutant}}{1 \text{ pollutant}}]$

= $\dfrac{(\text{ppm})(\text{g-mole pollutant}) \times 10^3}{24.45}$

The following examples illustrate some typical calculations
using the terms used in air pollution problems:

EXAMPLE 8.3   The exhaust gas from an automobile shows a CO concentration
of 1.5% by volume. What is the concentration of CO in
mg/m^3 at 25°C and 1 atmosphere?

SOLUTION 8.3

From definition of ppm

(ppm) = $\dfrac{(1.5)}{(100)}$ x 10^6 = 1.5×10^4

From equation 8.1 [Molecular Weight of CO is 28]

$\mu g/m^3$ = $(15.000)(28)/24.45$ = 17.1×10^6

mg/m^3 = 17.1×10^6 $[\mu g/m^3]$ x $1/1,000$ $[mg/\mu g]$

= 17.1×10^3

EXAMPLE 8.4   It has been reported that the SO_2 concentration is

400 [$\mu g/m^3$] at 25°C. What is the ppm value?

SOLUTION 8.4

From equation 8.1   (Molecular Weight SO_2 is 64)

400 [$\mu g/m^3$] = ppm x 64 x 10^3/24.45

ppm = 0.159

EXAMPLE 8.5   The "dust leading" has been reported as 0.000130
[grains/dry standard cubic ft]
(one lb mass is equivalent to 7.000 grains)

SOLUTION 8.5

0.00013 (grains/SCF) x 1/7.000 (lb/grain)

x 454 (g/lb) x 10^6 (μg/g) x 1/0.02833 (ft^3/m^3)

= 298 ($\mu g/m^3$)

A general list of primary air pollutants is presented in
Table 8.2. Of the large number of potentially dangerous substances, only a very few are present in the atmosphere in sufficient quantities to be of immediate concern. As of 1974, particulate, carbon monoxide, sulfur oxides, oxides of nitrogen,
unburned hydrocarbons, photochemical oxidants, asbestos, beryllium, and mercury were considered to be sufficiently hazardous
to health to warrant the establishment of either air quality
standards or emission standards by the federal government.

TABLE 8.2

General Primary Pollutants

Particulate matter
   Fine dust: less than 100 μ^{α} in diameter
   Coarse dust: above 100 μ in diameter
   Fumes: 0.001 - 1 μ in diameter
   Mist: 0.01 - 10 μ in diameter

Sulfur compounds
Organic compounds
Nitrogen compounds
Carbon compounds
Halogen compounds
Radioactive compounds

α Note: 1 μ = 10^{-4} cm.

8.3.1 Pollution Sources and Inventory

This section presents several tables which provide some
idea of the magnitude of the problem as well as identification
of the major sources of the pollutants. Tables 8.3 through
8.6 provide some estimates of the important emissions from
various sources.

TABLE 8.3

Estimated Nationwide Emissions for 1968 (tons/yr)

Pollutant	Transportation	Fuel Combustion, Stationary Sources	Industrial Processes	Solid Waste Disposal	Misc.	Total
CO	63,800,000	1,900,000	9,700,000	7,800,000	16,900,000	100,100,000
SO_x	800,000	24,400,000	7,300,000	100,000	600,000	33,200,000
NO_x	8,100,000	10,000,000	200,000	600,000	1,700,000	20,600,000
C_xH_x	16,600,000	700,000	4,600,000	1,600,000	8,500,000	32,000,000
Particulate	1,200,000	8,900,000	7,500,000	1,100,000	9,600,000	28,300,000
Total	90,500,000	45,900,000	29,300,000	11,200,000	37,300,000	214,200,000
Percent	42.1	21.0	13.6	5.2	17.5	

TABLE 8.4

Sources of Major Air Pollutants for 1966 (tons/yr)

Pollutant	Transportation	Industry	Electric Power	Space Heating	Refuse Disposal	Misc.[a]	Total
CO	66,000,000	2,000,000	1,000,000	2,000,000	1,000,000	16,900,000	88,900,000
SO_x	1,000,000	9,000,000	12,000,000	3,000,000	1,000,000	600,000	26,600,000
NO_x	6,000,000	2,000,000	3,000,000	1,000,000	1,000,000	1,700,000	14,700,000
C_xH_x	12,000,000	4,000,000	1,000,000	1,000,000	1,000,000	8,200,000	27,000,000
Particulate	1,000,000	6,000,000	3,000,000	1,000,000	1,000,000	9,600,000	21,600,000
Total	86,000,000	23,000,000	20,000,000	8,000,000	5,000,000	37,000,000	179,000,000
Percent	48	12.8	11.3	4.8	2.8	20.3	

[a]Misc includes forest and structural fires, and coal refuse and agricultural burning.

TABLE 8.5

Emissions from Fuel Combustion in
Stationary Sources (million tons/yr)

Source	1966			1968		
	PM	SO_2	NO_x	PM	SO_2	NO_x
Utilities	5.6	14.4	3.5	5.6	16.8	4.0
Industrial	3.0	5.5	2.4	2.7	5.1	5.1
Residential and Commercial	0.6	2.6	0.8	0.6	2.5	0.9
Total	9.2	22.5	6.7	8.9	24.4	10.0

There are significant differences between various sources
of tabulated data. It is not possible to measure these many
sources, and somewhat crude estimates are required. They
should be viewed and used with caution. Different sources do
not group the same data by the same classifications. In spite
of this, several trends can be noted.

1. Transportation contributes the greatest emissions.
2. Industry and electric generation are also large
    contributors.
3. The amount of NO_x and SO_2 has increased substan-
    tially.
4. The amount of particulates and hydrocarbons has
    increased.

This book emphasizes only those emissions which are the
result of combustion. The tables all show that this covers
most of the pollution which enters our atmosphere.

The damaging effects of air pollutants are not equivalent on an equal mass basis. For example, a given concentration of SO_2 may be more detrimental to health than an equi-

TABLE 8.6

Estimated Annual Emissions Without
Further Control in 1973 (thousnads of tons)

Emission Sources	Particu-late	SO$_2$	CO	HC	NO$_x$	%
All mobile sources	900	1,150	159,300	27,850	16,250	57.3
Steam-electric boilers	4,083	28,150			16,053	13.5
Fuels industry	296	4,943	10,490	2,228		5.0
Industrial boilers	6,867	7,774			2,069	4.7
Refuse disposal	2,993		5,320	7,283		4.3
Metals industry	1,118	5,021	4,678			3.0
Construction	3,323					0.9
Agriculture & Forests	2,583		275	23	2	0.8
Chemical industry	1,035	830			115	0.6
Sources not controlled	8,592	631	21,205	3,623	1.593	9.9
National Totals	31,790	48,499	201,268	41,007	36,082	100.0

Fuel industries: coal cleaning, petroleum refineries, petroleum production
   and storage, natural gas, solvent dry cleaning.
Agriculture & forests: grain handling, feed plants, forest products,
   Kraft pulp.
Construction: asbestos, asphalt batching, cement.
Chemical industry: lime, nitric acid, phosphate fertilizers, sulfuric acid.
Sources not controlled: coking, carbon black, accidental fires, sand & stone.

valent concentration of carbon monoxide. The data given in
the tables above does not indicate the real impact of the various species. Figure 8.6 shows one attempt to provide a more
meaningful comparison of the real impact. The mass values have
all been multiplied by a tolerance factor. It indicates that
sulfur oxides and particulates are the most serious pollutants.
These are followed closely by nitrogen oxides and hydrocarbons.
In terms of their source, combustion is listed as the major
contributor with industry and transportation making major contributions.

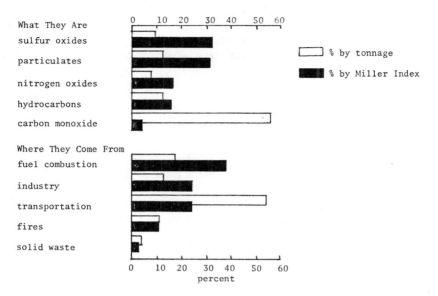

FIGURE 8.6 Air pollutants and their sources for the United
States in 1970. Two percentages are shown. One
is based on tonnage of emissions (Council on
Environmental Quality 1972)*, and the other is
based on a weighted index that takes into account
the relative health effects of each pollutant.

8.3.2 Spatial Effects

The major pollutants, in terms of their effect on the environment, have been briefly presented with the global average
level of pollutants distinguished from local levels. Many problems associated with the environment are a result of highly
concentrated man-made activity centers, rather than a problem
of total emissions. For example, 100,000 head of cattle on a
200,000 acre range present little abuse to the environment.
The same 100,000 head on a few acre feed-lot causes major environmental concern. It is not because more gas is passed by
the cattle, or more urine or solid waste is discharged. It is
the result of concentration of activity in a small area. The

local environment is overwhelmed and can no longer be cleansed
by natural forces.

The trend over the years has been away from a relatively
sparse distribution of activity to one of concentration.
Cities have grown at the expense of rural areas, with industrial
activity concentrated in cities. Many small industrial activities have been combined into larger plants, trending toward larger and larger industrial plants. The effect on the local environment has not gone unnoticed. It has often deteriorated!

This discussion has shown reasons for dispersing industrial activities into a number of small units. This may reduce
the pollution from extremely concentrated areas of emission but
does not reduce the total amount of pollution entering the environment. It will continue to add to the global problem.

Concentrating the activity into a small region makes the
treatment of the discharges a more achievable objective. In a
large facility more exotic clean-up systems must be considered
and more money may have to be spent.

Both dispersion of activities and concentration of activities along with treatment facilities will contribute to an improvement of the local environment. Both approaches have their
place and should be considered as options in any study.

8.3.3 Particulate Matter

From Figure 8.6, above, it can be seen that particulates
make up about 10% of the total tonnage but are considered to be
a major health hazard. Particulate matter is considered to present a health hazard to the lungs, enhance chemical reactions
in the atmosphere, reduce visibility, increase precipitation,
and change the radiation reaching the earth (changes the earth's

temperature and biological plant growth). It can be seen from
Tables 8.3 through 8.6 that the major source of particulates is
industrial and utility boilers. Because a large fraction of
the particulate comes from a reasonably small number of sources,
only a limited number of clean-up devices need be installed to
bring about a major reduction in the amount of particulate matter entering the atmosphere.

TABLE 8.7

Emission Factors for Selected Categories

of Uncontrolled Sources of Particulates

Emission Source	Emission Factor
Natural gas combustion	
Power plants	15 lb/million ft^3 of gas burned
Industrial boilers	18 lb/million ft^3 of gas burned
Domestic and commercial furnaces	19 lb/million ft^3 of gas burned
Distillate oil combustion	
Industrial and commercial furnaces	15 lb/thousand gal of oil burned
Domestic furnaces	8 lb/thousand gal of oil burned
Residual oil combustion	
Power plants	10 lb/thousand gal of oil burned
Industrial and commercial furnaces	23 lb/thousand gal of oil burned
Coal combustion	
Cyclone furnaces	(2 times ash %) lb/ton of coal burned
Other pulverized-coal furnaces	(13-17 times ash %) lb/ton of coal burned
Spreader stokers	(13 times ash %) lb/ton of coal burned
Other stokers	(2-5 times ash %) lb/ton of coal burned
Incineration	
Municipal, multiple chamber	17 lb/ton of refuse burned
Commercial, multiple chamber	3 lb/ton of refuse burned
Flue-fed incinerator	28 lb/ton of refuse burned
Domestic, gas-fired	15 lb/ton of refuse burned
Open burning of refuse	16 lb/ton of refuse burned
Motor Vehicles	
Gasoline-powered engines	12 lb/thousand gal of gasoline burned
Diesel-powered engines	110 lb/thousand gal of fuel burned
Cement manufacturing	38 lb/barrel of cement produced
Kraft pulp mills	
Lime kiln	94 lb/ton of dried pulp produced
Recovery furnaces*	150 lb/ton of dried pulp produced
Steel manufacturing	
Open-hearth furnaces	1.5-20 lb/ton of steel produced
Electric Arc furnaces	15 lb/ton of metal charged

*With primary stack gas scrubber
Source: Control Techniques for Particulate Air Pollutants.
    Washington, D.C.: HEW, December 1968

Table 8.7 gives emission factors for various utility and industrial furnaces. These values do not indicate what the specific emissions are. In addition, there is no indication as to the particle size range of the emissions. The behaviour of particulates in the atmosphere depends upon the particle size. Air borne particles range from 0.001 to 500 μ in diameter (μ is micron and is 10^{-6}m). Most of the particles are in the 0.1 to 10 μ diameter range. The general behaviour of particles, with respect to size, is listed below:

Size	Settling Velocity	Comment
<0.1 μ	4×10^{-5} cm/sec @0.1μ	Brownian movement -motion similar to molecules
1 -20 μ	4×10^{-3} cm/sec @1.0μ	Follows motion of gas in which it is borne
> 20 μ	30 cm/sec @100μ	Show significant settling velocity

The very small particles move in large random motion similar to those exhibited by gas molecules. They do not settle out. Large size particles have large settling velocities and are airborne for relatively short periods of time. From these simple observations it should be clear that a single collection device is not likely to be effective in removing all particles. For large size particles, the high settling velocity characteristic is used as the basis for cleaning a gas stream. For small particles, the settling velocity cannot be used. However, a single electron charge on a small particle will deflect the particle when passed through an electric field. This characteristic is used to remove small size particles.

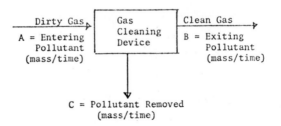

$$C = \text{Pollutant Removed} \\ \text{(mass/time)}$$

Referring to the sketch above, the efficiency of the gas clean-up device for particulate matter is given by

Efficiency: $\eta = (\frac{C}{A}) \times 100 = (\frac{A-B}{A})100$ (8.2)

EXAMPLE 8.6   The table provided belwo shows the size distribution of
a dust sample and the efficiency of removal. Calculate
the over-all collector efficiency?

Particle Diameter	Weight per 100g of Dust	Removal Efficiency	Particle Diameter	Weight per 100g of Dust	Removal Efficiency
(μ)	(g)	(%)	(μ)	(g)	(%)
<1	9	0	30–40	3	3
1–2	5	0	40–50	4	5
2–5	7	0	50–100	8	15
5–10	6	0	100–200	8	50
10–15	5	0	200–300	5	90
15–20	3	0	>300	32	100
20–30	5	1	TOTAL	100	

SOLUTION 8.6

Rearranging equation 8.2

$C = \eta \, A/100$

This equation is applied to each size material.
The results are tabulated below

Particle Diameter	Mass Retained	Particle Diameter	Mass Retained
(μ)	(g)	(μ)	(g)
<20	0.0	50-100	1.2
20-30	0.05	100-200	4.0
30-40	0.09	200-300	4.5
40-50	0.20	>300	320
		TOTAL	42.04

Over all efficiency = (42.04/100) x 100 = 42.04%

The example showed a gas clean-up device which was effective at high particle sizes (above 300 μ) and quite low for anything less than 100 (μ). In example problem 8.6 the dust was not all of one size. This is typical; the particulate matter is not a single size but varies over a range of sizes. Figure 8.7 is a typical plot used to represent the distribution of particle sizes. The ordinate is in units of weight percent per unit interval of particle size. The area under the curve is 100(%). Figure 8.7.b is the same curve, plotted on semi-log paper. If the distribution of particle sizes follows a Gaussian distribution, which is often the case, a plot of the log of the particle diameter versus log of the accumulative percent will yield a straight line.

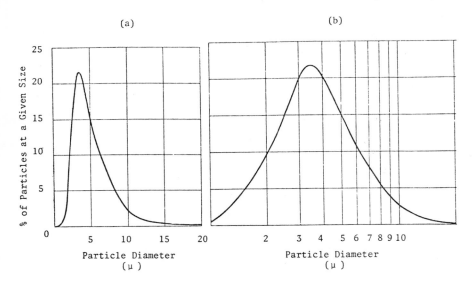

FIGURE 8.7 Plot of Particle Diameter Size Distribution
            (a) Linear Plot; (b) Semilog Plot

EXAMPLE 8.7   Using the weight distribution shown in Figure 8.7
                determine if the particle distribution follows a
                Gaussian distribution.

SOLUTION 8.7

Range Particle Size (μ)	Average Particle Size (μ)	Wt* (%)	Cumulative Weight (%)
<0.50	0.25	0.1	0.1
0.50-1.5	1.0	0.4	0.5
1.5- 2.5	2.0	9.5	10.0
2.5- 3.5	3.0	20.0	30.0
3.5- 4.5	4.0	20.0	50.0
4.5- 5.5	5.0	15.0	65.0
5.5- 6.5	6.0	11.0	76.0
6.5- 7.5	7.0	8.0	84.0
7.5- 8.5	8.0	5.5	89.5
8.5-11.5	10.0	5.5	95.0
11.5-16.5	14.0	4.0	99.0
16.5-23.5	20.0	0.8	99.8
>23.5		0.2	100.0

* Obtained from Figure 8.7

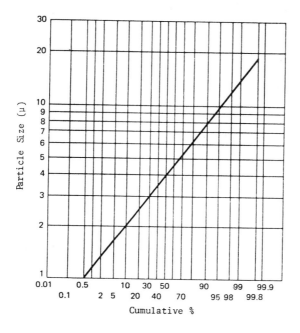

The values shown in the table are plotted above.
The straight line inidcates that a Gaussian distribution is followed.

EXAMPLE 8.8   The figure shown below provides the efficiency of
                  a particle collection system.

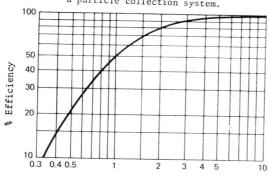

Particle Size
(μ)

a. Determine the efficiency of this unit on dust
    provided in example problem 8.7.

b. Determine if the dust passing through the unit
    Gaussian distribution.

SOLUTION 8.8

Particle Size Range (μ)	Weight (g)	Collection* Efficiency (%)	Weight Removed (g)	Weight Remaining (g)	Accumulative Weight (g)	Accumulative Weight% (%)
0.25	0.1	8.0	0.008	0.092	6.092	0.33
1.0	0.4	30.0	0.120	0.28	6.372	1.35
2.0	9.5	47.5	4.513	4.987	5.359	19.44
3.0	20.0	60.0	12.000	8.000	13.359	48.47
4.0	20.0	68.5	13.700	6.300	19.660	71.33
5.0	15.0	75.0	11.250	3.750	23.490	84.94
6.0	11.0	81.0	8.910	2.09	25.500	92.53
7.0	8.0	86.0	6.880	1.12	26.620	96.59
8.0	5.5	89.5	4.920	0.577	27.200	98.68
10.0	5.5	95.0	5.230	0.275	27.470	99.68
14.0	4.0	98.0	3.920	0.080	27.550	99.98
20.0	0.8	99	0.790	.0008	27.560	100
--	0.2	99	0.200	--	--	100

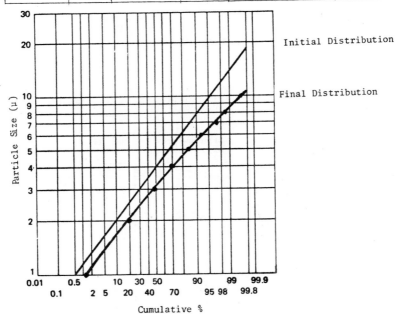

Initial Distribution

Final Distribution

b)   The data do not yield a straight line and does not represent a Gaussian Distribution

a)   Overall efficiency

$$= \frac{(100 - 27.56)}{100} \times 100 = 72.44\%$$

The simplest separation systems are based upon gravity. The important parameter in designing such systems is the settling, or terminal, velocity. This is the maximum downward velocity a particle attains in a direction parallel to earth's gravity field as it overcomes the buoyancy and frictional drag.

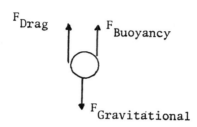

$$F_{Gravitational} = m_p g \quad \text{(Newton's Law)}$$

$$F_{Buoyancy} = m_p \left(\frac{\rho_g}{\rho_p}\right) \quad \text{(Archimedes' Law)}$$

$$F_{Drag} = \frac{\rho_g A v_t^2 Cd}{2}$$

Where: m_p = particle mass

g = acceleration due to gravity

ρ_g = gas density

ρ_p = particle density

v_t = terminal velocity

A = cross sectional area perpendicular to velocity vector

C_D = drag coefficient (dimensionless)

From the force balance

$$F_{Drag} + F_{Buoyancy} = F_{Gravity} \qquad (8.3)$$

After substitution of the relations given above

$$v_t = [\frac{2 \, m_p \, g(\rho_p - \rho_g)}{A \, C_D \, \rho_p \, \rho_g}]^{1/2} \qquad (8.4)$$

All of the properties in equation 8.4, except A and C_D, are physical properties and are usually known or may be estimated. The term A depends upon particle geometry. In most cases, spherical geometry may be assumed. In this case equation 8.4 becomes

$$v_t = [\frac{4g \, d_p \, (\rho_p - \rho_g)}{3\rho \, g \, C_D}]^{1/2} \qquad (8.5)$$

where d_p is particle diameter

The valuation of C_D is necessary to determine the terminal velocity. In stream line flow (Reynold's Number >0.5)

$$C_D = \frac{24}{(Re)} = \frac{24\mu}{d_p \, v_t \, \rho_g} \qquad (8.6)$$

where Re represents the Reynolds Number

$$\equiv \frac{d_p \, v_t \, \rho_g}{\mu}$$

μ - gas viscosity

Subsituting equation 6 into equation 5 yields

$$v_t = \frac{d_p^2 \, (\rho_p - \rho_g)}{18 \, \mu} \qquad \text{(Stokes' Law)} \qquad (8.7)$$

which is valid for Re less than 0.5. For larger Re values the
drag coefficient may be estimated from

$$C_D \doteq \frac{24}{Re} [1 = 0.15 \ Re^{0.687}] \tag{8.8}$$

which is valid up to Re values of 800. Where equation 8.7
allows for the terminal velocity v_t to be evaluated directly
for small Re values the evaluation of terminal velocity using
equation 8.8 and 8.5 requires a trial and error solution. Table
8.8 gives values of v_t for various particle sizes for spherical
particles settling in hot air.

From example 8.9 it can be seen that for any particles
less than 100 μ the time for the particles becomes prohibitive.
the gas volumes to be cleaned are large and a long residence
time requires extremely large settling systems.

8.3.4 Particulate Collection

There are six general types of collectors that will be
discussed:

1. Gravitational settling chambers
2. Centrifugal separators
3. Wet scrubbers
4. Filters
5. Electric precipitators
6. Ultrasonic agglomerators

Figure 8.8 shows a gravity separator. The particulate
laden gas stream enters at the left. This gas stream is al-

TABLE 8.8

Calculated Settling Velocities in Hot Air
(Spherical Particles)

Diameter (μ m)	Settling velocity (m/sec)			
	(Density = 2.0 g/cm^3)		(Density = 3.0 g/cm^3)	
	at 200°C	at 800°C	at 200°C	at 800°C
30	0.03	0.03	0.05	0.06
40	0.06	0.03	0.09	0.06
50	0.12	0.06	0.15	0.09
60	0.15	0.09	0.24	0.12
70	0.21	0.12	0.34	0.18
80	0.27	0.15	0.43*	0.24
90	0.37*	0.21	0.49	0.30
100	0.43	0.24*	0.55	0.37*
200	1.13	0.85	1.55	1.22
300	1.92	1.65	2.56	2.29
400	2.68	2.50	3.54	3.41
500	3.38	3.35	4.45	4.54
600	4.08	4.21	5.33	5.64
700	4.75	5.03	6.16	6.71
800	5.36	5.85	6.98	7.74
900	6.00	6.64	7.89	8.75
1000	6.58	7.41	8.50	9.72

*The asterisk denotes the largest particle for which Stokes' Law was used.

EXAMPLE 8.9   Estimate the terminal velocity for each element in
               the following table. Use values in Table 8.8.
               Determine time required to drop 5 m.

ρ(g/cc)/Size(μ)	30	100	300	1,000
1	0.015	0.29	1.28	4.39
2	0.03	0.43	1.92	6.58
3	0.045	0.55	2.56	8.50

T = 200°C

v = (m/sec)

SOLUTION 8.9   The values for ρ equal to 2 & 3 were obtained
               directly from Table 8.8. For ρ = 1.0 and μ = 30
               the particle follows Stoke's Law. Looking at
               equation 8.7 for the case, where $\rho_p \gg \rho_g$ it
               reduces to

$$v_t = d_p^2 \, \rho_p/18\mu$$

and the velocity is directly proportional to
density. The v_t for ρ = 1.0 is therefore 0.03/2
= 0.015. For larger particles the velocity increases only 1.5 times when the velocity is
doubled. This allows the other velocities to be
estimated.
Settling Time = 5(m)/v_t(m/sec) = ____(sec)

Values of settling time are given below

ρ(g/cc)/Size(μ)	30	100	300	1,000
1	333	17.2	2.9	1.1
2	166	11.6	2.6	0.8
3	111	9.0	2.0	0.6

lowed to expand in the chamber. The horizontal velocity drops
to U as shown in the figure. From the length L and this velocity the time it takes for the gas to pass through the settling
chamber becomes

$$\text{Residence Time:} \quad T = L/U \qquad (8.9)$$

The size of particles which are collected by such a system may
be estimated using the value of the terminal velocity determined

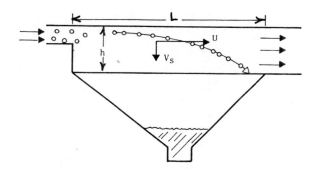

Fig. 8.8 A settling chamber.

earlier. Figure 8.8 shows that the maximum distance which a
particle must drop is given by h. If the residence time T is
known from the previous equation, the velocity which is necessary to allow a particle to settle out is given by

$$v = h/T \qquad\qquad (8.10)$$

In the ideal system all particles with a terminal velocity
greater than this value will be retained in the settling chamber.

EXAMPLE 8.10   Calculate the minimum size particle that will be
                collected with 100% efficiency for particles with
                a density of 3(g/cc) in air at 600°C. The cham-
                ber dimensions are

                    L = 30 (ft)
                    h = 6 (ft)

                The gas velocity is 5 (ft/sec).

                A gas cooler is installed before the settling
                chamber that reduces the temperature to 200°C.
                What is the minimum size under these conditions?

SOLUTION 8.10

        $T = 30(\text{ft})/5(\text{ft/sec}) = 6(\text{sec})$ Eq. 8.9

        $v_p = 6(\text{ft})/6(\text{sec}) = 1(\text{ft/sec})$ Eq. 8.10

        $1.0(\text{ft/sec}) \times 1/3.28(\text{m/ft}) = 0.305(\text{m/sec})$

        From Table 8.8 at v = 0.305 the particle diameter
        is about 100(μ).

        Cooling gas to 200°C reduces the gas volume. This
        reduces the velocity through the settling chamber.
        Assuming the ideal gas law holds the volume and
        therefore the velocity are directly proportional to
        the absolute temperature.

            $v_{200} = v_{600} \times 873/673(^{\circ}\text{K}/^{\circ}\text{K})$

                  $= 5(\text{ft/sec}) \times 873/673 = 2.71(\text{ft/sec})$

            $T_{200} = 30/2.71 = 11.1(\text{sec})$

            $v_p = 6/11.1 = 0.54(\text{ft/sec})$

        From Table 8.8 $d_p > 82(\mu)$ will be collected.

From Figure 8.8 it can be seen not all particles have to
travel a distance h to be collected. Some of the particles
with a terminal velocity less than the one calculated above
will be collected. Conversely, some of the larger particles
may become recaptured and be carried out.

Figure 8.9 shows a cyclone, or centrifugal separator.
The dirty gas enters on the left hand side. The gas enters
tangentially to the outer walls. The centrifugal action throws
the particles to the outer walls where they slide down the walls
to the bottom, where they are collected. The gas spirals downward in the main vortex, where it changes direction, spirals
upward in the vortex core and leaves at the top.

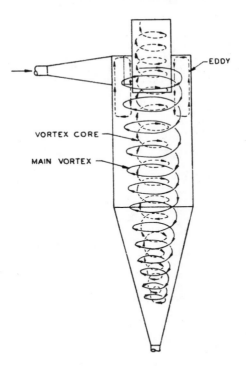

FIGURE 8.9 A Typical Cyclone Separator

The cyclone normally removes particles larger than 10 μ.

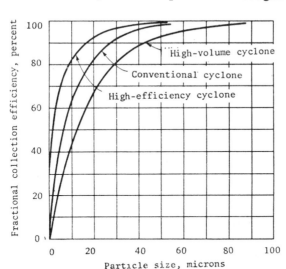

Figure 8.10   Fractional Collection Efficiency as
                a Function of Particle Size for
                Several Types of Cyclones

Figure 8.10 shows typical efficiencies for several centrifugal collectors. As a first approximation, the collection efficiency can be evaluated from

$$\eta = \frac{\text{Centrifugal Force}}{\text{Drag Force}} = \frac{v_p \, \rho_p \, d_p^2}{R \, \mu_g} \qquad (8.11)$$

where

v_p is tangential velocity

R is radius of rotation

the efficiency can be seen to vary linearly with both particle
density, ρ_p, and the tangential velocity. It decreases with
cyclone diameter R and gas viscosity. It is most sensitive

to particle diameter, and the efficiency changes as the square of the diameter. To increase the efficiency it is necessary to reduce R and increase v_p. These changes increase the pressure drop requiring more pumping costs. For air systems, the pressure drop ΔP can be evaluated from equation

$$\Delta P = 39.7 \; \frac{KQ^2P^2}{T^2} \qquad (8.12)$$

Where

Q is volume gas flow in ft^3/min

P is pressure in atm.

T is temperature in °R

K is a constant taken from Table 8.9

TABLE 8.9

Pressure Drop Parameter K Versus Cyclone Diameter

Cyclone diameter, in.	29	16	8.1	4.4
K	10^{-4}	10^{-3}	10^{-2}	10^{-1}

It is often necessary to estimate the performance for operating conditions other than the design conditions. For this purpose, the following equations may be useful.

For varying flow-rate:

$$\frac{100 - \eta_a}{100 - \eta_b} = (\frac{Q_b}{Q_a})^{0.5} \qquad \text{where Q is volume flow rate} \qquad (8.13a)$$

For varying gas viscosity:

$$\frac{100 - \eta_a}{100 - \eta_b} = (\frac{\mu_a}{\mu_b})^{0.5} \qquad \text{where } \mu \text{ is gas viscosity} \qquad (8.13b)$$

For varying density:

$$\frac{100 - n_a}{100 - n_b} = (\frac{\rho_p - \rho_{g,b}}{\rho_p - \rho_{g,a}})^{0.5}$$ where ρ_p is particle density

ρ_g is gas density

(8.13c)

There are several types of wet collectors. Several types are sketched in Figure 8.11. Removal of the particles results mainly from collisions between the liquid and solids. Fine particles in the range of 0.1 to 20 μ can be effectively removed.

Figure 8.11a shows a spray chamber. This is the simplest of all systems. The polluted gas flows upward and collides with liquid droplets formed by several spray nozzles. The mist eliminator is placed on top to remove most of the entrained liquid. The clean gas leaves the top, and the dirty water leaves the bottom.

The pressure (gas) is quite small, and efficiencies are high for sizes above 10 μ. High efficiencies can be obtained for particles as small as 1.0 μ using high pressure fog sprays. 2 - 10 gallons of water are normally used to clean 1,000 ft^3 of gas.

Figure 8.11b is a cyclone scrubber. The gas swirls tangentially as in the case of a dry cyclone. Water is introduced in the center. The water and the particulate are thrown to the walls by centrifugal action. The water rate is in the range of 1 to 8 gallons/1,000 ft^3 of gas. The pressure drop is 1 - 4 inches of water. In general efficiencies are over 100% for 100+ μ particles, 99% for 50 to 100 μ, and 90 to 98% for 5 to 50 μ. Often cyclonic collectors are used along with a Venturi scrubber.

The Venturi scrubber is shown in Figure 8.11c. Normally,

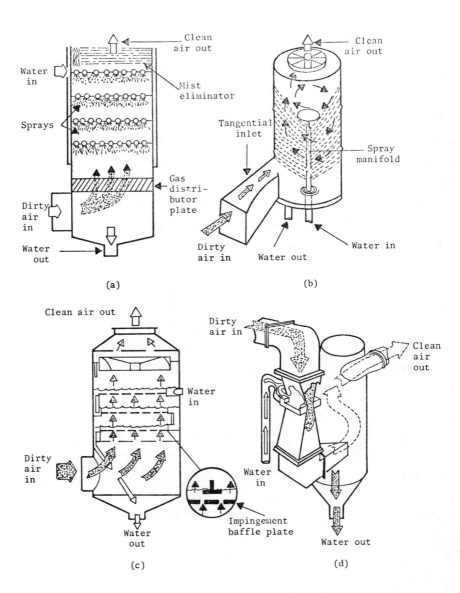

Figure 8.11   Wet collectors: (a) spray tower; (b) cyclone spray tower;
(c) impingement scrubber; (d) venturi scrubber.

the following conditions apply:

1. Gas velocity at Venturi - 50 to 180 (m/sec)
2. Area ration (Venturi throat/inlet) - 1/4
3. Angle of divergence - 5 to 7 degrees

Nozzles are located in the Venturi throat where scrubber liquor
is introduced. The high gas velocities help atomize the liquid.
The mechanism of collection is initial impact where the solid
particles, moving at the gas velocity, strike the liquor which
is introduced at right angles. The water droplets soon reach
the gas velocity, and the collection ceases.

To achieve the high velocity gas stream, the Venturi scrubber has a high pressure drop. Typically, the pressure drop is
3 - 100 inches of water. Efficiencies ranging above 99% for
sub-micron particles may be obtained. Figure 8.12 shows the
relationship between particle size, pressure drop, and collection efficiencies.

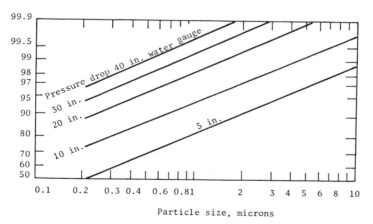

Figure 8.12 Relationship between collection efficiency and
particle size in venturi scrubbers.

The dust laden droplets must be removed from the gas stream.
This is typically done by passing it into a cyclonic separator
as shown in Figure 8.9c. About 2 - 12 gal/1,000 ft^3 of gas are
required.

   While high collection efficiencies may be obtained from a
wet scrubber, they have the disadvantage of producing a wet
sludge which must be treated. The gas must be cooled in order
to restrain the scrubber liquid from vaporizing. These may be
serious disadvantages.

   Several types of fabric filters are shown in Figure 8.13.
The fabric material must be compatible with both the carrier
gas and the particulate. The free space is usually over 95%.
The filters are usually formed into cylindrical tubes and hung
in multiple rows which provide a large surface area for the gas
to pass through. The gas can pass through the filter, and the
solids are retained. This solid layer, or cake, continues to
build up. This results in an increase in pressure drop. Period-
ically the cake is removed from the filter. Figure 8.13 shows
several systems to remove the cake.

   The pressure drop over the system is the sum of the pressure drop over the clean filter and the pressure drop of the
solid layer.

$$\Delta P \text{ (collector)} = \Delta P \text{ (filter)} + \Delta P \text{ (solid layer)} \qquad (8.14)$$

The pressure drop for both the clean filter and solid layer
may be expressed by Darcy's Law:

$$\frac{\Delta P}{x} = \frac{v\ \mu_g}{\kappa} \qquad\qquad (8.15)$$

   where x is thickness
          v is velocity of gas
          μ_g is viscosity of gas
          κ is permeability

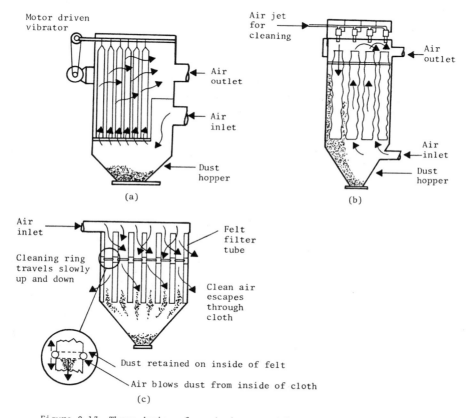

Figure 8.13 Three designs for a baghouse: (a) motor-driven vibrator; (b) air jet, and (c) cleaning ring for removing particles from fabric filters.

κ, the permeability, is hard to predict and is a function of porosity, specific surface area, size distribution, and other factors. It is normally determined experimentally. Example 8.11 shows this calculation.

Writing equation 8.15 for the solid layer:

$$\Delta P \text{ (solid layer)} = \frac{v \; \mu_g \; x}{\kappa} \tag{8.16}$$

The value of x may be obtained from a material balance on the solid.

$$\text{Total mass collected} = \rho_c \; (A \; x_c) \tag{8.17}$$

Where: ρ_c is cake density

A is total filter cross section

x is cake thickness

$$\text{Total mass collected} = (v \; A)(t)(L_d)(\eta) \tag{8.18}$$

Where: v is gas velocity

A same as above

t time of filtration

L_d dust loading (mass of solids/ft^3 of gas)

η removal efficiency

Equating 8.17 and 8.18 and solving for x_c

$$x_c = \frac{v \; L_d \; t}{\rho_c} \tag{8.19}$$

Substituting into equation 8.16

$$\Delta P \text{ (solid layer)} = \frac{v^2 \; L_d \; \mu_g \; t}{\kappa_c \; \rho_c} \tag{8.20}$$

or

$$\Delta P \text{ (solid layer)} = \kappa_1 \; v^2 \; L_d \; t \; [\kappa_1 = \frac{\mu_G}{\kappa_c \; \rho_c}] \tag{8.21}$$

In a normal operation v and L_d are constant, and the ΔP increases linearly with time. It can be seen from equation 8.21 that ΔP (solid layer) increases linearly with dust load and as the square of the gas velocity.

EXAMPLE 8.11   Air at 170 (^{o}F) passes through a fabric filter for 5.0 (hr.). At the beginning the pressure drop that resulted from the clean filter was 0.55 (inches) of water. The filter cake had a density of 1.28 (g/cc) and a dust loading of 14.0 (grain/ft^3). Estimate κ_c. The velocity is 4.2 (ft/min).

SOLUTION 8.11   Using equation 8.19

$$\kappa_c = \frac{(v^2)\,(L_d)\,(\mu_G)\,(t)}{(\rho_c)(\Delta P)}$$

$$= \frac{(4.7 \text{ ft/min})^2\,(14 \text{ grains/ft}^3)(0.0503 \text{ lb/ft-hr})(5.0 \text{ hr})}{1.28\,(\text{g/cc})(4.55 - 0.55) \text{ in } H_2O\,(32.2 \text{ ft-lb}_m/\text{lb}_f\text{-sec})}$$

$$= \frac{12.1\,(\text{gr})(\text{lb}_f)(\text{cm})^3\,(\text{sec})^2}{(\text{min})^2\,(\text{ft})^3\,(\text{g})\,(\text{in } H_2O)} \quad \rightarrow \quad 4.62 \times 10^{-11}\,(\text{ft})^2$$

A variety of fabrics are available for filters such as cotton, nylon, glass fibre, polyesters, etc. The cotton has an upper gas temperature of about 200°F, while the glass wool can go to 550°F. This is a limitation placed on fabric filter equipment.

The advantages of fabric filters include:

1. High collection efficiency over a broad range of particle sizes.

2. Extreme flexibility in design provided by the availability of various cleaning methods and filter media.

3. Volumetric capacities in a single installation which may range from 100 to 5 million ft^3/min.

4. Reasonable operating pressure drops and power requirements.

5. Ability to handle a diversity of solid materials.

Among acknowledged disadvantages are:

1. Space factors may prohibit consideration of bag houses.

2. Possibility of explosion if sparks are present in vicinity of a bag house.

3. Hygroscopic materials usually cannot be handled, owing to cloth cleaning problems.

Table 8.10 lists characteristics of fabric filters.

The final type of solid removal system is the electrostatic precipitator. Electrostatic precipitators are used to handle large volumes of gases from which aerosols must be removed. They have the advantage of low-pressure drop, on the order of 1 inch of water, high efficiency for small particle size, the ability to handle both gases and mists for high volume flow, and the relatively easy removal of the collected particulate. The four steps in the process are:

1. place a charge on the particle to be collected

2. migrate the particle to the collector

3. neutralize the charge at the collector

4. remove the collected particle.

Table 8.10 Filter fabric characteristics

Fiber	Operating exposure °F Long	Operating exposure °F Short	Supports combustion	Air permeability* cfm/ft²	Composition	Resistance† Abrasion	Resistance† Mineral acids	Resistance† Organic acids	Resistance† Alkali	Cost‡ rank
Cotton	180	225	yes	10-20	Cellulose	G	P	G	G	1
Wool	200	250	no	20-60	Protein	G	F	F	P	7
Nylon§	200	250	yes	15-30	Polyamide	E	P	F	G	2
Orlon§	240	275	yes	20-45	Polyacrylonitrile	G	G	G	F	3
Dacron§	275	325	yes	10-60	Polyester	E	G	G	G	4
Polypropylene	200	250	yes	7-30	Olefin	E	E	E	E	6
Nomex§	425	500	no	25-54	Polyamide	E	F	E	G	8
Fiberglass§	550	600	yes	10-70	Glass	P-F	E	E	P	5
Teflon§	450	500	no	15-65	Polyfluoroethylene	F	E	E	E	9

*cfm/ft² at 0.5 in. W.G.
†P = Poor, F = Fair, G = Good, E = Excellent.
‡Cost rank, 1 = lowest cost, 9 = highest cost.
§Dupont registered trademark.

A schematic of a precipitator is shown in Figure 8.14.

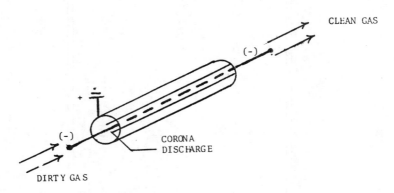

FIGURE 8.14   An Electrostatic Precipitator

The negatively charged center wire causes a discharge to the grounded, positive collector shown here as the outer cylinder. At the very high DC voltages used, 25 to 100 kv, a corona discharge occurs close to the negative electrode. The gas close to the negative electrode is thus ionized upon passing through the corona. As the negative ions and electrons migrate toward the collector electrode they, in turn, charge the passing particulate. The electric field then draws the particulate to the collector where it is deposited. The collected material is removed, usually by hitting the collector, a process which is called rapping. If a mist is being collected, then the material runs down the collectors and is removed at the bottom.

The theoretical efficiency, that is, weight fraction collected, of an electrostatic precipitator is given approximately by
the relation

$$\eta = 1 - e^{-2VL/RU} \tag{8.22}$$

where V is the particulate velocity towards the electrode,
    typically on the order of 0.1 to 0.7 ft/sec.
    L is the length of collector
    R is the radius of collector, if cylindrically
    shaped
    U is the bulk gas velocity, 2 to 8 ft/sec
Noting the collector area, A, is equal to $2\pi RL$ and the volumetric flow rate, Q is given as $\pi R^2 U$, we can rewrite this equation
as
$$\eta = 1 - e^{-AV/Q} \tag{8.23}$$
This expression, while approximate, indicates that higher efficiencies occur for low Q, high V, and large collector area. One
can obtain high efficiency, if required, by increasing the surface area. However, in order to increase efficiency from 90 to
99 percent one must approximately double the area; to go from
90 to 99.9 percent requires tripling the area.

An increase in flow rate reduces the efficiency and even
though only a small reduction in η might occur for a slight increase in Q, this means a large increase in emissions. A precipitator dropping from 99 percent efficiency to 97 percent triples
the emissions and it is the emissions which are the important
variable. In practice, the theoretical efficiency is never attained since some reentrainment of the collector material occurs.
The collectors are usually plates rather than cylinders. Plate-
type collectors are used for dry particulate collection as from
power plants while the tube type is often used for wet collection.

   The electrostatic precipitator system is not without problems. Ionization of the gas occurs only in a limited temperature range so that the device can suffer a loss in efficiency
if sudden changes occur in the operating conditions. Buildup
of collected material can cause "spark-over" between the electrodes which in turn causes a high current flow and excessive
power use. Buildup of material on the negative electrode can
suppress the corona discharge and reduce efficiency. The resistivity of the gas-particulate combination also affects the corona and the collection efficiency. Reentrainment of the collected material can interfere with the particle charging and result
in direct release of particulate up the stack.

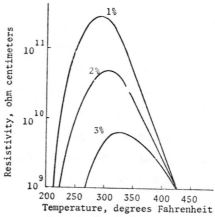

FIGURE 8.15 Variation of Resistivity
             of Fly Ash with Sulfur
             Content of Coal

The resistivity of the dust is a major parameter in the design of a precipitator. For effective collection, the resistivity should be in the range 10^4 to 10^{10} ohm-cm. This has serious implications in the cleanup of SO_2. Figure 8.15 shows the change in resistivity with the sulfur in the coal. When a utility shifts to a low sulfur coal, the resistivity increases beyond the effective range, and the effective removal of fly ash is reduced.

TABLE 8.11

Summary of Particulate Emission Control Techniques

Device	Minimum particle size†, μm	Efficiency % (mass basis)	Advantages	Disadvantages
Gravitational settler	>50	<50	Low pressure loss, simplicity of design & maintenance	Much space required. Low collection efficiency.
Cyclone	5-25	50-90	Simplicity of design & maintenance. Little floor space required. Dry continuous disposal of collected dusts. Low to moderate pressure loss. Handles large particles. Handles high dust loadings. Temperature independent.	Much head room required. Low collection efficiency of small particles. Sensitive to variable dust loadings and flow rates.
Wet collectors Spray tower Cyclonic Impingement Venturi	>10 >2.5 >2.5 >0.5	<80 <80 <80 <99	Simultaneous gas absorption and particle removal. Ability to cool and clean high-temperature, moisture-laden gases. Corrosive gases and mists can be recovered and neutralized. Reduced dust explosion risk. Efficiency can be varied.	Corrosion, erosion problems. Added cost of wastewater treatment and reclamation. Low efficiency on submicron particles. Contamination of effluent stream by liquid entrainment. Freezing problems in cold weather. Reduction in buoyancy and plume rise. Water vapor contributes to visible plume under some atmospheric conditions.

Table 8.11 (cont.)

Summary of Particulate Emission Control Techniques

Device	Minimum particle size†, μm	Efficiency % (mass basis)	Advantages	Disadvantages
Electrostatic precipitator	>1	95–99	99+ percent efficiency obtainable. Very small particles can be collected. Particles may be collected wet or dry. Pressure drops and power requirements are small compared with other high-efficiency collectors. Maintenance is nominal unless corrosive or adhesive materials are handled. Few moving parts. Can be operated at high temperatures(300 to 450 $^{\circ}$C)	Relatively high initial cost. Precipitators are sensitive to variable dust loadings or flow rates. Resistivity causes some material to be economically uncollectable. Precautions are required to safeguard personnel from high voltage. Collection efficiencies can deteriorate gradually and imperceptibly.
Fabric filtration	<1	>99	Dry collection possible. Decrease of performance is noticeable. Collection of small particles possible. High efficiencies possible.	Sensitivity to filtering velocity. High-temperature gases must be cooled to 100 to 450°C. Affected by relative humidity(condensation). Susceptibility of fabric to chemical attack.

Table 8.11 summarizes some of the major characteristics of particulate emission control techniques.

The removal of particulate matter is clearly an example of a tail-end cleanup system.

8.3.5 SO_2 Pollution

Referring to Figure 8.6 it can be seen that SO_2 pollution ranks as one of the major pollutants as far as health effects

are concerned. The major source of SO_2 is combustion in utility boilers (see Tables 8.3 - 8.6). As in the case of particulates, because it is generated at a limited number of sites,
control of SO_2 emissions must be considered.

EXAMPLE 8.12   Estimate the SO_2 generated from a 1,000 megawatt
                electric utility burning 2.0% sulfur coal.

SOLUTION 8.12   The following assumptions are made
                a) 10,000 (Btu) of fuel is required to produce
                    1 (Kw-hr) of electricity.

                b) Coal has a heating value of 15,000 (Btu/lb).

                1,000 {Mw} x 1,000 {Kw/Mw} x 10,000 {Btu/Kw-hr}

                x 1/15,000 {lb Coal/Btu} x 2/100 {lb S/lb Coal}

                = 13,333 {lb S/hr} or 58,000 {tons S/year}

                13,333 {lb S/hr} x 64/32 {lb SO_2/lb S}

                = 26,666 {lb SO_2/hr} or 116,000 {tons SO_2/year}

The total sulfur emissions have been estimated at 220×10^6
[tons/year]. The total man-made emissions are about 73×10^6
[tons/year] or about 33% on a global basis. The ambient SO_2 concentrations given in Table 8.1 show a concentration of 0.0002 ppm.
The concentration of SO_2 is much larger than this value. Values
approaching 1.0 ppm have been reported for 24 hour average concentrations in major metropolitan areas. This is 5,000 times
the ambient air concentration.

The sulfur appears in the atmosphere as SO_2, SO_3, H_2S and
marine sulfate compounds. It has been estimated that H_2S has a
lifetime of about 2 hours in urban environment and 2 days in remote areas. SO_2 has a life of about four days. The latter fig-

ure is difficult to determine since there are a number of removal mechanisms for SO_2. It can be washed out directly by
rain or deposited directly on vegetation. Under the action of
the nitrogen oxides and hydrocarbon compounds SO_2 can be converted into H_2SO_4, forming an aerosol haze. The presence of fine
metallic particles will aid in oxidizing SO_2 to SO_3 which sub-
sequently combines with a water molecule to form H_2SO_4. Lastly,
there is evidence that in foggy or very moist atmospheres, SO_2
can react with ammonia (present in the atmosphere in small a-
mounts) to form ammonium sulfate, $(NH_4)_2SO_4$, and this is subse-
quently removed from the atmosphere by washout. All these various removal mechanisms cause a short lifetime for SO_2 in the
atmosphere.

Sulfur dioxide is highly soluble and is absorbed in the
moist passages of the upper respiratory system. Exposure to
SO_2 levels on the order of 1 ppm leads to a constriction of
airways in the respiratory tract. In the presence of particulate matter, irritation is much greater.

8.3.6   Control of SO_2

As already described, the preferred way to control pollution is to prevent its generation. Gas and liquid fuels may
be desulfurized before being burned. The major problem with
SO_2 is the combustion of coal. Sulfur is found in coal as pyritic (FeS_2) sulfur, organic sulfur, and sulfates. A portion
of the pyritic sulfur may be removed by standard ore beneficia-
tion procedures. Figure 8.16 shows the amount of organic and
pyritic sulfur found in coal along with the fraction that may
be removed prior to combustion.

Major incentive for coal liquifaction and gasification is
that they allow for removal of sulfur prior to combustion.

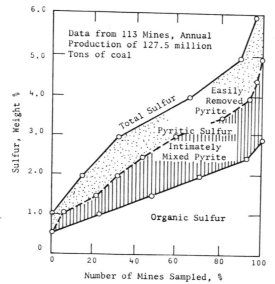

FIGURE 8.16   Maximum Sulfur Content versus
Percent of Mines Sampled. (AP-52)

Since only a portion of the sulfur can be removed by treatment prior to combustion, it will be necessary to provide additional treatment if SO_2 emissions to the atmosphere are to be
substantially reduced.   One method of reducing pollution levels
is to modify the process.   The fluid bed boiler represents a
"modification of process" which may allow high sulfur coal to
be burned with low SO_2 emissions.   The fluid bed characteristics
have been described earlier.   The fluidized bed boiler is shown
in Figure 8.17, on the next page.

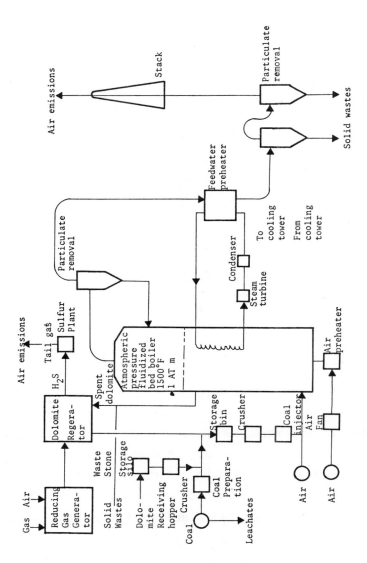

FIGURE 8.17   An atmospheric pressure fluidized-bed combustion system, including selected air and solid waste emission points.

The bed material is "reactive" toward SO_2 and removes SO_2 in the bed as soon as it is formed. The reactive material may be lime, limestone, or dolomite. These are solids which are crushed to the proper size and make up the bed material. The high sulfur coal is introduced to the bed. Combustion occurs, heat is released, and SO_2 is formed. The heat is removed by the water to generate steam. The SO_2 is removed by the lime, if lime is being used for bed material, to generate calcium sulfate. The chemical reaction is

$$CaO + 1/2\ O_2 + SO_2 \rightarrow CaSO_4$$

Figure 8.18 shows the formation of $CaSO_4$ on the surface of a lime particle.

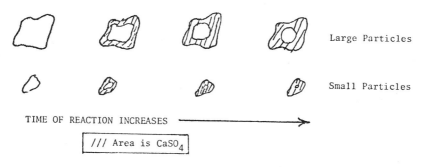

FIGURE 8.18   Calcium Sulfate Shell Builds Up On
             An Unreacted Core of Lime

The formation of the $CaSO_4$ layer restricts the rate at which SO_2 reaches the inner core of reactive CaO; the effect of a non-reactive layer on the rate of chemical reaction was discussed in Section 3.2.1.

In order to get a high conversion of CaO to $CaSO_4$, the
size of the particles should be small. In order to obtain high
through-put in a fluid bed, the particles should be large. A
compromise between high unit through-put and high CaO conversion
is necessary. Typical results show that about two times the
theoretical amount of lime is added to the fluid bed. The mixture of CaO and $CaSO_4$ must be disposed of.

The fluidized bed system did not retard the formation of
SO_2 but removed SO_2 once formed. The fluid bed carried out
two distinct functions: combustion and separation. A general
separation process is shown below:

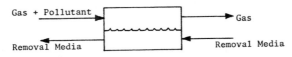

FIGURE 8.19 General Separation System

The stream containing the pollutant is contacted with a
separation media. The purpose of the media is to remove the
pollutant from the product gas stream. Several alternatives
may be envisioned. A few types often used are listed below:

1.   Chemical reaction--the pollutant is encouraged to
     react chemically. If the pollutant chemically
     reacts, its chemical identity is changed and the
     pollutant is destroyed. The chemical produced may
     itself be a pollutant.

2. Absorption--the pollutant is attracted to the surface of the removal media (a solid) and becomes attached. (This works like a molecular fly-paper.) When the surface is full, the media must be thrown away or regenerated.

3. Adsorption--the media will selectively take up the pollutant into the media. This scheme works on the basis of differences in solubility of different gases in liquids.

4. Molecular size--the media may be porous and will allow molecules of given shapes or sizes to become trapped in the media (molecular sieves). The media may be in the form of a membrane that will allow only the pollutant to diffuse (gaseous diffusion).

The first alternative removes the pollutant by changing its chemical identity while the others retain the pollutant but transfer it from the gas. It has not been eliminated. The result is the production of a <u>new waste stream</u>. It is usually desirable to regenerate the separation media for re-cycle. This is shown in Figure 8.20.

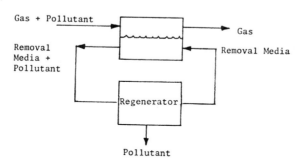

FIGURE 8.20 Regenerative Scrubber System

The pollutant has been separated and concentrated but it has not been eliminated. To eliminate the pollutant, it must be reacted chemically. It is fortunate when the pollutant

in its concentrated form has a beneficial use, or the concentrated pollutant could be reacted to provide a useful product.
Usually this is not the case. The disposal of the pollutant
has not been achieved by separation--only its concentration
increased. Cleaning up the gas did not solve the environmental
problem; it put the pollution "someplace else."

EXAMPLE 8.13 A separation scheme has been developed to remove the SO_2 from
the stack and produce H_2SO_4. The flow sheet is given below.

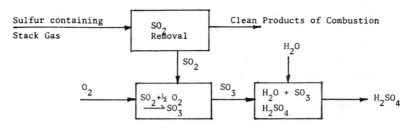

If 10% of all the utilities were fired with 2% S coal and the
separation scheme given above were installed to produce what
fraction of the H_2SO_4 production would be met?

SOLUTION 8.13

From Figure 1.6, 7.7 Q of coal used to
generate electricity

$7.7[Q/yr] \times 10^{15}[Btu/Q] \times 1/26 \times 10^6[ton\ coal/Btu]$

$\times 2/100[ton\ S/ton\ coal] \times 98/32[ton\ H_2SO_4/ton\ S]$

$= 1.8 \times 10^7[ton\ H_2SO_4/yr]$

For 10% utilization $1.8 \times 10^7 \times 0.1 = 1.8 \times 10^6[ton\ H_2SO_4/yr]$

Annual H_2SO_4 Production $\approx 4 \times 10^4[ton/yr]$

Fraction $= 1.8 \times 10^6/4 \times 10^4 = 45$

Forty-five times the annual use of H_2SO_4 would be produced.

The next few paragraphs provide some insight into the
design of a gas separation system.

The first problem faced is to provide a system where
the gas comes in contact with the separation media. The
separation media may be either a liquid or solid. Systems
for contacting a gas and solid have already been discussed

in Chapter 3. This section will emphasize the use of a
liquid separation media. Several methods are available, but
two methods are most used.

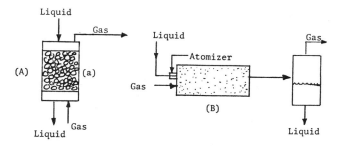

FIGURE 8.21 Basic Liquid-Gas Contacting System

1. Wetted Surface--this is shown in Figure 8.21(A). A
   vessel is packed with inert material. The liquid
   removal media is distributed over the top of the
   packed bed. As it flows downward, it coats (wets)
   the surface of this packing. This provides a large
   surface area to contact the gas. The gas may flow
   either co- or counter-current to the liquid.

2. Atomized Liquid--this is shown in Figure 8.21(B).
   The liquid is broken up into small size droplets and
   distributed in the gas. Extremely high surface
   areas are obtained. There is a wide variety of systems
   that fall into this class. The gas and liquid must be
   separated. The pollutant is transferred to the liquid
   stream.

To appreciate the design considerations for a separation
system without going into details, the discussion below makes
use of an analogy to heat transfer.

Figure 8.22 shows both a heat transfer and mass transfer
system. The pollutant stream is labeled with the subscript P.
For the heat transfer system the thermally polluted stream is
a hot liquid at temperature T_p. The pollutant, heat (q) is
removed by transferring it to a colder stream that enters
at T_M. The thermally polluted stream leaves with a lower

thermal pollution at T_1. The pollution removal media, L,
leaves at a higher thermal pollution level (T_2). The graph
shows the temperatures of the polluted stream (G) and the
removal media stream (L) as a function of distance.

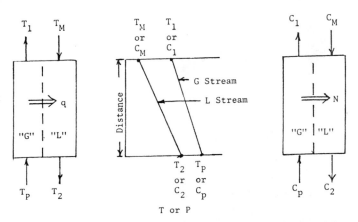

G - Polluted stream to be purified. L - Stream to accept pollutant.
Pollutant leaves the G stream and enters the L stream. Subscripts:
p = Polluted stream; M = Removal media stream.

FIGURE 8.22   Comparison of Heat and Mass Transfer Systems

For the mass transfer system the polluted stream (G) enters
at a pollution level C_p and is cleaned up by transferring mass
(N) to the liquid stream (L) that enters at concentration C_M
and leaves at a concentration C_2. The graph shows the concentration of both streams as a function of distance. Further
analogies are shown in Table 8.12.

For the transfer of heat, the area of contact between
the two streams is easy to evaluate because there is usually
a solid pipe wall separating the streams. For the packed
bed mass transfer system the area is the total wetted area
of the packing. The usual way to designate this area is in
terms of a constant "a" [surface area of packing/volume of
packing]. The mass transfer equation given in Table 8.12 may

be multiplied and divided by the total volume of the packed
system.

$$N = V \ Kg \ \frac{A}{V} \ \Delta C_{1n} = V \ K_{ga} \ \Delta C_{1n} \tag{8.24}$$

or

$$N/V = K_{ga} \ \Delta C_{1n}$$

TABLE 8.12
Analogy Between Heat Transfer and Mass Transfer Systems

	Heat Transfer	Mass Transfer
Entity Transferred	Heat - q(Energy)	Mass - N(Mass)
Driving Force	ΔT_{1n}	ΔC_{1n}
Evaluation of ΔT_{1n} & ΔC_{1n}	$(\Delta T_2 - \Delta T_1)/\ln(\Delta T_2/\Delta T_1)$	$(\Delta C_2 - \Delta C_1)/\ln(\Delta C_2/\Delta C_1)$
Transfer Equation	$q = hA\Delta T_{1n}$	$N = KgA\Delta C_{1n}$
Balance	$L(H_2 - H_M) = q$	$L(C_2 - C_M) = N$
	$V(H_1 - H_p) = q$	$V(C_1 - C_p) = N$

ΔT_2, ΔC_2 represent difference in concentration at one end of the system.
ΔT_1, ΔC_1 represent difference in concentration at the other end of the system.

The evaluation of ΔC is more complex than evaluation of
ΔT. Figure 8.23 shows a comparison of a heat transfer system
and a mass transfer system. For the case of thermal equilibrium, the two phases will reach the same temperature. For
chemical equilibrium the concentration in the two phases are
not equal. For dilute solutions, which is the case for most

pollutants, the concentration in the two phases at equilibrium
may be expressed by Henry's Law.

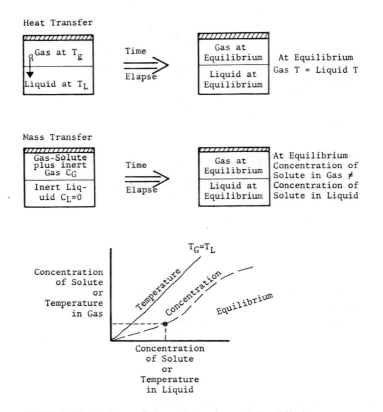

FIGURE 8.23 Analogy of Thermal to Chemical Equilibrium

$$x_i = (1/H_i)(p_i) \qquad\qquad (8.25)$$

p_i = equilibrium partial pressure

x_i = mole fraction in liquid at equilibrium

H_i = Henry's Law Constant

Figure 8.24 gives values of Henry's Law Constant.

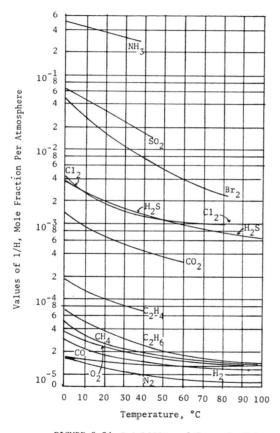

FIGURE 8.24 Solubility of Gases in Water

The value of ΔC in Equation 8.23 is often converted to ΔP (partial pressures). The ΔP represents the driving force

for mass transfer in the similar manner that ΔT provides the
driving force for heat transfer. The value of ΔP is

ΔP = Partial pressure of pollutant in gas - Partial
             pressure of pollutant in equilibrium with liquid

Substituting Equation 8.25

$$\Delta P = (pp_i - H_i\, x_i) \qquad\qquad\qquad (8.26)$$

EXAMPLE 8.14   A gas stream contains 0.35% SO_2. It is to be cleaned up
by washing with a liquid water stream to reduce the concentration to 0.035%. The liquid stream will leave at 95%
of its equilibrium value at 80°F. Assume 90% of the SO_2
is removed.

Determine

(a) The concentration of liquid leaving the scrubber
(b) Determine the liquid flow rate per 10^6 SCF of gas
     stream

SOLUTION 8.14

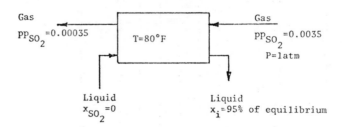

From Figure 8.24 $1/H = 0.025$

x_i = $(1/H)pp$ = $(0.025)(0.0035)$ = 8.75×10^{-5} @ equilibrium

x_i = $8.75 \times 10^{-5} \times 0.95 = 8.3 \times 10^{-5}$ @ of equilibrium

Evaluation of SO_2 transferred

SO_2(in)
   10^6[SCF]x1/359[lb-mole/SCF] x 0.0035[lb-mole SO_2/
      lb-mole gas] = 9.75[lb-mole]

SO_2(out)   [90% removed i.e 10% left]
   9.75[g/mole] x 0.1 = 0.975[g/mole]

SO_2 (Transferred)
   (9.75-0.975)[g/mole] = 8.77[g/mole]

Evaluation of Liquid (Material Balance)

In

$$L(x_{SO_2})_{in} + 9.75 = L(x_{SO_2})_{out} + 0.975$$

$$L(0) + 8.77 = L(8.3 \times 10^{-5})$$

$$L = 1.06 \times 10^5 \, [\text{lb-mole}]$$

$$1.06 \times 10^5 \, [\text{lb-moles H}_2\text{O}] \times 18 \, [\text{lbs/lb-mole}] \times 1/8.33 [\text{gal/lb}] = 2.28 \times 10^5 \, [\text{gal}]$$

The factors that influence the <u>size of the reactor system</u> according to Equation 8.20 are:

(1) K_{ga}--This is similar to heat transfer coefficient and is influenced by fluid flow characteristics and physical properties of the fluids and by the type of packing.

(2) C (or P)--The values of the concentration. The lower the concentrations the larger the scrubber volume. Also the lower the constant H the greater the ΔP and the lower the volume. The value of H increases with temperature.

EXAMPLE 8.15 Repeat Problem 8.14 but assume an initial concentration of 1/5 that in Problem 8.14. Determine the relative volume of the reactors. (H = 40; 1/H = 0.025)

SOLUTION 8.15

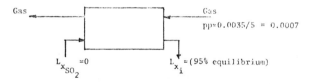

Repeating Problem 8.14 using the lower concentrations:

$$x_i = (0.025)(0.0007)(0.95) = 1.66 \times 10^{-5}$$

$$SO_2(\text{in}) = 9.75/5 \, [\text{lb-mole}] = 1.95 \, [\text{lb-mole}]$$

$$SO_2(\text{out}) = 1.95 \times 0.10 = 0.195 \, [\text{lb-moles}]$$

$$SO_2(\text{transferred}) = 1.755 \, [\text{lb-moles}]$$

$$L = 1.66/1.575 \times 10^{-5} = 1.05 \times 10^5$$

For conditions in Problem 8.14:

At the end of scrubber, the gas enters (from Equation 8.26)

$\Delta P_1 = 0.0035 - 40 \times 8.3 \times 10^{-5} = 1.8 \times 10^{-4}$

At the end where liquid enters (from Equation 8.26)

$\Delta P_2 = 3.5 \times 10^{-4}$ (End Water Enters)

$\Delta P_{1n} = [3.5 \times 10^{-4} - 1.8 \times 10^{-4}]/[\ln(3.5 \times 10^{-4}/1.8 \times 10^{-4})] = 2.6 \times 10^{-4}$

For lower concentrations given above

$\Delta P_1 = 0.0007 - 40 \times 1.66 \times 10^{-5} = 3.6 \times 10^{-5}$

$\Delta P_2 = (0.0007)(0.1) = 0.7 \times 10^{-4}$

$\Delta P_{1n} = [7 \times 10^{-5} - 3.6 \times 10^{-5}]/[\ln(7 \times 10^{-5}/3.6 \times 10^{-5})] = 5.15 \times 10^{-5}$

From Equation 8.9 $N/V = K_{ga}\Delta C_{1n}$

Since C is proportional to partial pressure, Equation 8.24 can be written

$N/V = K'_{ga} \Delta P_{1n}$

For High Concentration Case   For Lower Concentration Case

$(N/V)_H = K'_{ga}(3.5 \times 10^{-4})$ $(N/V)_L = K'_{ga}(0.7 \times 10^{-4})$

Dividing these two equations

$(N/V)_H/(N/V)_L = (K'_{ga}/K'_{ga})(2.6 \times 10^{-4}/5.15 \times 10^{-5}) = 5.0$

Since $N_H = 5 N_L$ (Five times the mass transferred)

$V_L/V_H = 1.0$

It takes the same water rate and reactor volume in both cases.
The first case, however, removes five times the amount of pollutant
in the same size system.

There is no attempt in this brief discussion to provide
the analytical tools necessary to determine the size of a gas
cleanup system. To do this would require more space than appropriate for a general survey text of this type. Most of the discussion has been directed toward a comparison of mass transfer
to heat transfer. The problem of mass transfer is far more
complex than indicated by the comparisons to heat transfer.

The scrubber systems described do not eliminate the pollutant but merely transfer it to the liquid. The liquid may be
cleaned up by removing the pollutant. This may be done by
(1) physical change and (2) chemical change, both shown below.

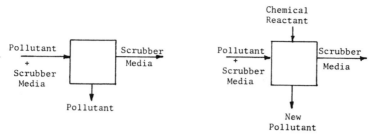

FIGURE 8.25 Scrubber Media Recovery

The first system shows a physical change and the pollutant
is the product. It is available in a concentrated form. In
the second system a chemical is added to react with the
pollutant. A new chemical is generated.

The need to remove impurities from a process stream
often arises for the protection of the process, the product
or the environment. In Chapter 5 it was required that several
catalytic reactors were used. The catalyst is extremely
sensitive to sulfur compounds, and their function to enhance
a desired reaction is destroyed by these sulfur compounds.
It is essential that all sulfur compounds be removed prior
to the catalytic reactor. The cleaning system is employed
to protect the process.

The product resulting from a conversion process is usually
specified to rather close tolerances, and it is often necessary to remove some chemical specie before it may be sold.
Industry clearly understands and is responsive to this problem.

The concern for the quality of our environment has
become a major concern in recent years. Previously, the
incentive to clean-up a stream or product was provided by the
market place. It was a matter of free enterprise economics.
The clean-up of the environment by contrast is being forced
upon industry by governmental legislation. It is met with

resistance by many of the industrial sectors because of the
lack of economic justification. It is being demanded by
various public interest groups. The battle will continue
over this environmental issue. It is too early to predict
the outcome; the conflict is likely to continue for decades.
What process or processes will ultimately be used to provide
for the desired (but not known) environmental quality is not
known. The current battle over SO_2 is an example. Widely
conflicting opinions are being circulated as to whether
there is known technology that can solve the problem in order
to meet EPA regulations. There is an abundance of evidence
presented on both sides of the issue.

A few systems that have been proposed or tried as a clean-
up system for SO_2 are discussed briefly. It is intended to
show that there are many alternative ways to effect a cleanup
of SO_2. Which of the systems described will be economically
and/or technologically feasible (if any) is not clear.

Table 8.13 shows some of the major processes under development in 1973 for removal of SO_2. It is divided into three
classes:

1. Throwaway processes where the separation media is dis-
    charged and may be a pollutant itself.

2. Regenerative processes where the sulfur in the SO_2 is
    concentrated to a level where it may be a useful pro-
    duct.

3. Dry processes where there is a reaction between a dry
    solid and SO_2.

Early work was done on injecting limestone into the combustion zone or into the flue gas. The limestone was to react
with the SO_2 to form $CaSO_4$ which would be removed in the electrostatic precipitators along with the fly ash. Unfortunately,
when over 100% more limestone than is theoretically needed was

added only 20 to 40% of the SO_2 was captured. This increased
the loading on the precipitator. Figure 8.26 shows a combination system.

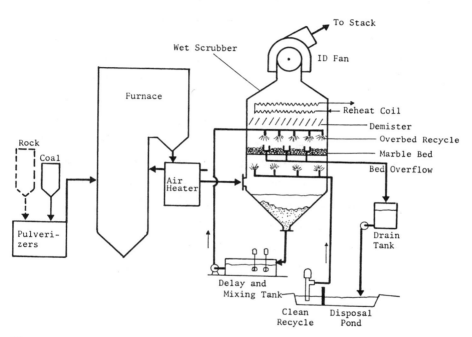

FIGURE 8.26 Schematic Diagram of Air Pollution Control System for Sulfur oxides
             and Particulate Removal, using Limestone Injection and a Scrubber.

In the system shown in Figure 8.26 a wet scrubber is added
to remove the SO_2 not captured by limestone injection. The wet
sludge contains both the $CaSO_4$ and ash and is not usable or sal-
able.   The addition of limestone increases chemical scaling.
corrosion, erosion, and provides a major sludge disposal problem.
An additional drawback to the wet process is the gas leaves at

$12n°F$. In order for the stack to work properly it must be re-
heated to about $300°F$.

The principal advantage of the lime-limestone scrubbing
technique is the relative ease with which the equipment can be
added to existing plants. Another advantage frequently cited
is that particulate as well as SO_2 is removed by the scrubber.
Hence no mechanical collector such as a precipitator is required.
This is a moot point. If the scrubber must be shut down for
maintenance, for example, the plant loses the capability for
particulate removal as well as SO_2 removal. Consequently, some
designers prefer to precede the absorber with an independent
particulate removal system. The wet scrubbing system also can
be used to remove SO_2 from industrial steam plants. In fact,
the wet scrubber is one of the few methods of removing SO_2 that
appear suitable for the smaller steam plants.

In the magnesium oxide (MgO) scrubbing process the slurry
functions to remove SO_2 in the same manner as lime. The important difference between these processes is the magnesium oxide
may be regenerated. The chemical process, as an example, scrubs
the flue gas with a slurry of MgO. Absorption takes place in a
Venturi scrubber by the reactions

$$MgO + SO_2 + 6H_2O \rightarrow MgSO_3 \cdot 6H_2O$$
$$MgO + SO_2 + 3H_2O \rightarrow MgSO_3 \cdot 3H_2O$$

The solution is centrifuged, and the hydrates are removed. The
$MgSO_3 \cdot xH_2O$ crystals are heated and the regeneration reaction
takes place.

$$MgSO_2 \cdot xH_2O + Heat \rightarrow MgO + SO_2 + H_2O$$

Typically the gas stream contains 10 - 15% SO_2. This is an
example of concentrating the pollutant. With high concentrations of SO_2 it may serve as a raw material to produce either

sulfuric acid or sulfur.

Theoretically the MgO is regenerated and will produce no new solid waste. It does have the disadvantage of requiring heat in the regeneration of MgO. This represents an energy loss.

Catalytic oxidation system is an approach used by Monsanto. The process is shown in Figure 8.20. Hot flue gas from the boiler passes into an electrostatic precipitator at high temperature (850°F) to remove the fly ash. The clean hot gas passes by a vanadium pentoxide catalyst which converts

$$SO_2 + {}^1/_2O_2 \rightarrow SO_3$$

The SO_3 reacts with the water vapor and forms sulfuric acid according to the reaction

$$SO_3 + H_2O \rightarrow H_2SO_4$$

The gas is cooled and sent to an absorption tower where the H_2SO_4 (sulfuric acid) is removed. Because of the high volume of gas to be treated, the equipment is large and there is no room to add it to many power plants. Corrosion problems are severe.

Scrubbing with either magnesium oxide or calcium oxide uses a slurry. Clear solutions of sodium or ammonia are excellent absorbers of SO_2. An example of clear solution absorption is one using Na_2SO_3. The basic reaction is

$$SO_2 + Na_2SO_3 + H_2O \rightarrow 2NaHSO_3$$

The sodium sulfite compound may be regenerated by the application of heat

$$2NaHSO_3 + Heat \rightarrow Na_2SO_3 + SO_2 + H_2O$$

There are many more systems to remove SO_2 than covered here.
This section only touched on a few of the systems being considered.

In summary, it can be stated that no single method of removing SO_2 from large quantities of flue gas has a clear-cut advantage over the others. All of the proposed systems require
extensive modification of the basic power plant design. Some
systems cannot be retrofitted to existing units. In some cases,
relatively large land areas are required to accommodate the
additional equipment for the SO_2 removal process. Additional
land area is required if a throwaway system is employed. It is
highly likely that a number of processes may prove fruitful,
tailored to a particular situation involving land availability,
the sulfur content of the fuel, presence of a local market for
recovery products such as sulfur or sulfuric acid, as well as
other factors.

The discussion of SO_2 removal focused on several avenues
to reduce the SO_2 pollution:

1. Change from high to low sulfur fuels
2. Use desulfurized fuels-See gasification and liqui-
    faction section
3. Improve combustion unit
4. Desulfurize the flue gas produced

With an abundance of coal and limits on gas and oil the first
choice of substitution, which has been the practice in the past,
is not a choice for the future. The remaining three avenues
must be more fully developed.

8.3.7 NO_x Pollution

Figure 8.6 shows that NO_x does not pose the same level of danger to health as either SO_2 or particulate. The effects are significant, however, and as control of SO_2 or particulates are adopted, it will become a major factor. Table 8.3 shows that the major contributors are a result of combustion from both stationary and mobile (transportation) sources.

On a global basis, man-made NO_x is not a serious problem. NO_x is formed naturally by processes of nature. It is estimated that 500×10^6 tons/year result from natural sources. This is over 15 times that released by man and leads to a background concentration of about 1 ppb. In major metropolitan areas concentrations are often 40 to 80 ppm. The average level in areas of California where smog is prevalent is about 0.25 ppm with maximum values of 3.5 ppm. The average residence time for NO is about 4 days, and NO_2 is about 5 days.

TABLE 8.14

Emission Factors for Nitrogen Oxides

Source	Average Emission Factor
Coal	
Household and commercial	8 lb/ton of coal burned
Industry and utilities	20 lb/ton of coal burned
Fuel oil	
Household and commercial	12 to 72 lb/1000 gal of oil burned
Industry	72 lb/1000 gal of oil burned
Utility	104 lb/1000 gal of oil burned
Natural gas	
Household and commercial	116 lb/million ft^3 of gas burned
Industry	214 lb/million ft of gas burned
Utility	390 lb/million ft of gas burned
Gas turbines	200 lb/million ft of gas burned
Waste disposal	
Conical incinerator	0.65 lb/ton of waste burned
Municipal incinerator	2 lb/ton of waste burned
Mobile source combustion	
Gasoline-powered vehicle	113 lb/1000 gal of gasoline burned
Diesel-powered vehicle	222 lb/1000 gal of oil burned
Aircraft: conventional	23 lb/flight per engine
fan-type jet	9.2 lb/flight per engine
Nitric acid manufacture	57 lb/ton of acid product

Source: NAPCA. Control Techniques for Nitorgen Oxides from Stationary Sources, AP-67. Washington, D.C.: HEW, 1970 (20).

The term NO_x refers to both NO and NO_2. On the basis of current medical knowledge, ambient levels of 0.5 ppm NO_x is not a health hazard. The real danger lies in the photochemical reactions leading to smog. These atmospheric reactions lead to compounds that have an adverse effect on health of man and on plants.

There are two sources of nitrogen which contribute to NO_x during combustion. One is the molecular nitrogen in the combustion air. The second one, perhaps the most important, is the nitrogen in the fuel itself.

Table 8.14 lists the NO_x emission levels for several types of combustion systems.

EXAMPLE 8.16   Determine the amount of NO_x emitted from the combustion of one million Btu of coal, oil and natural gas. Assume:
   i)   1 {lb} of coal releases 12,000 {Btu}
   ii) 1 {ft^3} of gas releases 1,000 {Btu}
   iii) 1 {gal} of fuel oil releases 150,000 {Btu}

SOLUTION 8.16   Coal -- 20 {lb NO_x/ton coal} x 1/2,000 {ton/lb} x
                       1/12,000 {lb/Btu} = 0.834 {lb NO_x/10^6 Btu}

                Gas -- 390/10^6 {lb NO_x/ft^3} x 1/1,000 {ft^3/Btu}
                       = 0.380 {lb NO_x/10^6 Btu}

                Oil -- 104/1,000 {lb NO_x/gal} x 1/150,000 {gal/Btu}
                       = 0.694 {lb/10^6 Btu}

The example shows that the fuels rank

                    coal > oil > natural gas

in order of decreasing NO_x per unit of energy released. There are two reasons for the higher NO_x emissions. (1) Coal and oil, to a lesser extent, contain nitrogen as part of the fuel (called fuel nitrogen). As previously mentioned, this nitrogen is much more likely to form NO_x than nitrogen that is in the air. (2) It is less likely for gas furnaces to have local hot spots (because of better mixing) where chemical equilibrium provides for little NO_x.

8.3.8 Control of NO_x

The chemical reactions of overall interest with regard to NO_x formation during combustion are

$$N_2 + O_2 \rightleftarrows 2NO$$

$$NO + ^1/_2O_2 \rightleftarrows NO_2$$

Thermodynamic equilibrium provides insight into those conditions which can result in the formation of oxides of nitrogen. Table 8.15 provides the equilibrium constant for the first reaction for various temperatures. It also gives the concentration of NO in parts per million for the initial ratio of N_2/O_2 of 4:1 and 40:1. The value of 4:1 is close to that of air (3.75:1) and the 40:1 is representative of products of combustion from a hydrocarbon burned in 10% excess air.

TABLE 8.15

Equilibrium Constant

$$K_p = \frac{(PP_{NO})}{(pp)_{N_2}^{1/2} \cdot (pp)_{O_2}^{1/2}}$$

for the reaction $1/2 N_2$ and $1/2 O_2$ NO and
Calculated Values of NO Concentration

Temperature		K_p	$N_2/O_2 = 4$	$N_2/O_2 = 40$
°K	°F		NO(ppm)	NO(ppm)
300	80	10^{-15}	--	--
1000	1340	8.7×10^{-5}	210	80
1200	1700	5.3×10^{-3}	1300	500
1500	2240	3.3×10^{-3}	4400	1650
2000	3140	2.0×10^{-2}	8000	2950

The adiabatic flame temperatures for most of the combustion reactions reach 2,000°K which allows for formation of NO. It is clear that little NO would be formed from N_2 if the temperature in the combustion zone remained low.

EXAMPLE 8.17   A waste stream is burned in a well insulated unit. The waste stream has the following composition

Component	Weight
Benzene	70
Carbon	8.2
Ash	9.3
Water	12.5

It is burned with 10% excess air heated to 300°C. There is a 5% loss from the furnace. Niessen, Combustion and Incineration Processes, Dekker New York (1978) provides a solution to this problem.

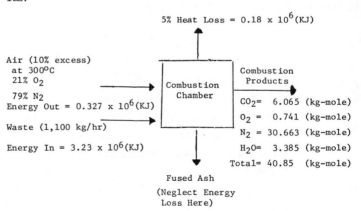

5% Heat Loss = 0.18×10^6 (KJ)

Air (10% excess) at 300°C
21% O_2
79% N_2
Energy Out = 0.327×10^6 (KJ)

Waste (1,100 kg/hr)

Energy In = 3.23×10^6 (KJ)

Combustion Chamber

Combustion Products
CO_2 = 6.065 (kg-mole)
O_2 = 0.741 (kg-mole)
N_2 = 30.663 (kg-mole)
H_2O = 3.385 (kg-mole)
Total = 40.85 (kg-mole)

Fused Ash
(Neglect Energy Loss Here)

A plot of the sensible heat of the product gas is plotted below.

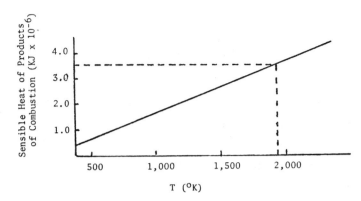

Determine the concentration of NO at equilibrium.

SOLUTION 8.17 The temperature of the gas must be determined to obtain an equilibrium constant. From an energy balance

$$\text{IN} \qquad\qquad = \qquad\qquad \text{OUT}$$
$$0.327 \times 10^6 (\text{KJ}) + 3.23 \times 10^6 (\text{KJ}) = 0.18 \times 10^6 (\text{KJ}) +$$
Energy in Combustion Products

ENERGY IN COMBUSTION PRODUCTS = 3.377×10^6 (KJ)

From diagram above

$\quad T = 1960$ ($^\circ$K)   or   1687 ($^\circ$C)

From Table 8.15, K_p = 0.16

Using the procedure described in section 3.2

Let α = kg-moles of NO formed

	IN	+	GEN	=	OUT	+	Consumed	Partial Pressure
N_2	30.66	+	0	=	P_{N2}	+	0.5 α	$(30.66-0.5\alpha)/40.854$
O_2	0.741	+	0	=	P_{O2}	+	0.5 α	$(0.741-0.5\alpha)/40.854$
NO	0	+	α	=	P_{NO}	+	0	$\alpha/40.854$
CO_2	6.065	+	0	=	P_{CO2}	+	0	
H_2O	3.385	+	0	=	P_{H2O}	+	0	
Total	40.854			=	P_{Total}			

$$K = 0.16 = \frac{(PP_{NO})}{(PP_{N_2})^{\frac{1}{2}} \cdot (PP_{O_2})^{\frac{1}{2}}} = \frac{\alpha}{(30.66-0.5\alpha)^{\frac{1}{2}}(0.741-0.5\alpha)^{\frac{1}{2}}}$$

$\alpha = 0.074$ (Kg-moles)

$$\frac{0.074 \text{ (Kg-moles NO)}}{40.854 \text{ (Kg-mole Combustion Products)}} = 0.0018$$

or 1800 (ppm)

Equilibrium calculations are not sufficient information to assure that NO will be produced. Equilibrium only indicates that NO could be formed. It is necessary for the rate of reaction to be high enough to provide an approach to equilibrium within the time the gas remains at a high temperature level if NO is to be generated. The actual time that gases are near their maximum temperature is about 2 seconds. Table 8.16 provides the time required to reach the 500 ppm NO concentration at different temperatures.

TABLE 8.16

Time Required to Form 500 (ppm) of NO
(Gas Contains 75% N_2; 3% O_2)

Temperature ($°F$)	Time (Sec)	Equilibrium Value (ppm)	% of Equilibrium
2400	1370	550	91
2800	16.2	1380	36
3200	1.1	2600	19
3600	0.11	4150	12

It can be seen that concentrations of 500 ppm NO should be anticipated if temperatures are greater than about 3100°F. Actual temperatures of 3,000°F occur in boilers and substantial concentrations of NO are predicted.

The discussion above considered only the formation of NO and did not consider further oxidation to NO_2. Table 8.17 shows the equilibrium concentrations which satisfy both of the reactions given earlier.

TABLE 8.17

Calculated Equilibrium Concentrations

Temperature ($^\circ$F)	NO (ppm)	NO_2 (ppm)	Ratio (NO_2/NO)
80	1.1×10^{-6}	3.3×10^{-5}	3.33
980	0.77	0.11	0.14
2060	250	0.87	0.0035
2912	2.000	1.8	0.0009

NO_2 concentrations remain at low levels. At low temperatures the ratio of NO_2/NO becomes large, but the concentration of both the oxides is small.

The discussion above showed that, according to the equilibrium, significant concentrations occur only at high temperatures. The gases do not leave the power station at high temperatures but are cooled and their sensible heat recovered. The equilibrium condition of this exit gas would indicate little NO should be found. Unfortunately, as soon as the gas leaving the heat zone is quenched (cooled rapidly), the reaction rate drops several orders of magnitude, and the approach to this new low (and desirable) concentration is not reached.

The control of nitrogen oxide is a good example of modifying the pollution generator. The discussion above suggests three ways to reduce the NO_x production:

1. Minimize the residence time at peak temperature
2. Reduce the peak temperature
3. Reduce the oxygen level

Several techniques to reduce NO_x are here described briefly:

1. Low Excess Air: Normal practice is to operate boilers
    at 10 to 20% excess air. This is done to assure com-
    plete combustion of the fuel but provides sufficient
    oxygen to react with nitrogen. For natural gas systems
    (and to a lesser extent, oil systems) designed to pro-
    vide extremely good fuel/air mixtures, the amount of
    NO_x is reduced and the thermal efficiency is increased.
    Extremely precise control is required to prevent CO
    and hydrocarbon emissions.

2. Two-stage Combustion: The fuel is burned in two stages.
    In the first stage, the furnace is fired with less than
    stoichiometric air. This both reduces the maximum
    temperature and eliminates any oxygen from the gas
    stream.

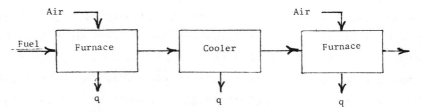

    The gas is cooled and then sent to a second stage,
    where additional oxygen is introduced to burn the re-
    maining fuel. The maximum temperature is lower and
    residence time in a zone containing O_2 is small.

3. Flue-gas Recirculation: To reduce the maximum tempera-
    ture, part of the flue gas is recycled. This reduces
    the maximum temperature in the furnace in the same man-
    ner as excess air, but does not contain oxygen to react
    with N_2.

EXAMPLE 8.18   Consider the recirculation system shown on the next
                page. By recirculation of the gas at 180°C the tem-
                perature in the combustion chamber reaches only 1,000°C.

                a. Determine gas recycle rate

                b. Determine NO concentration

                c. Comment on NO level

SOLUTION 8.18   The mixing chamber shown above does not exist. It is
                shown to simplify calculation procedure. The gas re-
                cycled has the same composition as that leaving the
                combustion chamber. The figure given in the previous
                example gave the energy content for 40.85 (Kg-mole) at
                various temperatures. From this information the energy
                content per mole at the temperatures of interest may be
                calculated.

Temperature		Total Energy	Molar Energy
(°C)	(°K)	(KJ)	(KJ/Kg)
180	453	0.4×10^6	9,800
1000	1273	2.0×10^6	49,000
1687	1960	3.4×10^6	83,200

From an energy balance around the mixer

Let n = Kg-moles recycled

$(40.85)(83,200) + n(9,800) = (40.85 + n)(49,000)$

$n(49,000 - 9,800) = 40.85(83,200 - 49,000)$

$n = 35.64$ (Kg-mole)

At 1,000°K $K_p = 8.7 \times 10^{-5}$

From example 8.16

$$K_p = \frac{\alpha}{(30.66 - 0.5\alpha)^{\frac{1}{2}} (0.741 - 0.5\alpha)^{\frac{1}{2}}} = 8.7 \times 10^{-5}$$

$\alpha = 0.0042$

Concentration $= \dfrac{0.0042}{40.85} = 0.000102$ or 102 (ppm)

The result of recycle air reduced the equilibrium
NO from 1800 (ppm) to 102 (ppm). The low temperature also reduces the rate of reaction so that
the concentration would be much less than 100 (ppm).

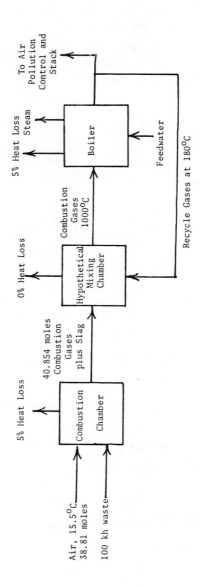

4. Fluidized Bed: The fluid bed furnace operates at a
    temperature which is less than 1900° F and because
    of the particle movement has no hot spots. Because
    of good gas-solid contact, it operates at a low ex-
    cess air rate.

All of these modifications reduce the NO_x emissions.

The previous example showed that by proper modification of
process, NO formation can be drastically reduced. This is fortunate, as NO_x is relatively stable and unreactive, and difficult to remove from a gas stream. All the problems associated
with SO_2 removal resulting from the treatment of large gas volumes and the generation of solid-liquid sludge are also present
in the cleanup of NO_x. In addition, the concentrations are lower and the NO_x less reactive. Little progress (relative to SO_2
removal) has been made to remove NO_x from flue gas.

The amount of NO_x formed from atmospheric nitrogen can be
controlled by modification of the combustion process. A major
portion of the NO_x may not come from atmospheric nitrogen, but
from "fuel nitrogen". This is nitrogen chemically bound to the
hydrocarbon fuel.

Oil: Heterocyclic rings of pyridine are common.

Coal: Contains both chain and ring nitrogen-bearing
       compounds.

Natural Gas: Essentially no nitrogen.

As much as 20 to 80% of the fuel nitrogen is converted to NO_x.
The theoretical mechanism is not known, but there is ample experimental evidence that fuel nitrogen plays a major role when
coal or oil is burned.

8.3.9 Carbon Monoxide, Hydrocarbon and Soot Pollution

All of these materials are products of incomplete combustion.

Not only do they contribute to atmospheric pollution, but they represent a loss of energy. A properly run combustion system should not produce these products. A look at Tables 8.3 through 8.6 shows that it is the transportation sector which is the major contributor to CO and hydrocarbon emissions.

CO is an important air pollutant. It is poisonous in high concentrations. The CO is produced by incomplete combustion, by char in the system, or as an intermediate combustion product. Hottel et al,"Combustion of Carbon Monoxide and Propane", 10th International Symposium on Combustion, Combustion Institute, Pittsburgh, Pa (1965) provided the following kinetic expression.

$$\frac{-dx_{CO}}{dt} = 12 \times 10^{10} \exp\left(\frac{-16,000}{1.99T}\right) x_{O_2}^{0.3} \; x_{CO} \; x_{H_2O}^{0.5} \cdot \left(\frac{P}{82.06 \; T}\right)^{1.8}$$

$$(8.27)$$

Where: x_{O_2} - mole fraction O_2

x_{CO} - "   "   CO

x_{H_2O} "   " H_2O

P - pressure in atm

T - $^\circ K$

Reviewing this expression it can be seen that

a) RATE is extremely sensitive to T
b) RATE is sensitive to pressure $(P)^{1.8}$
c) RATE is first order with respect to CO
d) RATE is not sensitive to $O_2 (x_{O_2})^{0.3}$
e) RATE is dependent on $H_2O \; (x_{H_2O})^{0.5}$

For dilute CO levels and for a constant temperature system the rate expression reduces to

$$\ln\frac{(x)_f}{(x)_i} = Kt \qquad\qquad (8.28)$$

Where: $(x)_f$ is final CO mole fraction

$(x)_i$ is initial CO mole fraction

$$K = 12 \times 10^{10} \exp\left(\frac{-16,000}{1.99T}\right) x_{O_2}^{0.3} \; x_{H_2O}^{0.5} \; \left(\frac{P}{82.06T}\right)^{1.8}$$

EXAMPLE 8.19 $470 \; m^3/min$ of a waste gas stream from a carbon char reactivation furnace is produced at 300°C. Its composition is:

Component	Mole percent
N_2	78.17
CO_2	18.71
CO	2.08
O_2	1.04
Total	100.0

The waste gas will be mixed with the flue gas from an atmospheric natural gas (methane) combustor operating at 50% excess air in a "fume incinerator" or after-burner. If CO combustion is assumed not to begin until the two gas streams are intimately mixed, plot the required residence time to reduce CO concentration to 10 ppm (neglecting dilution effects) against the mixed-gas temperature. Note that residence time is directly related to the incinerator volume (capital cost) and mixed-gas temperature to methane consumption rate (operating cost). For ease in computation, assume the heat capacity of the two gas streams are constant, identical, and equal to about 8 kcal/kg mol °C. Neglect heat release from CO combustion. Let the datum for sensible heat content be 15.5°C. The system diagram is shown below:

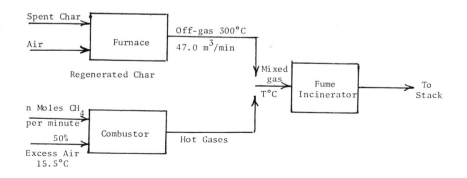

SOLUTION 8.19 For 1 mol of CH_4, at 50% excess air, the hot combustor
gas produced has the composition:

Component	Moles
N_2	11.29
CO_2	1.00
O_2	1.00
H_2O	2.00
Total	15.29

The molal flow rate of gases from the furnace is

$$470\left(\frac{273}{300+273}\right)\left(\frac{1}{22.4}\right)= 10 \text{ mol/min}$$

Basis: 1 min of operation with a methane firing rate
of n moles per minute.

Mixed Gas Temperature

1. Total heat content

    a. Sensible heat in furnace gas = $10 \overline{Mc_p}$
        $(300 - 15.5)= 22,760$ kcal

    b. Heat of combustion of methane = $(191,910)n$ kcal

2. Mixed-gas temperature (T °C)

$$T = \frac{\text{Total heat content}}{(\overline{Mc_p})(\text{no. moles})} + 15.5 = \frac{22,760+191,910n}{8(10+15.29n)} +15.5$$

Mixed Gas Composition

Component	Moles		
	From furnace	From methane	Total
N_2	7.82	11.29n	7.82 + 11.29n
CO_2	1.87	n	1.87 + n
O_2	0.10	n	0.1 + n
H_2O	0.0	2.0 n	2.0 n
CO	0.21	0	0.21
	10.0	15.29n	10 + 15.29n

Oxygen mole fraction $= f_{O_2}^* = \dfrac{0.1 + n}{10 + 15.29n}$

Water mole fraction $= f_{H_2O} = \dfrac{2n}{10 + 15.29n}$

CO Reaction Requirement

$(f_{CO})_i$ = Initial CO mole fraction $= 2.08 \times 10^{-2}$

$(f_{CO})_f$ = Final CO mole fraction $= 10^{-5}$

$\dfrac{(f_{CO})_f}{(f_{CO})_i} = \dfrac{10^{-5}}{2.08 \times 10^{-2}} = 4.81 \times 10^{-4} = \exp(-Kt)$

$Kt = -\ln(4.81 \times 10^{-4})$

$Kt = 7.64$

$t = \dfrac{7.64}{K}$ sec

Reaction Time and Methane Usage. From the relationships developed above, the reaction rate parameter K and the required residence time t can be calculated as a function of n, the methane consumption rate. Then, for a given value of n, and the associated total mole flow rate and absolute temperature (T'), the volume flow rate of the mixed gas $\dot{V}$ can be calculated:

$\dot{V} = (\text{mol/min}) \dfrac{22.4}{6} \dfrac{T'}{273}$ m^3/sec

The chamber volume V is then given by:

$V = \dot{V}t$ m^3

The calculations are shown in the Table below and the results are plotted

---

*Neglect, for simplicity, the change in O_2 concentration due to CO combustion and changes in the total moles in the system.

Calculations for Example 8.19					
n (Moles CH_4/min)	0.2	0.4	0.6	0.8	1.0
Total moles	13.06	16.12	19.17	22.23	25.29
Temperature (°C)	600	787	899	991	1076
T' (K)	874	1060	1172	1264	1350
f_{O_2}	0.023	0.031	0.037	0.040	0.043
f_{H_2O}	0.031	0.050	0.063	0.072	0.079
$K(sec^{-1})$	88	529	1187	2038	3095
$t(sec)$	0.087	0.014	0.006	0.004	0.002
$\dot{V}$ (m^3/sec)	15.6	23.4	30.7	38.4	46.7
V (m^3)	1.360[a]	0.337	0.198	0.144	0.115

[a]Note that if the oxygen mole fraction corresponding to complete combustion
of CO had been used (f_{O_2} = 0.09) instead of f_{O_2} = 0.023, the volume would
have increased to 1.95 m^3.

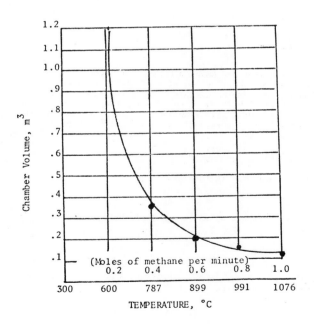

Several points become clear:

The strong influence of temperature, typical of kinetically controlled reactions

It would appear that around 650°C, the combustion rate of CO increases sharply. This could be considered an "ignition temperature" for the low oxygen concentration mixture existing in the afterburner. At normal atmospheric oxygen concentrations, the CO ignition temperature is approximately 600°C (note that for $f_{O_2} = 0.2$, the chamber volume is only about 0.7 m^3 at 600°C and a V-n curve such as shown in the figure would just be breaking sharply upward).

Example 8.19 is taken from "Combustion and Incineration Processes" by W.R. Niesson, Marbel Dekker (1978).

The formation of soot is undesirable from many points of view:

1. It represents an energy loss.

2. It is unsightly.

3. Swing from oxidizing to reducing conditions that are likely in systems producing soot leads to high corrosion of fire-side boiler tubes.

4. Lowers resistivity in electrostatic precipitation.

As described in section 3.2, the combustion of a solid depends upon both the chemical reaction rate and the diffusional rate of O_2.

One expression for the time it takes to burn up a solid soot sphere is

$$t_b (\text{sec}) = \frac{1}{pp_{O_2}} \left[\frac{d_o \ (\text{cm})}{0.13 \ \exp\left[\frac{-35,700}{R}\left(\frac{1}{T} - \frac{1}{1600}\right)\right]} + \frac{d_o^2 \ (\text{cm})^2}{8.67 \times 10^{-6} \ T^{0.75}} \right]$$

(Kinetic term)   (Diffusion term)

(8.29)

Where: pp_o^2 is partial pressure of oxygen.

The first term is a chemical reaction rate term and the second results from a diffusional resistance.

EXAMPLE 8.20   Determine the time it takes to burn out soot with diameters of 2, 20, 200 (μ) in a stream of gas at $1273^\circ K$ and a pp_{O_2} = 0.0172 atm.

SOLUTION 8.20 $t_b = \dfrac{1}{0.0172} \left[\dfrac{d_o}{0.13 \ \exp\left[\dfrac{-35,700}{1.99}\right](\dfrac{1}{1273} - \dfrac{1}{1600})} + \right.$

$$\left. \dfrac{d_o^2}{8.67 \times 10^{-6} \ T^{0.75}} \right]$$

$$t_b = 58 \left[\dfrac{d_o}{2.33} + \dfrac{d_o^2}{0.00185} \right]$$

μ	$\dfrac{58d_o}{2.33}$	$\dfrac{58d_o^2}{0.00185}$	t_b
2	0.0052	0.00125	0.0067
20	0.052	0.125	0.175
200	0.52	12.5	13.02

For the small size particle the first term controlled. In the case of the largest particle the diffusional term controlled.

8.4   THERMAL POLLUTION

The Laws of Thermodynamics have shown that any conversion of energy to work results in the need to discharge
wasted energy to a low temperature sink (our biosphere).
There is no way to avoid this dumping of heat. It is a
product of our industrial society. Eventually if man continues to use more and more energy and dump more and more
heat the effect of thermal pollution will limit man's
activities. The thermal pollution of our environment has
been termed "the ultimate pollution". If all other environmental pollution problems were solved and infinite sources
of energy become available (fusion), the use of energy would
remain finite, bounded by the need to dump heat into our
biosphere.   .

Just as the problem of chemical pollution was a spatial
problem and related to a concentration of activities the thermal
pollution problem is a local problem at present. The electrical
energy of a power plant is sent out into the electric grid
where it is degraded to heat by an electric consumer. This
portion of the energy is distributed.

Most of the energy from the power plant is lost as heat
in the area immediately adjacent to the generating facility.
The trend toward larger and larger power plant size provides for
larger and larger energy point sources. In the case of a fossil
(or nuclear) power plant this wasted energy is produced where
little existed before its construction.

In contrast hydroelectric power provides no new thermal
production, but rather a redistribution of thermal energy. The

water going over a dam converts potential energy to thermal
energy at the bottom of the dam. When a turbine is installed,
part of the energy is converted to electricity and distributed
in regions away from the dam. The amount released at the bottom
of the dam is reduced.

The difference between the burning of fuel for power and
hydroelectric power is that burning fuel releases new energy
into the biosphere while use of hydroelectric power only alters
the spatial distribution of energy.

In any consideration of thermal pollution two factors are
important:

(a) The total energy released.

(b) The spatial distribution of energy.

The trend in energy is not to distribute but to concentrate the
energy generation into small areas or point sources.

The energy sources most discussed to provide additional
energy contribute more to thermal pollution than the use of
fossil fuels. Table 8.18 provides the amount of energy that is
lost to the biosphere to produce 1,000 Mw of electricity.

TABLE 8.18
Heat Lost to Biosphere for 1000 MW Electrical Capacity
(Typical)

System	Efficiency	Heat Loss (MW)
Coal, Oil, Coal	40%	1,500
Nuclear	36%	1,777
Solar (Flat Plate)	1%	99,000
Ocean Gradient	3%	32,333
Solar (Focusing)	30%	2,333
Geothermal	20%	5,000

What can be seen from Table 8.18 is that nuclear power contributes
significantly more wasted heat than coal and oil. Geothermal
and ocean gradients provide for a substantial increase in
wasted energy.

It will be essential in any long-term future planning to
be able to assess the effect of discharging the heat to the
biosphere. There is not enough information known to make an
accurate assessment, but it is essential that the problem
be considered. The discussion that is provided below takes
a simplistic approach to a complex problem. It approaches the
problem in the manner often used in teaching economics. [Only
one factor will be varied and the effect noted "with all other
factors held constant."]

In Chapter 7 the energy balance for the earth biosphere
was developed. Under present conditions the amount of energy
released by man is a small fraction of the amount received
from the sun; but as the demand increases and new nations join
the industrial society, man's contribution will soon become
important.

Chapter 7 has already discussed the factors

(a) Solar radiation constant.

(b) The fraction of solar radiation penetrating the
     earth's atmosphere.

(c) The earth's albedo.

At the present time, any major disturbance on the sun's surface
(sun spots) would have a much larger effect upon the earth's
temperature than any of man's activities and is likely to
be a dominant factor.

The fraction of the solar energy reaching the earth's
surface is affected by the dust in the atmosphere and the
cloud cover. The dust particles tend to scatter (divert) the

solar radiation away from the earth. The same is true of clouds,
and it has been suggested that the water vapor trail resulting
from jet planes will influence the earth's energy balance and
the effect on the earth temperature.

A change in the earth albedo (the ratio of energy reflected
to energy absorbed) will influence the earth's equilibrium
temperature. The polar ice caps reflect almost all of the
energy reaching it. A melting of the ice cover to expose new
surfaces that will absorb more of the radiant energy will result in a warming of the surface. The cultivation of large
area of green plant life in desert areas will also change the
albedo. This affects the earth's temperature.

The earth loses its heat by radiation to space. This
radiation is infrared and CO_2 does not allow this radiation
to pass. Space becomes opaque to radiation rather than clear.
This means that CO_2 acts as a blanket and does not allow the
earth to lose heat at the same rate as it does without the
CO_2 present. This effect causes the temperature of the earth
to rise. Certainly a rapid increase in the amount of fossil
fuel burned could lead to a noticeable greenhouse effect.

The discussion that has been presented does not consider
that the effects described are not interconnected and are independent. It is not that simple. An incomplete picture can be
misleading. For example, if for some reason the solar constant would
increase, what might happen? Most assured the earth would warm-up,
and the radiation leaving the earth would increase. If only the
value of S were changed in the energy balance, the earth would heat
up until a new temperature would be established. It is not that
simple. The warmer temperature would result in a melting of the
ice cap. This would give a larger albedo. The earth will

absorb a greater fraction of solar energy, increase the earth's
temperature even more, melting more snow; thereby, absorbing more
energy and so on. The system is unstable and if no other factors
come in to play will result in destruction.

The point to be learned from this discussion is that the
earth is in a state of thermal equilibrium with the solar system.
Until recently the contribution of energy resulting from man's
activities has been minimal. But this is no longer the situation,
man is now beginning to contribute a significant fraction of
the energy provided by the sun. There is no assurance what
the total effects will be. It is a complex matter and is most
often ignored in making plans for future energy resource
utilization.   Alternatives such as geothermal, fission, fusion,
solar satelites all add energy to the biosphere and all are
ultimately limited by thermal energy release. The Second Law
has stated that the entropy of the world is increasing, and all
activities ultimately result in release of heat. New alternative
energy sources cannot get around this limitation.

The discussion above considered global averages. In many
localities the man-produced energy already exceeds the solar
flux. Table 8.19 shows values for several regions. The
average solar flux is 250 watt/m^2. This table shows that
Manhattan is already 2.5 times this value, the U.S. is about
0.1% of this value while the megalopolis stretching from
Boston to Washington is about 2% of the incident solar energy.

The higher levels in smaller regions are less important
first because the effect is dispersed by winds and natural
movement of the heat released to the atmosphere and second a
local effect will not have much influence on the overall
relationship between the earth and its thermal balance with
the sun.

TABLE **8.19**

Man-Generated Energy Flux (1971)

	Area (10^6 Km^2)	Flux Watts/M^2
Global average	500	0.016
Land surface	150	0.054
U.S.	7.8	0.24
Eastern U.S.	.9	1.1
U.S.S.R	22.4	0.05
Central Russia	0.3	0.85
Western Europe	1.7	0.74
West Germany	0.25	1.36
	Area (10^3 Km^2)	
Boston-Washington	87	4.4
Los Angeles County	10	7.5
Moscow	0.88	127
Manhattan	0.06	630

The question as to how much man's contribution can be tolerated is open to question. If one rather arbitrarily assumes "the rule of one," i.e., a one percent contribution from man represents the top limit and it is assumed that this may be off by a factor of ten, the flux that may be tolerated ranges between 2.5 to 0.025 watts/m^2. The total U.S. value is already in this range.

It has been pointed out many times that all thermal power cycles must have a low temperature energy sink to dump waste heat. In almost all cases rivers and streams provide these sinks. In fact, the term "thermal pollution" is usually used to describe the heating up of these water bodies by waste energy from power plants. The amount of heat that must be disposed of depends upon the plant size and plant efficiency. Modern fossil fuel plants will dump about 40% of the fuel

energy to water bodies and nuclear plants about 67%. This
increase in waste heat from nuclear plants to water bodies
is due to two factors.

    (a)   Lower thermal efficiency as a result of lower
          temperature power cycles.

    (b)   All the waste heat goes to cooling water in a nuclear
          plant whereas part of the heat is lost to stack gas
          in fossil fuel plants.

The effect of dumping heat into the rivers and streams
raising the temperature level will:

1. Result in fish kill. If temperatures are changed
   significantly from normal temperatures, fish will be
   killed. This depends upon the specie of fish and the
   abruptness of change.

2. Lower the oxygen level. An increase in temperature
   lowers the solubility of oxygen and increases the
   metabolic activity. The fish needs more oxygen but
   has less available.

3. Behavior patterns. Many behavior patterns such as
   migration, breeding, etc., are triggered by temperature
   changes that are altered.

The question remains as to the level of temperature that
may be tolerated. This is certainly site specific and depends
upon the specie of life in the river, their tolerance to
temperature, the size of area effected, the other activities
on the water body. To obtain some idea of the temperature
rise that may be tolerated the very general guidelines are:

    1-2°F Conservative ΔT

    3-5°F Moderate ΔT

    10°F   Liberal ΔT

These are for guidelines only. Each situation must be considered
separately.

EXAMPLE 8.21   The average discharge rate for all streams is
2×10^6 [ft^3/sec]. Estimate the temperature rise
of the water.

SOLUTION **8.21**

From Figure 1.6, 5.5 Q of electricity are consumed.
For every Btu of electrical energy produced, about
1.5 are dumped to condensor.

$q = 5.5[Q/yr] \times 10^{15}[Btu/Q] \times 1.5[Btu \text{ to condensor/Btu}$

$\text{electrical}] \times 1/8.76 \times 10^3[yr/hr] \times 1/3600[hr/sec]$

$= 2.6 \times 10^8 [Btu/sec]$

$q = 2.6 \times 10^8 = m\ C_p\Delta T = 2 \times 10^6[ft^3/sec] \times 62.4[lb/ft^3] \times \Delta T$

$\Delta T = 2.6 \times 10^8/1.25 \times 10^8 = 2.1[°F]\ (1.2°C)$

The value from Example 8.21 was based upon average flow and
generation rates. The peak electrical generation rates are
in the heat of summer when the water flow rates are minimal.
This would indicate that the temperature rise would far
exceed the value calculated. Add to this the fact that the
generating plants are not distributed on streams in the direct
ratio to the water flow, and it becomes clear that the thermal
pollution of rivers is at hand.

Fortunately, there are factors that are at work to minimize the temperature increase. The river loses heat by a
combination of radiation, evaporation, and conduction. When
the temperature rises, the amount of energy lost bv radiation
and the amount of energy lost by evaporation increases.

EXAMPLE **8.22**   Estimate the fraction of heat lost from a stream to air as a result
of increasing the temperature from 70 to 76°F. Assume that the
rate that water vaporized is evaporated is proportional to the
vapor pressure.

SOLUTION **8.22**

@ 70°F   v.p. = 0.36 psia
@ 76°F   v.p. = 0.47 psia
% increase =(0.47 - 0.36)/0.36 ⅄ 100 = 31%

This example shows that the amount of heat lost to the
air because of evaporation significantly increases with a

small change in temperature. Fortunately, the effect is one of self-regulation and the effect of the added heat is reduced.

The use of lakes for dumping heat is often used. The analysis of the heating of lakes is complicated by thermal gradients in lakes and where water is removed and returned to the lake. To obtain a feel for the magnitude of the temperature changes that might be expected the following equation estimates the effect of dumping heat into lakes:

$$T(°C) = \text{Heat Added } (\text{watts/m}^2)/22$$

EXAMPLE **8.23**   Estimate the mean temperature increase by adding heat equivalent to 5% of the thermal flux.

SOLUTION **8.23**

$$\Delta T(°C) = (0.05)(250)/22 = 0.6°C(1°F)$$

The discussion of thermal pollution was directed toward power plants where fuel is burned to produce energy. No mention was made of energy released for heat only, such as home heating or cooling. No mention was made of the vast amounts of heat that must be dissipated from large coal conversion systems. No mention was made of the contribution of the transportation industry.

EXAMPLE **8.24**   Estimate the heat loss that must be dissipated by a coal gasification plant to produce 250×10^9 SCF/Day synthetic gas plant. Plant 70% effective in converting coal to gas. Determine:

    (a) Flow rate sufficient to accept this heat and not increase by more than 3.0°C.

    (b) Lake surface area sufficient to accept this plant and not increase in temperature by more than 1°C.

SOLUTION **8.24**

(a) 250×10^6 [SCF/Day of CH_4] x 1000[Btu/SCF] x .3/.7[Btu(lost)/Btu(fuel)]

$= 1.1 \times 10^{11}$ [Btu/day]

$q = m\ C_p \Delta T = 1.1 \times 10^{11}$ [Btu/day]

$= m$ x 1[Btu/lb°F] x 3.0[°F]

$m = 3.67 \times 10^{10}$ [lbs/day] or 4.4×10^9 [gal/day] or 5.9×10^8 [ft^3/day]

or 6.82×10^3 [ft^3/sec]

(b) $q = 1.1 \times 10^{11}$ [Btu/day] x 1/3413[Kw-hr/Btu]

x 1/24[day/hr] x 1000[W/Kw] = 1/34x10^9 [watts]

$A = 1.34 \times 10^9/22 = 6.09 \times 10^7$ [m^2] or 230[mi]^2 or 1.47×10^5 [acre]

The problem of thermal pollution must ultimately be taken
into account. The amount of dust pollution and chemical
pollution that can be emitted may be controlled by man and
discharge limits may be set by man. Not so with thermal pollution. The amount of heat pumped into our biosphere is set
by the laws of thermodynamics, a law of nature, and we shall
ultimately learn to live with these laws.

A solution to our energy problem that receives much
attention is that the U.S. imbarked on a program similar to
the effort in space. There is one basic difference. The
laws of nature had predicted that a man could be put on the
moon. The only problem was technological. The laws of
nature tell us we cannot continue to dump more energy on our
biosphere and no amount of American ingenuity will change this.

Problems for Chapter 8.0

1. Howard K. Smith of ABC News in an editorial August 25, 1976, commented that the drought in Europe may be the result of shifting climatic patterns resulting from the release of man-produced energy. Estimate the increase in temperature that could result in these activities.

2. Compare the size of two scrubber units. One cleans up SO_2 from a flue gas resulting from combustion of coal (98% C, 2% S) in 50% excess air. The second first gasifies the solid. All S goes to H_2S.

$$C + H_2O \longrightarrow CO + H_2$$
$$S + H_2 \longrightarrow H_2S$$

The sulfur is removed and the gas then burns in 50% excess air.
   (a) Compare the volumes of gas in each case.
   (b) Determine the volume % of H_2S and SO_2.
   (c) Assuming 90% removal of S, compare the volumes of scrubber for each case. Assume water is used as scrubber media and water leaves 90% saturated.
   (d) Assuming 90% removal of S, compare the volumes of scrubber for each case. Assume a basic solution is used as scrubber media. For H_2S the partial pressure at equilibrium = 0.

3. The dust particles entering the atmosphere act as reflectors of solar energy. Assume that the dust level increases to a point where 7 units of energy rather than 6 units shown in Figure 7.5 are reflected. What effect would it have on the earth's temperature.

4. Assume that the fraction of "holes" in the thermal shield is inversely proportional to the concentration of CO_2. What would the effect of increasing the equilibrium concentration in the earth's atmosphere have on the earth's average temperature (assume 50% holes at present CO_2 levels)? Continuous increase in fossil fuel use will increase CO_2 levels.

5. The increased combustion of fossil fuel will result in an increase in the CO_2 level. This has the effect of increasing the amount of the earth's radiation that is absorbed in the atmosphere. Assume that the present CO_2 level provides for 50% of the radiation to be absorbed

(i.e., 50% holes, 50% cover) and the doubling of the
CO_2 level would halve the area of the holes that the
atmosphere presently provides. What change in temperature would the increase in CO_2 cause?

6.

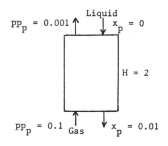

The diagram above shows a system for the removal of a
pollutant P from a gas stream. pp_p is the partial pressure of the pollutant in a gas stream. x_p is the mole
fraction of pollutant in a liquid stream. What would
happen to the volume of an absorber system if the pressure of the system were doubled? Doubling the pressure
doubles the partial pressure but has no effect on H.
Assume that the liquid is adjusted so that the concentrations in the liquid phase do not change and the effect on the value of K_g may be neglected.

7. The Bulletin of Atomic Scientists, February 1978, discussed the build-up of CO_2 from burning fossil fuel.
The level of CO_2 has grown gradually at a rate of about
4.3%/yr. In 1860 the CO_2 level was about 285 ppm and
is presently about 325 ppm. The article suggests levels
of 400 ppm by the year 2000 and 650 ppm in 2050. Using
the model provided in the text that explains the average
earth's temperature by assuming a shell that intercepts
50% of the radiation leaving the earth, determine the
earth's temperature in 2000 and 2050. Assume the fraction of holes to be inversely proportional to the CO_2
concentration.

8. The impact of strip-mining coal for a power plant is
to be determined. Consider an electrical power plant
(40% efficient) with an average output power of 1000 Mv.
Determine the quantity of coal necessary to operate the
plant for a 50-year period. Assume the coal is produced
by strip mining a vein with an average thickness of 30 ft
and a density 1.5 times that of water. Determine the land
area that would need to be stripped over the 50-year period.

9. See Chapter 3, Problem 15.

10. 1 ppm SO_2 = 2620 $\mu g/m^3$ at 25°C, 760 mm H_g

   1 ppm SO_2 = 2860 $\mu g/m^3$ at 0°C, 760 mm H_g

   Derive these relations and therefore a general conversion from ppm to $\mu g/m^3$

11. Consider a particle with an apparent density of 1.7 g/cm^3 falling through ambient air at 80°F (27°C). (a) Determine the terminal speed in feet per minute and centimeters per second for a 10 μ diameter particle. (b) What is the ratio of vertical distance to horizontal distance the particle travels if the horizontal air speed is 33 ft/s? (c) Determine the terminal speed in feet per minute and centimeters per second for a 1200 μ diameter particle. (c) Repeat part (b) for the 1200 μ particle.

12. A 1000-MW pulverized coal-fired steam power plant of 40 percent thermal efficiency uses coal with an ash content of 12 percent and a heating value of 26,700 kJ/kg. Assume that 50 percent of the ash goes up the stack as particulate in the flue gas. The particulate emission is controlled by an electrostatic precipitator which has the following removal efficiencies and weight distribution in the given size ranges.

Particle size, μ	0-5	5-10	10-20	20-40	>40
Efficiency, %	70	92.5	96	99	100
Wt., %	14	17	21	23	25

Determine the amount of fly ash emitted, in kilograms per second, with the flue gas.

13. A particulate sample in an air stream has the following weight percent distribution: 1 μ, 10 percent; 10 μ, 20 percent; 50 μ, 40 percent; 100 μ, 20 percent; 200 μ, 10 percent. A cyclone separator is employed on the basis of the following data: inlet width, 12 in.; effective turns, 5; inlet gas velocity, 20 ft/s; particle density, 1.6 g/cm^3; gas density, 0.074 lb/ft^3; gas viscosity, 0.045 lb/ft-hr. Using Figure 8.10 estimate the percentage of the total weight that would be removed in the cyclone separator.

14. Settling chambers are commonly used in a sinter plant to remove large particles of quartz and iron oxide from effluent gas streams. A settling chamber 3 m high and wide and 6 m long is available. The volumetric flow rate of gas through the chamber is 5000 m^3/hr. The densities of quartz and iron oxide particles are 2.6 and 4.5, respectively. Compute and plot efficiency curves for this unit at the given gas flow rate for both quartz and iron oxide particles as a function of particle diameter.

15. A gravity settling chamber is used to collect 70 μ particles having a specific gravity of 1.5. If the particles are carried in standard air, what is the maximum gas velocity that can be used in a collector 3 m high and 5 m long, neglecting recirculation effects within the chamber?

16. A particulate sample in a gas stream has the following particle size and weight distribution:

Avg. particle size, μ	1	2	5	10	15	40	60
Cumul, wt., %	2	14	50	80	92	99.5	99

Using Figure 8.10 as representative efficiency data for a collector, determine the over-all expected collection efficiency, in percent. Is the sample distributed log-normal?

17. Air at 150°F with a dust loading of 5 gr/ft^3 passes through a fabric filter with a superficial velocity of 3 ft/min (air-to-cloth ratio). The dust permeability is 2.50×10^{-11} ft^2, and the filter cake density averages 1.4 g/cm^3. (a) If the residual pressure drop across the cleaned cloth is 0.40 in. water, estimate the time required for the total pressure drop to reach 3.5 in. water, in hours, on the basis of Equation 8.20   (b) What is the value of the filter drag, S, in in. water/ft/min?

18. Air at 200°F with a dust loading of 10 gr/ft^3 passes through a fabric filter. After a period of 7 hr the total pressure drop across the filter is measured as 3.60 in. water. The filter cake density is 1.10 g/cm^3 and the residual pressure drop across the cleaned filter is 0.45 in. water. If the air velocity is maintained at 4 ft/min during the test, estimate the dust permeability, K_p, in units of (a) lb-in./in. water-min^2 and (b) square feet, on the bases of Equation

19. An effluent air stream contains 5.4 percent SO_2 by volume. It is desired to reduce this concentration to 1000 ppm by passing the gas stream through a packed tower before releasing the flow to the atmosphere. The inlet-gas flow rate is 1000 gt^3/min measured at 1 atm and 68°F, and the absorber is counterflow. The equilibrium of SO_2 in water is assumed to be reasinably represented by the relation $y = 30x$. Estimate the minimum amount of water required as a solvent, in cubic feet per minute.

20. A power plant of 1,000 megawatt electrical capacity is to run on residual fuel oil of 1.5 percent sulfur content. Estimate the NO_x and SO_x emissions per year from such a plant if the oil produces 18,000 Btu/lb and has a density of 1 g/cm^3. The plant is on line 90 percent of the time.

21. When burning Utah King mine coal (ultimate analysis: carbon 72.3 percent wt; hydrogen 5.8 percent wt; nitrogen 1.3 percent wt; sulfur 0.5 percent wt; oxygen 14.9 percent wt; ash 5.2 percent wt), what will be the ppm (dry) SO_2 in the stack at 3 percent O_2 in the flue gas? What will be the ppm NO_x as NO if we assume all the chemically bound nitrogen is converted to NO? Calculate these emissions in lb/million Btu. The heating value of the coal is 12,000 Btu/lb.

22. The effluent from a drying oven at 475°K contains 75 ppm carbon monoxide and 1750 ppm hydrocarbons. After the effluent is heated to a temperature of 960°K, the conversion efficiency based solely on hydrocarbons in 98 percent; based on CO and HC it is 90 percent. (a) What is the CO content at the afterburner outlet, in parts per million? (b) What is the CO content in parts per million at the outlet if the CO-HC efficiency is 92 percent at 1000°K?

23. A semi-anthracite coal has the following ultimate analysis of basic elements in percent by weight: sulfur, 0.6; hydrogen, 3.7; carbon, 79.5; nitrogen, 0.9; oxygen, 4.7; and an ash content of 10.6 percent. The coal is burned with 20 percent excess air beyond that required to change S, H, and C to SO_2, H_2O, and CO_2. If (a) 50 percent and (b) 20 percent of the bound nitrogen is converted to NO, and none of the atmospheric nitrogen is oxidized to NO, what would be the parts per million of NO in the flue gas?

24. Estimate by means of the emission factor table in this chapter the NO_x emissions, in tons, from the burning of waste for a city of 100,000 population. A municipal incinerator without controls will be used. Assume that each person generates 5 lb of waste per day, and that industrial sources within the city generate another 2.5 lb of waste per capita per day.

25. The quantities of various exhaust products of internal combustion can be estimated on the basis of chemical equilibrium considerations. In particular, one can estimate the mole fractions of CO, NO, and NO_2 at equilibrium under different conditions. (a) Compute the equilibrium mole fraction of CO in a system of CO, CO_2, and O_2 at 1 atm pressure at 2000, 3000, and 4000°K according to

$$CO_2 \rightleftarrows CO + \frac{1}{2} O_2$$

under conditions in which the ratio of the number of
oxygen atoms to the number of carbon atoms is 2, 3.125
(stoichiometric combustion of octane), and 5. The
equilibrium constant is

$$K = 3 \times 10^4 e^{-67,000/RT}$$

(b) Compute the equilibrium mole fraction of NO in a system
of N_2, O_2, and NO at 1 atm pressure at 2000, 3000, and
4000°K according to

$$\frac{1}{2}N_2 + \frac{1}{2}O_2 \underset{\leftarrow}{\rightarrow} NO$$

under conditions in which the ratio of the number of nitrogen
atoms to the number of oxygen atoms is 4 (air) and 40 (combustion flue gases at 10 percent excess air).
(c) It is of interest to estimate the quantity of NO_2 that
can be formed during combustion. Determine the mole fraction
of NO_2 that is attained at equilibrium in the system

$$\frac{1}{2}N_2 + \frac{1}{2}O_2 \underset{\leftarrow}{\rightarrow} NO$$

$$NO + \frac{1}{2}O_2 \underset{\leftarrow}{\rightarrow} NO_2$$

at 25, 1000, and 1600°C both in air and in flue gas of composition
3.3 percent O_2 and 76 percent N_2. The equilibrium constant for
the second reaction is

$$K = 2.5 \times 10^{-4} e^{13,720/RT}$$

26. Small quantities of NO_2 can be formed in combustion exhaust gases
by the third-order reaction

$$2NO + O_2 \xrightarrow{\quad k \quad} 2NO_2$$

You wish to estimate the amount of NO_2 that can be formed by
this route under conditions typical of those in the exhause
of the automobile. Assume that the exhaust system of a car
can be represented as a straight pipe through which the
exhaust gases flow in so-called plug flow (each element
travels through the pipe independently of the other elements).
Assume that the concentration of NO at the beginning of the
exhaust system is 2000 ppm. For initial O_2 concentrations
of 10^2, 10^3, and 10^4 ppm, compute the concentration of NO_2
formed for a residence time of 2 sec if the temperature of
the exhaust gases (a) is constant at 800°C; (b) is constant
at 300°C; (c) decreases linearly from 800 to 300°C.

The following rate constant data are available:

T°C	k, 1^2 mole^{-2} sec^{-1}
0	17.9 x 10^3
40	13.1 x 10^3
200	6.6 x 10^3
390	5.1 x 10^3

9.0 ECONOMICS, POLITICS, AND SOCIAL IMPLICATIONS

Any discussion of economics in this text has been avoided. Avoided because it would be obsolete before the text reaches the market. Avoided because the simplistic approach that would have been provided would be inadequate and misleading.

If this book were written two years ago, almost all energy costs would have ranged less than $1.00/10^6$ Btu; today the range could go upward to $3.00/10^6$ and in the next decade values of $6-\$10/10^6$ have been projected. The change is so rapid and the forces for change so unpredictable that any analysis would not likely be meaningful.

If an economic analysis were attempted it would have been limited to engineering economics. The engineer in his economics only measures the cost of goods, materials, and the time value of money. They become quite proficient at determining the dollars that would be returned on a certain dollar investment. This is certainly of interest to the company management and the stockholders.

The economics of the engineer is not sufficient to deal

with the problems that affect national security, international
diplomacy, balance of payments, depletion of natural non-
renewable resources, effects of new industry on the social
structure. None of these factors even enter into the analysis.

The engineering analysis would show that the larger
you build a process the lower the per unit cost of the product
and the higher the output per man. These are the types of
analysis provided by the engineer. The concept is, in the
limit self-destructive, because a plant size approaching
infinity would produce an infinite amount of a product at
zero price and require few employees (per unit of product).
Of course to keep up full employment, we would need an
infinite number of these plants.

It may be shown that on a "reasonable" scale that increasing size will reduce unit costs. All of the discussion
on thermal pollution, air pollution, etc., indicate that the
problem is one of concentration of activity. Many smaller
plants are more advantageous than a single large plant. For
almost all industrial products, the size is limited. There
is only a limited market for the product, and the plant size
is restricted.

In the case of energy the need for energy is so large
that the normal constraints do not apply. Almost any amount
of energy that can be generated can be consumed provided the
price is reasonable. Therefore, the general approach in
attacking the energy problem is to construct super size
plants. This follows the trend of super size tankers, super
size jets, super size power generators, super sized cities,
etc. It has already been learned that super sized tankers
cause super sized problems on the environment when an accident
occurs. One can contemplate the super sized tragedy when
super sized jets collide. We have learned to some extent

what it means when super sized cities are mismanaged. One
may legitimately ask what might occur from an accident in a
super sized coal conversion complex or in a super sized
nuclear energy park. The economics of the engineer do not
factor this into his return on investment.

Almost the entire thrust in approaching the energy problem is toward achieving "economies of size," large nuclear
plants, large gasification plants, large energy farms. The
place in the country for the small business man, the entre-
preneur is to be reduced. Independence and creativity will
be decreased as man becomes an employee of a super sized
corporation over which he has no control. Instead of a con-
troller, most will become controlees. Consider the 35 mile
by 35 mile energy plantation. This is not a "mom and pop"
operation. The family farm and the free farmer is replaced
by the company employee and union. It is these mom and pop
farms that produce our doctors, lawyers, leaders to a greater
extent than do factory employees. Is this change of social
structure to be considered?

It is a difficult task to attempt to compare energy costs
because they are not free market costs and subject to arbitrary
changes. Electric utility costs, natural gas prices, and
imported oil prices are controlled. New gas costs more than
old gas. What about the oil depletion allowance? How does
the U.S. price natural resources that are being depleted and
are not?

It would seem that the energy prices may be more dependent
upon a decision by one (or a few men) over which the U.S. has
little control than any other single factor. Couple this
with the fact that the U.S. has no long range energy policy,
the future remains ominous. It appears the better part of
valor to retreat from any effort to provide costs or economics

associated with such a complex interconnecting maze of intrigue.

For this reason the text has dealt with implications imposed on any solution by the conservation of mass and laws of thermodynamics. What ever the future policies may be, these laws will remain invariant.

APPENDIX A

THERMODYNAMIC DATA

TABLE 1
Saturation Temperatures for Steam

NOTE: The following steam tables and Mollier chart have been
abstracted from Thermodynamic Properties of Steam (Copyright,
1937, by Keenan and Keyes) published by John Wiley & Sons, Inc.
Reference to this book is necessary for values outside the range
of the abstract and for other data.

Temp. F t	Abs. Pressure Lb. Sq. In. P	Specific Volume Sat. Liquid V_f	Evap. V_{fg}	Sat. Vapor V_g	Enthalpy Sat. Liquid h_f	Evap. h_{fg}	Sat. Vapor h_g	Entropy Sat. Liquid s_f	Evap. s_{fg}	Sat. Vapor s_g	Temp. F t
32	0.08854	0.01602	3306	3306	0.00	1075.8	1075.8	0.0000	2.1877	2.1877	32
35	0.09995	0.01602	2947	2947	3.02	1074.1	1077.1	0.0061	2.1709	2.1770	35
40	0.12170	0.01602	2444	2444	8.05	1071.3	1079.3	0.0162	2.1435	2.1597	40
45	0.14752	0.01602	2036.4	2036.4	13.06	1068.4	1081.5	0.0262	2.1167	2.1429	45
50	0.17811	0.01603	1703.2	1703.2	18.07	1065.6	1083.7	0.0361	2.0903	2.1264	50
60	0.2563	0.01604	1206.6	1206.7	28.06	1059.9	1088.0	0.0555	2.0393	2.0948	60
70	0.3631	0.01606	867.8	867.9	38.04	1054.3	1092.3	0.0745	1.9902	2.0647	70
80	0.5069	0.01608	633.1	633.1	48.02	1048.6	1096.6	0.0932	1.9428	2.0360	80
90	0.6982	0.01610	468.0	468.0	57.99	1042.9	1100.9	0.1115	1.8972	2.0087	90
100	0.9492	0.01613	350.3	350.4	67.97	1037.2	1105.2	0.1295	1.8531	1.9826	100
110	1.2748	0.01617	265.3	265.4	77.94	1031.6	1109.5	0.1471	1.8106	1.9577	110
120	1.6924	0.01620	203.25	203.27	87.92	1025.8	1113.7	0.1645	1.7694	1.9339	120
130	2.2225	0.01625	157.32	157.34	97.90	1020.0	1117.9	0.1816	1.7296	1.9112	130
140	2.8886	0.01629	122.99	123.01	107.89	1014.1	1122.0	0.1984	1.6910	1.8894	140
150	3.718	0.01634	97.06	97.07	117.89	1008.2	1126.1	0.2149	1.6537	1.8685	150
160	4.741	0.01639	77.27	77.29	127.89	1002.3	1130.2	0.2311	1.6174	1.8485	160
170	5.992	0.01645	62.04	62.06	137.90	996.3	1134.2	0.2472	1.5822	1.8293	170
180	7.510	0.01651	50.21	50.23	147.92	990.2	1138.1	0.2630	1.5480	1.8109	180
190	9.339	0.01657	40.94	40.96	157.95	984.1	1142.0	0.2785	1.5147	1.7932	190
200	11.526	0.01663	33.62	33.64	167.99	977.9	1145.9	0.2938	1.4824	1.7762	200
210	14.123	0.01670	27.80	27.82	178.05	971.6	1149.7	0.3090	1.4508	1.7598	210
212	14.696	0.01672	26.78	26.80	180.07	970.3	1150.4	0.3120	1.4446	1.7566	212
220	17.186	0.01677	23.13	23.15	188.13	965.2	1153.4	0.3239	1.4301	1.7440	220
230	20.780	0.01684	19.365	19.382	198.23	958.8	1157.0	0.3387	1.3901	1.7288	230
240	24.969	0.01692	16.306	16.323	208.34	952.2	1160.5	0.3531	1.3609	1.7140	240
250	29.825	0.01700	13.804	13.821	218.48	945.5	1164.0	0.3675	1.3323	1.6998	250
260	35.429	0.01709	11.746	11.763	228.64	938.7	1167.3	0.3817	1.3043	1.6860	260
270	41.858	0.01717	10.044	10.061	238.84	931.8	1170.6	0.3958	1.2769	1.6727	270
280	49.203	0.01726	8.628	8.645	249.06	924.7	1173.8	0.4096	1.2501	1.6597	280
290	57.556	0.01735	7.444	7.461	259.31	917.5	1176.8	0.4234	1.2238	1.6472	290
300	67.013	0.01745	6.449	6.466	269.59	910.1	1179.7	0.4369	1.1980	1.6350	300
320	89.66	0.01765	4.896	4.914	290.28	894.9	1185.2	0.4637	1.1478	1.6115	320
340	118.01	0.01787	3.770	3.788	311.13	879.0	1190.1	0.4900	1.0992	1.5891	340
360	153.04	0.01811	2.939	2.957	332.18	862.2	1194.4	0.5158	1.0519	1.5677	360
380	195.77	0.01836	2.317	2.335	353.45	844.6	1198.1	0.5413	1.0059	1.5471	380
400	247.31	0.01864	1.8447	1.8633	374.97	826.0	1201.0	0.5664	0.9608	1.5272	400
420	308.83	0.01894	1.4811	1.5000	396.77	806.3	1203.1	0.5912	0.9166	1.5078	420
440	381.59	0.01926	1.1979	1.2171	418.90	785.4	1204.3	0.6158	0.8730	1.4887	440
460	466.9	0.0196	0.9748	0.9944	441.7	763.2	1204.6	0.6402	0.8298	1.4700	460
480	566.1	0.0200	0.7972	0.8172	464.4	739.4	1203.7	0.6645	0.7868	1.4513	480
500	680.8	0.0204	0.6545	0.6749	487.8	713.9	1201.7	0.6887	0.7438	1.4325	500
520	812.4	0.0209	0.5385	0.5594	511.9	686.4	1198.2	0.7130	0.7006	1.4136	520
540	962.5	0.0215	0.4434	0.4649	536.6	656.6	1193.2	0.7374	0.6568	1.3942	540
560	1133.1	0.0221	0.3647	0.3868	562.2	624.2	1186.4	0.7621	0.6121	1.3742	560
580	1325.8	0.0228	0.2989	0.3217	588.9	588.4	1177.3	0.7872	0.5659	1.3532	580
600	1542.9	0.0236	0.2432	0.2668	617.0	548.5	1165.5	0.8131	0.5176	1.3307	600
620	1786.6	0.0247	0.1955	0.2201	646.7	503.6	1150.3	0.8398	0.4664	1.3062	620
640	2059.7	0.0260	0.1538	0.1798	678.6	452.0	1130.5	0.8679	0.4110	1.2789	640
660	2365.4	0.0278	0.1165	0.1442	714.2	390.2	1104.4	0.8987	0.3485	1.2472	660
680	2708.1	0.0305	0.0810	0.1115	757.3	309.9	1067.2	0.9351	0.2719	1.2071	680
700	3093.7	0.0369	0.0392	0.0761	823.3	172.1	995.4	0.9905	0.1484	1.1389	700
705.4	3206.2	0.0503	0	0.0503	902.7	0	902.7	1.0580	0	1.0580	705.4

A-1

TABLE 2
Saturation Pressures for Steam

Abs. Press. Lb. Sq. In. p	Temp. F t	Specific Volume		Enthalpy			Entropy			Internal Energy			Abs. Press. Lb. Sq. In. p
		Sat. Liquid V_f	Sat. Vapor V_g	Sat. Liquid h_f	Evap. h_{fg}	Sat. Vapor h_g	Sat. Liquid S_f	Evap. S_{fg}	Sat. Vapor S_g	Sat. Liquid U_f	Evap. U_{fg}	Sat. Vapor U_g	
1.0	101.74	0.01614	333.6	69.70	1036.3	1106.0	0.1326	1.8456	1.9782	69.70	974.6	1044.3	1.0
2.0	126.08	0.01623	173.73	93.99	1022.2	1116.2	0.1749	1.7451	1.9200	93.98	957.9	1051.9	2.0
3.0	141.48	0.01630	118.71	109.37	1013.2	1122.6	0.2008	1.6855	1.8863	109.36	947.3	1056.7	3.0
4.0	152.97	0.01636	90.63	120.86	1006.4	1127.3	0.2198	1.6427	1.8625	120.85	939.3	1060.2	4.0
5.0	162.24	0.01640	73.52	130.13	1001.0	1131.1	0.2347	1.6094	1.8441	130.12	933.0	1063.1	5.0
6.0	170.06	0.01645	61.98	137.96	996.2	1134.2	0.2472	1.5820	1.8292	137.94	927.5	1065.4	6.0
7.0	176.85	0.01649	53.64	144.76	992.1	1136.9	0.2581	1.5586	1.8167	144.74	922.7	1067.4	7.0
8.0	182.86	0.01653	47.34	150.79	988.5	1139.3	0.2674	1.5383	1.8057	150.77	918.4	1069.2	8.0
9.0	188.28	0.01656	42.40	156.22	985.2	1141.4	0.2759	1.5203	1.7962	156.19	914.6	1070.8	9.0
10	193.21	0.01659	38.42	161.17	982.1	1143.3	0.2835	1.5041	1.7876	161.14	911.1	1072.2	10
14.696	212.00	0.01672	26.80	180.07	970.3	1150.4	0.3120	1.4446	1.7566	180.02	897.5	1077.5	14.696
15	213.03	0.01672	26.29	181.11	969.7	1150.8	0.3135	1.4415	1.7549	181.06	896.7	1077.8	15
20	227.96	0.01683	20.089	196.16	960.1	1156.3	0.3356	1.3962	1.7319	196.10	885.8	1081.9	20
30	250.33	0.01701	13.746	218.82	945.3	1164.1	0.3680	1.3313	1.6993	218.73	869.1	1087.8	30
40	267.25	0.01715	10.498	236.03	933.7	1169.7	0.3919	1.2844	1.6763	235.90	856.1	1092.0	40
50	281.01	0.01727	8.515	250.09	924.0	1174.1	0.4110	1.2474	1.6585	249.93	845.4	1095.3	50
60	292.71	0.01738	7.175	262.09	915.5	1177.6	0.4270	1.2168	1.6438	261.90	836.0	1097.9	60
70	302.92	0.01748	6.206	272.61	907.9	1180.6	0.4409	1.1906	1.6315	272.38	827.8	1100.2	70
80	312.03	0.01757	5.472	282.02	901.1	1183.1	0.4531	1.1676	1.6207	281.76	820.3	1102.1	80
90	320.27	0.01766	4.896	290.56	894.7	1185.3	0.4641	1.1471	1.6112	290.27	813.4	1103.7	90
100	327.81	0.01774	4.432	298.40	888.8	1187.2	0.4740	1.1286	1.6026	298.08	807.1	1105.2	100
120	341.25	0.01789	3.728	312.44	877.9	1190.4	0.4916	1.0962	1.5878	312.05	795.6	1107.6	120
140	353.02	0.01802	3.220	324.82	868.2	1193.0	0.5069	1.0682	1.5751	324.35	785.2	1109.6	140
160	363.53	0.01815	2.834	335.93	859.2	1195.1	0.5204	1.0436	1.5640	335.39	775.8	1111.2	160
180	373.06	0.01827	2.532	346.03	850.8	1196.9	0.5325	1.0217	1.5542	345.42	767.1	1112.5	180
200	381.79	0.01839	2.288	355.36	843.0	1198.4	0.5435	1.0018	1.5453	354.68	759.0	1113.7	200
250	400.95	0.01865	1.8438	376.00	825.1	1201.1	0.5675	0.9588	1.5263	375.14	740.7	1115.8	250
300	417.33	0.01890	1.5433	393.84	809.0	1202.8	0.5879	0.9225	1.5104	392.79	724.3	1117.1	300
350	431.72	0.01913	1.3260	409.69	794.2	1203.9	0.6056	0.8910	1.4966	408.45	709.6	1118.0	350
400	444.59	0.0193	1.1613	424.0	780.5	1204.5	0.6214	0.8630	1.4844	422.6	695.9	1118.5	400
450	456.28	0.0195	1.0320	437.2	767.4	1204.6	0.6356	0.8378	1.4734	435.5	683.2	1118.7	450
500	467.01	0.0197	0.9278	449.4	755.0	1204.4	0.6487	0.8147	1.4634	447.6	671.0	1118.6	500
550	476.94	0.0199	0.8424	460.8	743.1	1203.9	0.6608	0.7934	1.4542	458.8	659.4	1118.2	550
600	486.21	0.0201	0.7698	471.6	731.6	1203.2	0.6720	0.7734	1.4454	469.4	648.3	1117.7	600
700	503.10	0.0205	0.6554	491.5	709.7	1201.2	0.6925	0.7371	1.4296	488.8	627.5	1116.3	700
800	518.23	0.0209	0.5687	509.7	688.9	1198.6	0.7108	0.7045	1.4153	506.6	607.8	1114.4	800
900	531.98	0.0212	0.5006	526.6	668.8	1195.4	0.7275	0.6744	1.4020	523.1	589.0	1112.1	900
1000	544.61	0.0216	0.4456	542.4	649.4	1191.8	0.7430	0.6467	1.3897	538.4	571.0	1109.4	1000
1100	556.31	0.0220	0.4001	557.4	630.4	1187.8	0.7575	0.6205	1.3780	552.9	553.5	1106.4	1100
1200	567.22	0.0223	0.3619	571.7	611.7	1183.4	0.7711	0.5956	1.3667	566.7	536.3	1103.0	1200
1300	577.46	0.0227	0.3293	585.4	593.2	1178.6	0.7840	0.5719	1.3559	580.0	519.4	1099.4	1300
1400	587.10	0.0231	0.3012	598.7	574.7	1173.4	0.7963	0.5491	1.3454	592.7	502.7	1095.4	1400
1500	596.23	0.0235	0.2765	611.6	556.3	1167.9	0.8082	0.5269	1.3351	605.1	486.1	1091.2	1500
2000	635.82	0.0257	0.1878	671.7	463.4	1135.1	0.8619	0.4230	1.2849	662.2	403.4	1065.6	2000
2500	668.13	0.0287	0.1307	730.6	360.5	1091.1	0.9126	0.3197	1.2322	717.3	313.3	1030.6	2500
3000	695.36	0.0346	0.0858	802.5	217.8	1020.3	0.9731	0.1885	1.1615	783.4	189.3	972.7	3000
3206.2	705.40	0.0503	0.0503	902.7	0	902.7	1.0580	0	1.0580	872.9	0	872.9	3206.2

A-2

TABLE 3
Superheated Vapor

Abs. Press. Lb./Sq. In. (Sat Temp)		Temperature F 200	300	400	500	600	700	800	900	1000	1200	1400
1 (101.74)	v	392.6	452.3	512.0	571.6	631.2	690.8	750.4	809.9	869.5	988.7	1107.8
	h	1150.4	1195.8	1241.7	1288.3	1335.7	1383.8	1432.8	1482.7	1533.5	1637.7	1745.7
	s	2.0512	2.1153	2.1720	2.2233	2.2702	2.3137	2.3542	2.3923	2.4283	2.4952	2.5566
5 (162.24)	v	78.16	90.25	102.26	114.22	126.16	138.10	150.03	161.95	173.87	197.71	221.6
	h	1148.8	1195.0	1241.2	1288.0	1335.4	1383.6	1432.7	1482.6	1533.4	1637.7	1745.7
	s	1.8718	1.9370	1.9942	2.0456	2.0927	2.1361	2.1767	2.2148	2.2509	2.3178	2.3792
10 (193.21)	v	38.85	45.00	51.04	57.05	63.03	69.01	74.98	80.95	86.92	98.84	110.77
	h	1146.6	1193.9	1240.6	1287.5	1335.1	1383.4	1432.5	1482.4	1533.2	1637.6	1745.6
	s	1.7927	1.8595	1.9172	1.9689	2.0160	2.0596	2.1002	2.1383	2.1744	2.2413	2.3028
14.696 (212.00)	v		30.53	34.68	38.78	42.86	46.94	51.00	55.07	59.13	67.25	75.37
	h		1192.8	1239.9	1287.1	1334.8	1383.2	1432.3	1482.3	1533.1	1637.5	1745.5
	s		1.8160	1.8743	1.9261	1.9734	2.0170	2.0576	2.0958	2.1319	2.1989	2.2603
20 (227.96)	v		22.36	25.43	28.46	31.47	34.47	37.46	40.45	43.44	49.41	55.37
	h		1191.6	1239.2	1286.6	1334.4	1382.9	1432.1	1482.1	1533.0	1637.4	1745.4
	s		1.7808	1.8396	1.8918	1.9392	1.9829	2.0235	2.0618	2.0978	2.1648	2.2263
40 (267.25)	v		11.040	12.628	14.168	15.688	17.198	18.702	20.20	21.70	24.69	27.68
	h		1186.8	1236.5	1284.8	1333.1	1381.9	1413.3	1481.4	1532.4	1637.0	1745.1
	s		1.6994	1.7608	1.8140	1.8619	1.9058	1.9467	1.9850	2.0212	2.0883	2.1498
60 (292.71)	v		7.259	8.357	9.403	10.427	11.441	12.449	13.452	14.454	16.451	18.446
	h		1181.6	1233.6	1283.0	1331.8	1380.9	1430.5	1480.8	1531.9	1636.6	1744.8
	s		1.6492	1.7135	1.7678	1.8162	1.8605	1.9015	1.9400	1.9762	2.0434	2.1049
80 (312.03)	v			6.220	7.020	7.797	8.562	9.322	10.077	10.830	12.332	13.830
	h			1230.7	1281.1	1330.5	1379.9	1429.7	1480.1	1531.3	1636.2	1744.5
	s			1.6791	1.7346	1.7836	1.8281	1.8694	1.9079	1.9442	2.0115	2.0731
100 (327.81)	v			4.937	5.589	6.218	6.835	7.446	8.052	8.656	9.860	11.060
	h			1227.6	1279.1	1329.1	1378.9	1428.9	1479.5	1530.8	1635.7	1744.2
	s			1.6518	1.7085	1.7581	1.8029	1.8443	1.8829	1.9193	1.9867	2.0484
120 (341.25)	v			4.081	4.636	5.165	5.683	6.195	6.702	7.207	8.212	9.214
	h			1224.4	1277.2	1327.7	1377.8	1428.1	1478.8	1530.2	1635.3	1743.9
	s			1.6287	1.6869	1.7370	1.7822	1.8237	1.8625	1.8990	1.9664	2.0281
140 (353.02)	v			3.468	3.954	4.413	4.861	5.301	5.738	6.172	7.035	7.895
	h			1221.1	1275.2	1326.4	1376.8	1427.3	1478.2	1529.7	1634.9	1743.5
	s			1.6087	1.6683	1.7190	1.7645	1.8063	1.8451	1.8817	1.9493	2.0110
160 (363.53)	v			3.008	3.443	3.849	4.244	4.631	5.015	5.396	6.152	6.906
	h			1217.6	1273.1	1325.0	1375.7	1426.4	1477.5	1529.1	1634.5	1743.2
	s			1.5908	1.6519	1.7033	1.7491	1.7911	1.8301	1.8667	1.9344	1.9962
180 (373.06)	v			2.649	3.044	3.411	3.764	4.110	4.452	4.792	5.466	6.136
	h			1214.0	1271.0	1323.5	1374.7	1425.6	1476.8	1528.6	1634.1	1742.9
	s			1.5745	1.6373	1.6894	1.7355	1.7776	1.8167	1.8534	1.9212	1.9831
200 (381.79)	v			2.361	2.726	3.060	3.380	3.693	4.002	4.309	4.917	5.521
	h			1210.3	1268.9	1322.1	1373.6	1424.8	1476.2	1528.0	1633.7	1742.6
	s			1.5594	1.6240	1.6767	1.7232	1.7655	1.8048	1.8415	1.9094	1.9713
220 (389.86)	v			2.125	2.465	2.772	3.066	3.352	3.634	3.913	4.467	5.017
	h			1206.5	1266.7	1320.7	1372.6	1424.0	1475.5	1527.5	1633.3	1742.3
	s			1.5453	1.6117	1.6652	1.7120	1.7545	1.7939	1.8308	1.8987	1.9607
240 (397.37)	v			1.9276	2.247	2.533	2.804	3.068	3.327	3.584	4.093	4.597
	h			1202.5	1264.5	1319.2	1371.5	1423.2	1474.8	1526.9	1632.9	1742.0
	s			1.5319	1.6003	1.6546	1.7017	1.7444	1.7839	1.8209	1.8889	1.9510

A-3

TABLE 3 (Continued)
Superheated Vapor

Abs. Press. Lb./Sq. In. (Sat Temp)		Temperature F							
		500	600	700	800	900	1000	1200	1400	1600
260 (404.42)	v	2.063	2.330	2.582	2.827	3.067	3.305	3.776	4.242	4.707
	h	1262.3	1317.7	1370.4	1422.3	1474.2	1526.3	1632.5	1741.7	1854.2
	s	1.5897	1.6447	1.6922	1.7352	1.7748	1.8118	1.8799	1.9420	1.9995
280 (411.05)	v	1.9047	2.156	2.392	2.621	2.845	3.066	3.504	3.938	4.370
	h	1260.0	1316.2	1369.4	1421.9	1473.5	1525.8	1632.1	1741.5	1854.0
	s	1.5796	1.6354	1.6834	1.7265	1.7662	1.8033	1.8716	1.9337	1.9912
300 (417.33)	v	1.7675	2.005	2.227	2.442	2.652	2.859	3.269	3.674	4.078
	h	1257.6	1314.7	1368.3	1420.6	1472.8	1525.2	1631.7	1741.0	1853.7
	s	1.5701	1.6268	1.6751	1.7184	1.7582	1.7954	1.8638	1.9260	1.9835
350 (431.72)	v	1.4923	1.7036	1.8980	2.084	2.266	2.445	2.798	3.147	3.493
	h	1251.5	1310.9	1365.5	1418.5	1471.1	1523.8	1630.7	1740.3	1853.1
	s	1.5481	1.6070	1.6563	1.7002	1.7403	1.7777	1.8463	1.9086	1.9663
400 (444.59)	v	1.2851	1.4770	1.6508	1.8161	1.9767	2.134	2.445	2.751	3.055
	h	1245.1	1306.9	1362.7	1416.4	1469.4	1522.4	1629.6	1739.5	1852.5
	s	1.5281	1.5894	1.6398	1.6842	1.7247	1.7623	1.8311	1.8936	1.9513
450 (456.28)	v	1.1231	1.3005	1.4584	1.6074	1.7516	1.8928	2.170	2.443	2.714
	h	1238.4	1302.8	1359.9	1414.3	1467.7	1521.0	1628.6	1738.7	1851.9
	s	1.5095	1.5735	1.6250	1.6699	1.7108	1.7486	1.8177	1.8803	1.9381
500 (467.01)	v	0.9927	1.1591	1.3044	1.4405	1.5715	1.6996	1.9504	2.197	2.442
	h	1231.3	1298.6	1357.0	1412.1	1466.0	1519.6	1627.6	1737.9	1851.3
	s	1.4919	1.5588	1.6115	1.6571	1.6982	1.7363	1.8056	1.8683	1.9262
550 (476.94)	v	0.8852	1.0431	1.1783	1.3038	1.4241	1.5414	1.7706	1.9957	2.219
	h	1223.7	1294.3	1354.0	1409.9	1464.3	1518.2	1626.6	1737.1	1850.6
	s	1.4751	1.5451	1.5991	1.6452	1.6868	1.7250	1.7946	1.8575	1.9155
600 (486.21)	v	0.7947	0.9463	1.0732	1.1899	1.3013	1.4096	1.6208	1.8279	2.033
	h	1215.7	1289.9	1351.1	1407.7	1462.5	1516.7	1625.5	1736.3	1850.0
	s	1.4586	1.5323	1.5875	1.6343	1.6762	1.7174	1.7846	1.8476	1.9056
700 (503.10)	v		0.7934	0.9077	1.0108	1.1082	1.2024	1.3853	1.5641	1.7405
	h		1280.6	1345.0	1403.2	1459.0	1513.9	1623.5	1734.8	1848.8
	s		1.5084	1.5665	1.6147	1.6573	1.6963	1.7666	1.8299	1.8881
800 (518.23)	v		0.6779	0.7833	0.8763	0.9633	1.0470	1.2088	1.3662	1.5214
	h		1270.7	1338.6	1398.6	1455.4	1511.0	1621.4	1733.2	1847.5
	s		1.4863	1.5476	1.5972	1.6407	1.6801	1.7510	1.8146	1.8729
900 (531.98)	v		0.5873	0.6863	0.7716	0.8506	0.9262	1.0714	1.2124	1.3509
	n		1260.1	1332.1	1393.9	1451.8	1508.1	1619.3	1731.6	1846.3
	s		1.4653	1.5303	1.5814	1.6257	1.6656	1.7371	1.8009	1.8595
1000 (544.61)	v		0.5140	0.6084	0.6878	0.7604	0.8294	0.9615	1.0893	1.2146
	h		1248.8	1325.3	1389.2	1448.2	1505.1	1617.3	1730.0	1845.0
	s		1.4450	1.5141	1.5670	1.6121	1.6525	1.7245	1.7886	1.8474
1100 (556.31)	v		0.4532	0.5445	0.6191	0.6866	0.7503	0.8716	0.9885	1.1031
	h		1236.7	1318.3	1384.3	1444.5	1502.2	1615.2	1728.4	1843.8
	s		1.4251	1.4989	1.5535	1.5995	1.6405	1.7130	1.7775	1.8363
1200 (567.22)	v		0.4016	0.4909	0.5617	0.6250	0.6843	0.7967	0.9046	1.0101
	h		1223.5	1311.0	1379.3	1440.7	1499.2	1613.1	1726.9	1842.5
	s		1.4052	1.4843	1.5409	1.5879	1.6293	1.7025	1.7672	1.8263
1400 (587.10)	v		0.3174	0.4062	0.4714	0.5281	0.5805	0.6789	0.7727	0.8640
	h		1193.0	1295.5	1369.1	1433.1	1493.2	1608.9	1723.7	1840.0
	s		1.3639	1.4567	1.5177	1.5666	1.6093	1.6836	1.7489	1.8083

A-4

TABLE 3 (Continued)
Superheated Vapor

Abs. Press. Lb./Sq. In. (Sat Temp)		Temperature F							
		600	700	800	900	1000	1200	1400	1600
1600 (604.90)	v		0.3417	0.4034	0.4553	0.5027	0.5906	0.6738	0.7545
	h		1278.7	1358.4	1425.3	1487.0	1604.6	1720.5	1837.5
	s		1.4303	1.4964	1.5476	1.5914	1.6669	1.7328	1.7926
1800 (621.03)	v		0.2907	0.3502	0.3986	0.4421	0.5218	0.5968	0.6693
	h		1260.3	1347.2	1417.4	1480.8	1600.4	1717.3	1835.0
	s		1.4044	1.4765	1.5301	1.5752	1.6520	1.7185	1.7786
2000 (635.82)	v		0.2489	0.3074	0.3532	0.3935	0.4668	0.5352	0.6011
	h		1240.0	1335.5	1409.2	1474.5	1596.1	1714.1	1832.5
	s		1.3783	1.4576	1.5139	1.5603	1.6384	1.7055	1.7660
2500 (668.13)	v		0.1686	0.2294	0.2710	0.3061	0.3678	0.4244	0.4784
	h		1176.8	1303.6	1387.8	1458.4	1585.3	1706.1	1826.2
	s		1.3073	1.4127	1.4772	1.5273	1.6088	1.6775	1.7389
3000 (695.36)	v		0.0984	0.1760	0.2159	0.2476	0.3018	0.3505	0.3966
	h		1060.7	1267.2	1365.0	1441.8	1574.3	1698.0	1819.9
	s		1.1966	1.3690	1.4439	1.4984	1.5837	1.6540	1.7163
3206.2 (705.40)	v			0.1583	0.1918	0.2288	0.2806	0.3267	0.3703
	h			1250.5	1355.2	1434.7	1569.8	1694.6	1817.2
	s			1.3508	1.4309	1.4874	1.5742	1.6452	1.7080
3500	v		0.0306	0.1364	0.1762	0.2058	0.2546	0.2977	0.3381
	h		780.5	1224.9	1340.7	1424.5	1563.3	1689.8	1813.6
	s		0.9515	1.3241	1.4127	1.4723	1.5615	1.6336	1.6968
4000	v		0.0287	0.1052	0.1462	0.1743	0.2192	0.2581	0.2943
	h		763.8	1174.8	1314.4	1406.8	1552.1	1681.7	1807.2
	s		0.9347	1.2757	1.3827	1.4482	1.5417	1.6154	1.6795
4500	v		0.0276	0.0798	0.1226	0.1500	0.1917	0.2273	0.2602
	h		753.5	1113.9	1286.5	1388.4	1540.8	1673.5	1800.9
	s		0.9235	1.2204	1.3529	1.4253	1.5235	1.5990	1.6640
5000	v		0.0268	0.0593	0.1036	0.1303	0.1696	0.2027	0.2329
	h		746.4	1047.1	1256.5	1369.5	1529.5	1665.3	1794.5
	s		0.9152	1.1622	1.3231	1.4034	1.5066	1.5839	1.6499
5500	v		0.0262	0.0463	0.0880	0.1143	0.1516	0.1825	0.2106
	h		741.3	985.0	1224.1	1349.3	1518.2	1657.0	1788.1
	s		0.9090	1.1093	1.2930	1.3821	1.4908	1.5699	1.6369

Mollier Chart for Steam

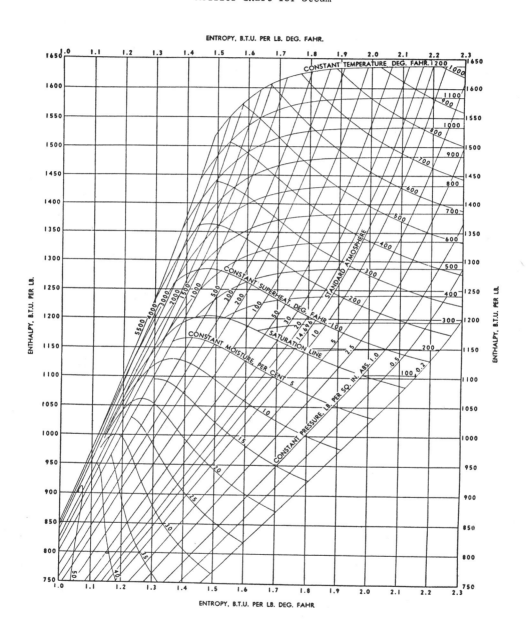

A-6

TABLE 4 $\bar{H}_T$ (J/kg-mole) $\times 10^{-3}$ For Common Gases (T in °K)

Temperature	CO_2	C_2H_4	Air	SO_2	NH_3	SO_3	C_2H_6	NO	H_2O	N_2	CO	O_2	H_2	CH_4
4.00×10^2	4.04×10^3	5.05×10^3	2.98×10^3	4.26×10^3	3.77×10^3	5.68×10^3	6.09×10^3	3.04×10^3	3.44×10^3	2.96×10^3	2.97×10^3	3.03×10^3	2.96×10^3	3.88×10^3
5.00×10^2	8.31×10^3	1.09×10^4	5.96×10^3	8.73×10^3	7.79×10^3	1.18×10^4	1.33×10^4	6.09×10^3	6.93×10^3	5.92×10^3	5.94×10^3	6.10×10^3	5.87×10^3	8.31×10^3
6.00×10^2	1.29×10^4	1.76×10^4	9.00×10^3	1.35×10^4	1.21×10^4	1.85×10^4	2.16×10^4	9.21×10^3	1.05×10^4	8.93×10^3	8.99×10^3	9.27×10^3	8.80×10^3	1.33×10^4
7.00×10^2	1.77×10^4	2.50×10^4	1.21×10^4	1.84×10^4	1.67×10^4	2.57×10^4	3.10×10^4	1.24×10^4	1.42×10^4	1.20×10^4	1.21×10^4	1.25×10^4	1.17×10^4	1.89×10^4
8.00×10^2	2.27×10^4	3.31×10^4	1.53×10^4	2.36×10^4	2.16×10^4	3.32×10^4	4.13×10^4	1.56×10^4	1.81×10^4	1.51×10^4	1.53×10^4	1.58×10^4	1.47×10^4	2.49×10^4
9.00×10^2	2.79×10^4	4.18×10^4	1.85×10^4	2.89×10^4	2.67×10^4	4.11×10^4	5.25×10^4	1.89×10^4	2.20×10^4	1.83×10^4	1.85×10^4	1.92×10^4	1.77×10^4	3.15×10^4
1.00×10^3	3.33×10^4	5.11×10^4	2.18×10^4	3.43×10^4	3.21×10^4	4.94×10^4	6.45×10^4	2.23×10^4	2.61×10^4	2.15×10^4	2.18×10^4	2.27×10^4	2.07×10^4	3.84×10^4
1.10×10^3	3.89×10^4	6.08×10^4	2.51×10^4	3.99×10^4	3.77×10^4	5.78×10^4	7.71×10^4	2.57×10^4	3.02×10^4	2.48×10^4	2.51×10^4	2.61×10^4	2.37×10^4	4.58×10^4
1.20×10^3	4.45×10^4	7.09×10^4	2.85×10^4	4.55×10^4	4.35×10^4	6.65×10^4	9.03×10^4	2.92×10^4	3.45×10^4	2.81×10^4	2.85×10^4	2.97×10^4	2.68×10^4	5.35×10^4
1.30×10^3	5.02×10^4	8.14×10^4	3.19×10^4	5.12×10^4	4.95×10^4	7.53×10^4	1.04×10^5	3.27×10^4	3.89×10^4	3.15×10^4	3.19×10^4	3.32×10^4	2.99×10^4	6.16×10^4
1.40×10^3	5.60×10^4	9.20×10^4	3.54×10^4	5.69×10^4	5.57×10^4	8.41×10^4	1.18×10^5	3.62×10^4	4.34×10^4	3.50×10^4	3.54×10^4	3.68×10^4	3.31×10^4	6.99×10^4
1.50×10^3	6.18×10^4	1.03×10^5	3.89×10^4	6.25×10^4	6.21×10^4	9.30×10^4	1.33×10^5	3.98×10^4	4.81×10^4	3.85×10^4	3.89×10^4	4.04×10^4	3.63×10^4	7.85×10^4
1.60×10^3	6.76×10^4	1.14×10^5	4.24×10^4	6.81×10^4	6.86×10^4	1.02×10^5	1.47×10^5	4.34×10^4	5.28×10^4	4.20×10^4	4.25×10^4	4.40×10^4	3.95×10^4	8.73×10^4
1.70×10^3	7.34×10^4	1.25×10^5	4.60×10^4	7.36×10^4	7.53×10^4	1.11×10^5	1.62×10^5	4.71×10^4	5.77×10^4	4.56×10^4	4.61×10^4	4.76×10^4	4.29×10^4	9.62×10^4
1.80×10^3	7.91×10^4	1.36×10^5	4.96×10^4	7.90×10^4	8.21×10^4	1.19×10^5	1.77×10^5	5.07×10^4	6.27×10^4	4.92×10^4	4.97×10^4	5.12×10^4	4.62×10^4	1.05×10^5
1.90×10^3	8.47×10^4	1.47×10^5	5.33×10^4	8.42×10^4	8.89×10^4	1.28×10^5	1.91×10^5	5.44×10^4	6.78×10^4	5.29×10^4	5.34×10^4	5.48×10^4	4.97×10^4	1.14×10^5
2.00×10^3	9.02×10^4	1.57×10^5	5.70×10^4	8.92×10^4	9.59×10^4	1.36×10^5	2.06×10^5	5.81×10^4	7.30×10^4	5.66×10^4	5.71×10^4	5.83×10^4	5.32×10^4	1.24×10^5
2.10×10^3	9.55×10^4	1.68×10^5	6.07×10^4	9.40×10^4	1.03×10^5	1.44×10^5	2.20×10^5	6.19×10^4	7.83×10^4	6.04×10^4	6.08×10^4	6.18×10^4	5.68×10^4	1.33×10^5
2.20×10^3	1.01×10^5	1.78×10^5	6.44×10^4	9.86×10^4	1.10×10^5	1.51×10^5	2.34×10^5	6.56×10^4	8.38×10^4	6.42×10^4	6.46×10^4	6.53×10^4	6.04×10^4	1.42×10^5
2.30×10^3	1.06×10^5	1.88×10^5	6.82×10^4	1.03×10^5	1.17×10^5	1.58×10^5	2.48×10^5	6.94×10^4	8.93×10^4	6.81×10^4	6.84×10^4	6.87×10^4	6.41×10^4	1.51×10^5
2.40×10^3	1.10×10^5	1.97×10^5	7.20×10^4	1.07×10^5	1.25×10^5	1.65×10^5	2.61×10^5	7.31×10^4	9.50×10^4	7.20×10^4	7.22×10^4	7.20×10^4	6.80×10^4	1.60×10^5
2.50×10^3	1.15×10^5	2.06×10^5	7.58×10^4	1.11×10^5	1.32×10^5	1.71×10^5	2.73×10^5	7.69×10^4	1.01×10^5	7.59×10^4	7.61×10^4	7.54×10^4	7.19×10^4	1.69×10^5
2.60×10^3	1.19×10^5	2.14×10^5	7.96×10^4	1.14×10^5	1.39×10^5	1.76×10^5	2.85×10^5	8.07×10^4	1.07×10^5	7.99×10^4	7.99×10^4	7.86×10^4	7.59×10^4	1.78×10^5
2.70×10^3	1.23×10^5	2.21×10^5	8.35×10^4	1.17×10^5	1.46×10^5	1.81×10^5	2.96×10^5	8.44×10^4	1.13×10^5	8.40×10^4	8.38×10^4	8.17×10^4	7.99×10^4	1.87×10^5
2.80×10^3	1.26×10^5	2.28×10^5	8.74×10^4	1.19×10^5	1.53×10^5	1.85×10^5	3.06×10^5	8.82×10^4	1.19×10^5	8.81×10^4	8.77×10^4	8.47×10^4	8.41×10^4	1.95×10^5
2.90×10^3	1.30×10^5	2.34×10^5	9.21×10^4	1.21×10^5	-	1.89×10^5	3.15×10^5	9.19×10^4	1.25×10^5	9.22×10^4	9.16×10^4	8.76×10^4	8.84×10^4	2.03×10^5

TABLE 5 $\bar{S}_T$ [J/kg-mole°K] x 10^{-3} For Common Gases (T in °K)

Temperature	CO_2	C_2H_4	Air	SO_2	NH_3	SO_3	C_2H_6	NO	H_2O	N_2	CO	O_2	H_2	CH_4
4.00×10^2	1.16×10^1	1.45×10^1	8.58×10^0	1.23×10^1	1.09×10^1	1.64×10^1	1.75×10^1	8.76×10^0	9.93×10^0	8.54×10^0	8.56×10^0	8.74×10^0	8.55×10^0	1.11×10^1
5.00×10^2	2.12×10^1	2.75×10^1	1.52×10^1	2.22×10^1	1.98×10^1	3.00×10^1	3.35×10^1	1.56×10^1	1.77×10^1	1.51×10^1	1.52×10^1	1.56×10^1	1.50×10^1	2.10×10^1
6.00×10^2	2.95×10^1	3.97×10^1	2.08×10^1	3.08×10^1	2.77×10^1	4.22×10^1	4.87×10^1	2.12×10^1	2.45×10^1	2.06×10^1	2.07×10^1	2.14×10^1	2.04×10^1	3.01×10^1
7.00×10^2	3.69×10^1	5.11×10^1	2.56×10^1	3.85×10^1	3.48×10^1	5.33×10^1	6.31×10^1	2.62×10^1	3.00×10^1	2.54×10^1	2.55×10^1	2.64×10^1	2.49×10^1	3.87×10^1
8.00×10^2	4.36×10^1	6.19×10^1	2.98×10^1	4.54×10^1	4.13×10^1	6.33×10^1	7.69×10^1	3.05×10^1	3.51×10^1	2.95×10^1	2.97×10^1	3.08×10^1	2.89×10^1	4.68×10^1
9.00×10^2	4.97×10^1	7.22×10^1	3.36×10^1	5.16×10^1	4.73×10^1	7.27×10^1	9.01×10^1	3.44×10^1	3.97×10^1	3.33×10^1	3.35×10^1	3.48×10^1	3.24×10^1	5.45×10^1
1.00×10^3	5.54×10^1	8.19×10^1	3.70×10^1	5.74×10^1	5.30×10^1	8.13×10^1	1.03×10^2	3.79×10^1	4.40×10^1	3.67×10^1	3.70×10^1	3.84×10^1	3.55×10^1	6.18×10^1
1.10×10^3	6.07×10^1	9.12×10^1	4.02×10^1	6.27×10^1	5.83×10^1	8.94×10^1	1.15×10^2	4.12×10^1	4.80×10^1	3.98×10^1	4.02×10^1	4.17×10^1	3.84×10^1	6.88×10^1
1.20×10^3	6.56×10^1	10.0×10^1	4.31×10^1	6.76×10^1	6.34×10^1	9.69×10^1	1.26×10^2	4.42×10^1	5.17×10^1	4.27×10^1	4.31×10^1	4.48×10^1	4.11×10^1	7.55×10^1
1.30×10^3	7.02×10^1	1.08×10^2	4.59×10^1	7.21×10^1	6.82×10^1	1.04×10^2	1.37×10^2	4.70×10^1	5.52×10^1	4.54×10^1	4.59×10^1	4.76×10^1	4.36×10^1	8.20×10^1
1.40×10^3	7.45×10^1	1.16×10^2	4.85×10^1	7.63×10^1	7.28×10^1	1.10×10^2	1.48×10^2	4.96×10^1	5.86×10^1	4.80×10^1	4.84×10^1	5.03×10^1	4.59×10^1	8.81×10^1
1.50×10^3	7.85×10^1	1.24×10^2	5.09×10^1	8.02×10^1	7.72×10^1	1.17×10^2	1.58×10^2	5.21×10^1	6.18×10^1	5.04×10^1	5.09×10^1	5.28×10^1	4.82×10^1	9.41×10^1
1.60×10^3	8.22×10^1	1.31×10^2	5.32×10^1	8.38×10^1	8.14×10^1	1.22×10^2	1.67×10^2	5.44×10^1	6.48×10^1	5.26×10^1	5.32×10^1	5.51×10^1	5.03×10^1	9.97×10^1
1.70×10^3	8.57×10^1	1.38×10^2	5.53×10^1	8.71×10^1	8.54×10^1	1.28×10^2	1.76×10^2	5.66×10^1	6.78×10^1	5.48×10^1	5.54×10^1	5.73×10^1	5.23×10^1	1.05×10^2
1.80×10^3	8.90×10^1	1.44×10^2	5.74×10^1	9.02×10^1	8.93×10^1	1.33×10^2	1.84×10^2	5.87×10^1	7.06×10^1	5.69×10^1	5.74×10^1	5.93×10^1	5.42×10^1	1.10×10^2
1.90×10^3	9.20×10^1	1.50×10^2	5.94×10^1	9.30×10^1	9.30×10^1	1.37×10^2	1.92×10^2	6.07×10^1	7.34×10^1	5.89×10^1	5.94×10^1	6.13×10^1	5.61×10^1	1.15×10^2
2.00×10^3	9.48×10^1	1.55×10^2	6.13×10^1	9.56×10^1	9.66×10^1	1.41×10^2	2.00×10^2	6.26×10^1	7.61×10^1	6.08×10^1	6.13×10^1	6.31×10^1	5.79×10^1	1.20×10^2
2.10×10^3	9.74×10^1	1.60×10^2	6.31×10^1	9.80×10^1	1.00×10^2	1.45×10^2	2.07×10^2	6.44×10^1	7.87×10^1	6.26×10^1	6.32×10^1	6.48×10^1	5.96×10^1	1.25×10^2
2.20×10^3	9.98×10^1	1.65×10^2	6.48×10^1	1.00×10^2	1.03×10^2	1.49×10^2	2.13×10^2	6.62×10^1	8.12×10^1	6.44×10^1	6.49×10^1	6.64×10^1	6.13×10^1	1.29×10^2
2.30×10^3	1.02×10^2	1.69×10^2	6.65×10^1	1.02×10^2	1.07×10^2	1.52×10^2	2.19×10^2	6.79×10^1	8.37×10^1	6.61×10^1	6.66×10^1	6.79×10^1	6.30×10^1	1.33×10^2
2.40×10^3	1.04×10^2	1.73×10^2	6.81×10^1	1.04×10^2	1.10×10^2	1.55×10^2	2.25×10^2	6.95×10^1	8.61×10^1	6.78×10^1	6.82×10^1	6.94×10^1	6.46×10^1	1.37×10^2
2.50×10^3	1.06×10^2	1.77×10^2	6.97×10^1	1.05×10^2	1.13×10^2	1.57×10^2	2.30×10^2	7.10×10^1	8.85×10^1	6.94×10^1	6.98×10^1	7.07×10^1	6.62×10^1	1.40×10^2
2.60×10^3	1.08×10^2	1.80×10^2	7.12×10^1	1.06×10^2	1.15×10^2	1.59×10^2	2.34×10^2	7.25×10^1	9.08×10^1	7.10×10^1	7.13×10^1	7.20×10^1	6.77×10^1	1.44×10^2
2.70×10^3	1.09×10^2	1.83×10^2	7.26×10^1	1.08×10^2	1.18×10^2	1.61×10^2	2.39×10^2	7.39×10^1	9.31×10^1	7.25×10^1	7.28×10^1	7.31×10^1	6.93×10^1	1.47×10^2
2.80×10^3	1.10×10^2	1.85×10^2	7.40×10^1	1.08×10^2	1.21×10^2	1.63×10^2	2.42×10^2	7.52×10^1	9.53×10^1	7.40×10^1	7.42×10^1	7.42×10^1	7.08×10^1	1.50×10^2
2.90×10^3	1.11×10^2	1.87×10^2	7.54×10^1	1.09×10^2	-	1.64×10^2	2.45×10^2	7.66×10^1	9.75×10^1	7.54×10^1	7.56×10^1	7.53×10^1	7.23×10^1	1.53×10^2

TABLE 6 $\bar{B}_T$ [J/Kg-mole] $\times 10^{-3}$ For Common Gases (T in °K)

Temperature	CO_2	C_2H_4	Air	SO_2	NH_3	SO_3	C_2H_6	NO	H_2O	N_2	CO	O_2	H_2	CH_4
4.00×10^2	5.73×10^2	7.27×10^2	4.18×10^2	6.03×10^2	5.35×10^2	8.09×10^2	8.80×10^2	4.27×10^2	4.84×10^2	4.16×10^2	4.17×10^2	4.27×10^2	4.15×10^2	5.57×10^2
5.00×10^2	2.01×10^3	2.70×10^3	1.42×10^3	2.10×10^3	1.88×10^3	2.88×10^3	3.31×10^3	1.45×10^3	1.65×10^3	1.41×10^3	1.42×10^3	1.46×10^3	1.39×10^3	2.05×10^3
6.00×10^2	4.09×10^3	5.75×10^3	2.81×10^3	4.27×10^3	3.86×10^3	5.93×10^3	7.12×10^3	2.87×10^3	3.30×10^3	2.78×10^3	2.80×10^3	2.90×10^3	2.73×10^3	4.34×10^3
7.00×10^2	6.69×10^3	9.77×10^3	4.49×10^3	6.95×10^3	6.35×10^3	9.80×10^3	1.22×10^4	4.60×10^3	5.31×10^3	4.44×10^3	4.48×10^3	4.66×10^3	4.31×10^3	7.35×10^3
8.00×10^2	9.72×10^3	1.47×10^4	6.39×10^3	1.01×10^4	9.29×10^3	1.44×10^4	1.84×10^4	6.55×10^3	7.61×10^3	6.32×10^3	6.39×10^3	6.65×10^3	6.10×10^3	1.10×10^4
9.00×10^2	1.31×10^4	2.03×10^4	8.48×10^3	1.35×10^4	1.26×10^4	1.95×10^4	2.57×10^4	8.70×10^3	1.02×10^4	8.39×10^3	8.48×10^3	8.85×10^3	8.03×10^3	1.52×10^4
1.00×10^3	1.68×10^4	2.67×10^4	1.07×10^4	1.72×10^4	1.63×10^4	2.51×10^4	3.39×10^4	1.10×10^4	1.30×10^4	1.06×10^4	1.07×10^4	1.12×10^4	1.01×10^4	2.00×10^4
1.10×10^3	2.08×10^4	3.36×10^4	1.31×10^4	2.12×10^4	2.03×10^4	3.12×10^4	4.29×10^4	1.31×10^4	1.59×10^4	1.30×10^4	1.31×10^4	1.37×10^4	1.23×10^4	2.53×10^4
1.20×10^3	2.50×10^4	4.11×10^4	1.56×10^4	2.54×10^4	2.46×10^4	3.76×10^4	5.27×10^4	1.60×10^4	1.91×10^4	1.54×10^4	1.56×10^4	1.63×10^4	1.46×10^4	3.10×10^4
1.30×10^3	2.93×10^4	4.91×10^4	1.82×10^4	2.97×10^4	2.92×10^4	4.43×10^4	6.31×10^4	1.87×10^4	2.25×10^4	1.80×10^4	1.82×10^4	1.90×10^4	1.69×10^4	3.71×10^4
1.40×10^3	3.38×10^4	5.74×10^4	2.09×10^4	3.41×10^4	3.40×10^4	5.12×10^4	7.41×10^4	2.14×10^4	2.60×10^4	2.07×10^4	2.09×10^4	2.18×10^4	1.94×10^4	4.36×10^4
1.50×10^3	3.84×10^4	6.60×10^4	2.37×10^4	3.86×10^4	3.91×10^4	5.82×10^4	8.56×10^4	2.43×10^4	2.97×10^4	2.34×10^4	2.37×10^4	2.47×10^4	2.19×10^4	5.04×10^4
1.60×10^3	4.31×10^4	7.49×10^4	2.66×10^4	4.31×10^4	4.44×10^4	6.54×10^4	9.74×10^4	2.72×10^4	3.35×10^4	2.63×10^4	2.66×10^4	2.76×10^4	2.46×10^4	5.75×10^4
1.70×10^3	4.78×10^4	8.39×10^4	2.95×10^4	4.76×10^4	4.98×10^4	7.26×10^4	1.09×10^5	3.02×10^4	3.75×10^4	2.92×10^4	2.96×10^4	3.05×10^4	2.73×10^4	6.49×10^4
1.80×10^3	5.26×10^4	9.30×10^4	3.25×10^4	5.21×10^4	5.54×10^4	7.97×10^4	1.22×10^5	3.32×10^4	4.16×10^4	3.23×10^4	3.26×10^4	3.35×10^4	3.01×10^4	7.24×10^4
1.90×10^3	5.73×10^4	1.02×10^5	3.56×10^4	5.65×10^4	6.12×10^4	8.68×10^4	1.34×10^5	3.63×10^4	4.59×10^4	3.53×10^4	3.57×10^4	3.65×10^4	3.30×10^4	8.01×10^4
2.00×10^3	6.19×10^4	1.11×10^5	3.87×10^4	6.07×10^4	6.71×10^4	9.37×10^4	1.46×10^5	3.95×10^4	5.03×10^4	3.85×10^4	3.88×10^4	3.95×10^4	3.59×10^4	8.79×10^4
2.10×10^3	6.65×10^4	1.20×10^5	4.19×10^4	6.48×10^4	7.32×10^4	1.00×10^5	1.59×10^5	4.27×10^4	5.49×10^4	4.17×10^4	4.21×10^4	4.25×10^4	3.90×10^4	9.58×10^4
2.20×10^3	7.10×10^4	1.29×10^5	4.51×10^4	6.88×10^4	7.93×10^4	1.07×10^5	1.71×10^5	4.59×10^4	5.96×10^4	4.50×10^4	4.53×10^4	4.55×10^4	4.21×10^4	1.04×10^5
2.30×10^3	7.53×10^4	1.37×10^5	4.84×10^4	7.25×10^4	8.56×10^4	1.13×10^5	1.82×10^5	4.92×10^4	6.44×10^4	4.84×10^4	4.86×10^4	4.85×10^4	4.54×10^4	1.12×10^5
2.40×10^3	7.94×10^4	1.45×10^5	5.17×10^4	7.60×10^4	9.19×10^4	1.19×10^5	1.94×10^5	5.21×10^4	6.94×10^4	5.18×10^4	5.19×10^4	5.14×10^4	4.87×10^4	1.20×10^5
2.50×10^3	8.33×10^4	1.53×10^5	5.50×10^4	7.92×10^4	9.82×10^4	1.24×10^5	2.05×10^5	5.57×10^4	7.45×10^4	5.53×10^4	5.53×10^4	5.43×10^4	5.21×10^4	1.28×10^5
2.60×10^3	8.70×10^4	1.60×10^5	5.84×10^4	8.21×10^4	1.05×10^5	1.29×10^5	2.15×10^5	5.91×10^4	7.97×10^4	5.88×10^4	5.87×10^4	5.71×10^4	5.57×10^4	1.35×10^5
2.70×10^3	9.04×10^4	1.67×10^5	6.18×10^4	8.46×10^4	1.11×10^5	1.33×10^5	2.25×10^5	6.21×10^4	8.51×10^4	6.24×10^4	6.21×10^4	5.99×10^4	5.93×10^4	1.43×10^5
2.80×10^3	9.36×10^4	1.73×10^5	6.53×10^4	8.68×10^4	1.17×10^5	1.37×10^5	2.34×10^5	6.57×10^4	9.06×10^4	6.60×10^4	6.56×10^4	6.26×10^4	6.30×10^4	1.50×10^5
2.90×10^3	9.64×10^4	1.78×10^5	6.88×10^4	8.85×10^4	-	1.40×10^5	2.42×10^5	6.91×10^4	9.62×10^4	6.97×10^4	6.91×10^4	6.52×10^4	6.69×10^4	1.58×10^5

TABLE 7 $\bar{C}_p$ [J/Kg-mole] x 10^{-3} For Evaluation of $\bar{H}_T$ by $\bar{H}_T = C_p(T_2 - 298)$ (T in °K)

Temperature	CO_2	C_2H_4	Air	SO_2	NH_3	SO_3	C_2H_6	NO	H_2O	N_2	CO	O_2	H_2	CH_4
4.00×10^2	3.96×10^1	4.95×10^1	2.92×10^1	4.18×10^1	3.70×10^1	5.57×10^1	5.97×10^1	2.98×10^1	3.38×10^1	2.90×10^1	2.91×10^1	2.97×10^1	2.90×10^1	3.80×10^1
5.00×10^2	4.12×10^1	5.40×10^1	2.95×10^1	4.32×10^1	3.86×10^1	5.86×10^1	6.58×10^1	3.01×10^1	3.43×10^1	2.93×10^1	2.94×10^1	3.02×10^1	2.91×10^1	4.11×10^1
6.00×10^2	4.26×10^1	5.82×10^1	2.98×10^1	4.46×10^1	4.01×10^1	6.13×10^1	7.16×10^1	3.05×10^1	3.49×10^1	2.96×10^1	2.98×10^1	3.07×10^1	2.91×10^1	4.41×10^1
7.00×10^2	4.40×10^1	6.22×10^1	3.01×10^1	4.58×10^1	4.16×10^1	6.39×10^1	7.71×10^1	3.08×10^1	3.54×10^1	2.99×10^1	3.01×10^1	3.11×10^1	2.92×10^1	4.70×10^1
8.00×10^2	4.52×10^1	6.60×10^1	3.04×10^1	4.70×10^1	4.30×10^1	6.62×10^1	8.23×10^1	3.11×10^1	3.60×10^1	3.01×10^1	3.04×10^1	3.15×10^1	2.93×10^1	4.97×10^1
9.00×10^2	4.64×10^1	6.95×10^1	3.07×10^1	4.80×10^1	4.44×10^1	6.83×10^1	8.72×10^1	3.15×10^1	3.66×10^1	3.04×10^1	3.07×10^1	3.19×10^1	2.94×10^1	5.23×10^1
1.00×10^3	4.75×10^1	7.28×10^1	3.10×10^1	4.89×10^1	4.57×10^1	7.03×10^1	9.18×10^1	3.18×10^1	3.71×10^1	3.07×10^1	3.10×10^1	3.23×10^1	2.95×10^1	5.48×10^1
1.10×10^3	4.85×10^1	7.58×10^1	3.13×10^1	4.98×10^1	4.70×10^1	7.21×10^1	9.61×10^1	3.21×10^1	3.77×10^1	3.09×10^1	3.13×10^1	3.26×10^1	2.95×10^1	5.71×10^1
1.20×10^3	4.93×10^1	7.86×10^1	3.16×10^1	5.05×10^1	4.82×10^1	7.37×10^1	1.00×10^2	3.23×10^1	3.83×10^1	3.12×10^1	3.16×10^1	3.29×10^1	2.96×10^1	5.93×10^1
1.30×10^3	5.01×10^1	8.12×10^1	3.18×10^1	5.11×10^1	4.94×10^1	7.51×10^1	1.04×10^2	3.26×10^1	3.88×10^1	3.15×10^1	3.18×10^1	3.32×10^1	2.97×10^1	6.15×10^1
1.40×10^3	5.08×10^1	8.35×10^1	3.21×10^1	5.16×10^1	5.06×10^1	7.63×10^1	1.07×10^2	3.29×10^1	3.94×10^1	3.17×10^1	3.21×10^1	3.34×10^1	2.99×10^1	6.34×10^1
1.50×10^3	5.14×10^1	8.56×10^1	3.23×10^1	5.20×10^1	5.17×10^1	7.74×10^1	1.10×10^2	3.31×10^1	4.00×10^1	3.20×10^1	3.24×10^1	3.36×10^1	3.00×10^1	6.53×10^1
1.60×10^3	5.19×10^1	8.75×10^1	3.26×10^1	5.23×10^1	5.27×10^1	7.82×10^1	1.13×10^2	3.33×10^1	4.06×10^1	3.22×10^1	3.26×10^1	3.38×10^1	3.02×10^1	6.70×10^1
1.70×10^3	5.23×10^1	8.91×10^1	3.28×10^1	5.25×10^1	5.37×10^1	7.89×10^1	1.15×10^2	3.36×10^1	4.11×10^1	3.25×10^1	3.29×10^1	3.40×10^1	3.04×10^1	6.86×10^1
1.80×10^3	5.26×10^1	9.05×10^1	3.30×10^1	5.26×10^1	5.46×10^1	7.94×10^1	1.18×10^2	3.38×10^1	4.17×10^1	3.28×10^1	3.31×10^1	3.41×10^1	3.06×10^1	7.01×10^1
1.90×10^3	5.29×10^1	9.16×10^1	3.33×10^1	5.25×10^1	5.55×10^1	7.97×10^1	1.19×10^2	3.40×10^1	4.23×10^1	3.30×10^1	3.33×10^1	3.42×10^1	3.08×10^1	7.15×10^1
2.00×10^3	5.30×10^1	9.25×10^1	3.35×10^1	5.24×10^1	5.64×10^1	7.98×10^1	1.21×10^2	3.42×10^1	4.29×10^1	3.33×10^1	3.35×10^1	3.43×10^1	3.10×10^1	7.27×10^1
2.10×10^3	5.30×10^1	9.32×10^1	3.37×10^1	5.22×10^1	5.72×10^1	7.97×10^1	1.22×10^2	3.43×10^1	4.35×10^1	3.35×10^1	3.38×10^1	3.43×10^1	3.12×10^1	7.38×10^1
2.20×10^3	5.29×10^1	9.36×10^1	3.39×10^1	5.19×10^1	5.79×10^1	7.95×10^1	1.23×10^2	3.45×10^1	4.40×10^1	3.38×10^1	3.40×10^1	3.43×10^1	3.15×10^1	7.48×10^1
2.30×10^3	5.28×10^1	9.38×10^1	3.41×10^1	5.14×10^1	5.86×10^1	7.91×10^1	1.24×10^2	3.47×10^1	4.46×10^1	3.40×10^1	3.42×10^1	3.43×10^1	3.18×10^1	7.56×10^1
2.40×10^3	5.25×10^1	9.38×10^1	3.43×10^1	5.09×10^1	5.92×10^1	7.84×10^1	1.24×10^2	3.48×10^1	4.52×10^1	3.42×10^1	3.44×10^1	3.43×10^1	3.20×10^1	7.63×10^1
2.50×10^3	5.22×10^1	9.35×10^1	3.44×10^1	5.02×10^1	5.98×10^1	7.76×10^1	1.24×10^2	3.49×10^1	4.58×10^1	3.45×10^1	3.45×10^1	3.42×10^1	3.23×10^1	7.70×10^1
2.60×10^3	5.17×10^1	9.29×10^1	3.46×10^1	4.94×10^1	6.04×10^1	7.66×10^1	1.24×10^2	3.50×10^1	4.64×10^1	3.47×10^1	3.47×10^1	3.41×10^1	3.26×10^1	7.74×10^1
2.70×10^3	5.12×10^1	9.22×10^1	3.48×10^1	4.86×10^1	6.09×10^1	7.55×10^1	1.23×10^2	3.51×10^1	4.70×10^1	3.50×10^1	3.49×10^1	3.40×10^1	3.30×10^1	7.78×10^1
2.80×10^3	5.05×10^1	9.12×10^1	3.49×10^1	4.76×10^1	6.13×10^1	7.41×10^1	1.22×10^2	3.52×10^1	4.76×10^1	3.52×10^1	3.51×10^1	3.39×10^1	3.33×10^1	7.80×10^1
2.90×10^3	4.98×10^1	8.99×10^1	3.51×10^1	4.65×10^1	-	1.26×10^1	1.21×10^2	3.55×10^1	4.81×10^1	3.54×10^1	3.52×10^1	3.37×10^1	3.36×10^1	7.81×10^1

TABLE 8 $\bar{C}_p$ [J/Kg-mole°K] x 10^{-3} For Evaluation of $\bar{S}_T$ by $\bar{S}_T = \bar{C}_p \ln(T_2/298)$ (T in °K)

Temperature	CO_2	C_2H_4	Air	SO_2	NH_3	SO_3	C_2H_6	NO	H_2O	N_2	CO	O_2	H_2	CH_4
4.00×10^2	3.95×10^1	4.92×10^1	2.92×10^1	4.17×10^1	3.69×10^1	5.56×10^1	5.93×10^1	2.97×10^1	3.37×10^1	2.90×10^1	2.91×10^1	2.97×10^1	2.90×10^1	3.79×10^1
5.00×10^2	4.09×10^1	5.32×10^1	2.94×10^1	4.30×10^1	3.83×10^1	5.81×10^1	6.47×10^1	3.01×10^1	3.42×10^1	2.93×10^1	2.94×10^1	3.01×10^1	2.91×10^1	4.06×10^1
6.00×10^2	4.21×10^1	5.67×10^1	2.97×10^1	4.41×10^1	3.95×10^1	6.04×10^1	6.95×10^1	3.04×10^1	3.47×10^1	2.95×10^1	2.96×10^1	3.05×10^1	2.91×10^1	4.30×10^1
7.00×10^2	4.32×10^1	5.98×10^1	2.99×10^1	4.51×10^1	4.07×10^1	6.24×10^1	7.30×10^1	3.06×10^1	3.51×10^1	2.97×10^1	2.99×10^1	3.09×10^1	2.92×10^1	4.53×10^1
8.00×10^2	4.41×10^1	6.27×10^1	3.02×10^1	4.59×10^1	4.18×10^1	6.41×10^1	7.79×10^1	3.09×10^1	3.55×10^1	2.99×10^1	3.01×10^1	3.12×10^1	2.92×10^1	4.74×10^1
9.00×10^2	4.50×10^1	6.53×10^1	3.04×10^1	4.67×10^1	4.28×10^1	6.57×10^1	8.15×10^1	3.11×10^1	3.60×10^1	3.01×10^1	3.03×10^1	3.15×10^1	2.93×10^1	4.93×10^1
1.00×10^3	4.58×10^1	6.77×10^1	3.06×10^1	4.74×10^1	4.38×10^1	6.72×10^1	8.48×10^1	3.13×10^1	3.64×10^1	3.03×10^1	3.06×10^1	3.17×10^1	2.94×10^1	5.10×10^1
1.10×10^3	4.65×10^1	6.98×10^1	3.08×10^1	4.80×10^1	4.47×10^1	6.84×10^1	8.78×10^1	3.15×10^1	3.67×10^1	3.05×10^1	3.08×10^1	3.19×10^1	2.94×10^1	5.27×10^1
1.20×10^3	4.71×10^1	7.18×10^1	3.10×10^1	4.85×10^1	4.55×10^1	6.96×10^1	9.06×10^1	3.17×10^1	3.71×10^1	3.07×10^1	3.10×10^1	3.22×10^1	2.95×10^1	5.42×10^1
1.30×10^3	4.76×10^1	7.36×10^1	3.11×10^1	4.89×10^1	4.63×10^1	7.06×10^1	9.31×10^1	3.19×10^1	3.75×10^1	3.08×10^1	3.11×10^1	3.24×10^1	2.96×10^1	5.56×10^1
1.40×10^3	4.81×10^1	7.51×10^1	3.13×10^1	4.93×10^1	4.70×10^1	7.14×10^1	9.54×10^1	3.21×10^1	3.79×10^1	3.10×10^1	3.13×10^1	3.25×10^1	2.97×10^1	5.70×10^1
1.50×10^3	4.86×10^1	7.66×10^1	3.15×10^1	4.96×10^1	4.78×10^1	7.22×10^1	9.75×10^1	3.22×10^1	3.82×10^1	3.12×10^1	3.15×10^1	3.27×10^1	2.98×10^1	5.82×10^1
1.60×10^3	4.89×10^1	7.79×10^1	3.16×10^1	4.99×10^1	4.84×10^1	7.28×10^1	9.93×10^1	3.24×10^1	3.86×10^1	3.13×10^1	3.16×10^1	3.28×10^1	2.99×10^1	5.93×10^1
1.70×10^3	4.92×10^1	7.90×10^1	3.18×10^1	5.00×10^1	4.91×10^1	7.33×10^1	1.01×10^2	3.25×10^1	3.89×10^1	3.15×10^1	3.18×10^1	3.29×10^1	3.00×10^1	6.04×10^1
1.80×10^3	4.95×10^1	8.00×10^1	3.19×10^1	5.02×10^1	4.97×10^1	7.37×10^1	1.02×10^2	3.27×10^1	3.93×10^1	3.16×10^1	3.19×10^1	3.30×10^1	3.01×10^1	6.14×10^1
1.90×10^3	4.97×10^1	8.08×10^1	3.21×10^1	5.02×10^1	5.02×10^1	7.40×10^1	1.03×10^2	3.28×10^1	3.96×10^1	3.18×10^1	3.21×10^1	3.31×10^1	3.03×10^1	6.22×10^1
2.00×10^3	4.98×10^1	8.15×10^1	3.22×10^1	5.02×10^1	5.07×10^1	7.42×10^1	1.05×10^2	3.29×10^1	4.00×10^1	3.19×10^1	3.22×10^1	3.31×10^1	3.04×10^1	6.31×10^1
2.10×10^3	4.99×10^1	8.21×10^1	3.23×10^1	5.02×10^1	5.12×10^1	7.43×10^1	1.06×10^2	3.30×10^1	4.03×10^1	3.21×10^1	3.23×10^1	3.32×10^1	3.05×10^1	6.38×10^1
2.20×10^3	4.99×10^1	8.25×10^1	3.24×10^1	5.01×10^1	5.17×10^1	7.44×10^1	1.07×10^2	3.31×10^1	4.06×10^1	3.22×10^1	3.25×10^1	3.32×10^1	3.07×10^1	6.45×10^1
2.30×10^3	4.99×10^1	8.29×10^1	3.25×10^1	4.99×10^1	5.21×10^1	7.43×10^1	1.07×10^2	3.32×10^1	4.09×10^1	3.24×10^1	3.26×10^1	3.32×10^1	3.08×10^1	6.51×10^1
2.40×10^3	4.99×10^1	8.31×10^1	3.27×10^1	4.97×10^1	5.26×10^1	7.41×10^1	1.08×10^2	3.33×10^1	4.13×10^1	3.25×10^1	3.27×10^1	3.32×10^1	3.10×10^1	6.56×10^1
2.50×10^3	4.98×10^1	8.31×10^1	3.28×10^1	4.95×10^1	5.29×10^1	7.39×10^1	1.08×10^2	3.34×10^1	4.16×10^1	3.26×10^1	3.28×10^1	3.32×10^1	3.11×10^1	6.60×10^1
2.60×10^3	4.96×10^1	8.31×10^1	3.29×10^1	4.92×10^1	5.33×10^1	7.35×10^1	1.06×10^2	3.35×10^1	4.19×10^1	3.28×10^1	3.29×10^1	3.32×10^1	3.13×10^1	6.64×10^1
2.70×10^3	4.94×10^1	8.30×10^1	3.30×10^1	4.88×10^1	5.36×10^1	7.31×10^1	1.08×10^2	3.35×10^1	4.22×10^1	3.29×10^1	3.30×10^1	3.32×10^1	3.14×10^1	6.68×10^1
2.80×10^3	4.92×10^1	8.27×10^1	3.30×10^1	4.84×10^1	5.39×10^1	7.26×10^1	1.08×10^2	3.36×10^1	4.26×10^1	3.30×10^1	3.31×10^1	3.31×10^1	3.16×10^1	6.70×10^1
2.90×10^3	4.89×10^1	8.23×10^1	3.31×10^1	4.80×10^1	-	7.20×10^1	1.08×10^2	3.36×10^1	4.29×10^1	3.32×10^1	3.32×10^1	3.31×10^1	3.18×10^1	6.73×10^1

APPENDIX B-MATERIAL BALANCES

Material Balances

The purpose of this section is to provide those without experience with solving material balance problems the necessary background to perform simple calculations. The material will be developed in two sections. In Section A, no chemical reactions will take place. In section B, problems involving chemical reactions will be considered.

Section A - Material Balances Without Chemical Reactions

Figure 1 represents a basic calculation unit or module where two streams of different compositions are mixed.

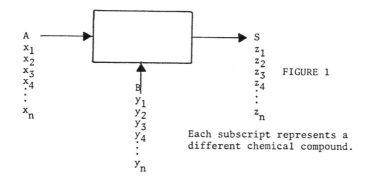

FIGURE 1

Each subscript represents a different chemical compound.

A, B, and S represent total flow rates and x, y, and z the weight fractions (or mole fractions). Since there is no chemical reaction, the amount (weight) of each chemical compound is conserved.

$$\left\{ \begin{array}{l} \text{Wt. of compound 1 in} \\ \text{all incoming streams} \end{array} \right. = \left. \begin{array}{l} \text{Wt. of compound 1 out} \\ \text{in all exit streams} \end{array} \right\} \quad \begin{array}{l} \text{Material Balance} \\ \text{on Component 1} \end{array} \quad \text{(A-1)}$$

For every chemical compound a similar balance equation may be written (for an N component system N equations may be written). In addition to these material balance relationships the Σ of the weight fractions = 1 for each entering and exiting stream. For the general system shown above there are

    a) N-independent component balance relationships

    b) $3 - \Sigma$ weight fractions = 1 equations $(\Sigma x=1, \Sigma y=1, \Sigma z=1)$

These equations may be used to solve all material balances involved in this text that do not involve chemical reactions.

Example B-1: Two streams containing oil, coal, and ash are mixed. One stream is 70% oil, 28% coal, and 2% ash. The second stream is 50% oil, 40% coal, and the remaining (% by weight), ash. The flow rates of the two streams are 1,000 and 2,000 lbs/hr respectively. What is the flow rate and composition of the mixed stream?

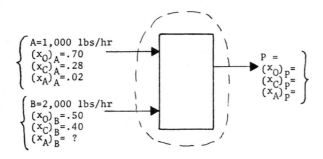

$$A = 1,000 \text{ lbs/hr}$$
$$(x_O)_A = .70$$
$$(x_C)_A = .28$$
$$(x_A)_A = .02$$

$$B = 2,000 \text{ lbs/hr}$$
$$(x_O)_B = .50$$
$$(x_C)_B = .40$$
$$(x_A)_B = ?$$

$$P =$$
$$(x_O)_P =$$
$$(x_C)_P =$$
$$(x_A)_P =$$

Unknowns: P, $(x_O)_P$, $(x_C)_P$, $(x_A)_P$, $(x_A)_B$ - 5 unknown values

Relations:

Mass Balance on Oil $(1,000)(.7) + 2,000(.5) = P(x_C)_P$ (1)

Mass Balance on Coal $(1,000)(.28) + 2,000(.40) = P(x_C)_P$ (2)

Mass Balance on Ash $(1,000)(.02) + 2,000(x_A)_B = P(x_A)_P$ (3)

Overall Mass Balance $1,000 + 2,000 = P$ (4)

Equation 4 is not independent but is the sum of equations 1, 2, and 3. Any three of these equations may be used to solve the problem. In addition to these material balance relations are two Σ equations.

Σx on Stream B: $.5 + .4 + (x_A)_B = 1.0$ (5)

Σx on Stream P: $(x_C)_P + (x_C)_P + (x_N)_P = 1.0$ (6)

There are 5 Equations Independent and 5 Unknowns and a solution is possible.

Solution:

From Equation 4 $P = 3,000$

From Equation 5 $(x_A)_B = 1.0$

From Equation 1 $700 + 1,000 = 3,000 (x_C)_P$

$(x_C)_D = .567$

From Equation 2 $280 + 800 = 3,000(x_C)_D$

$(x_C)_D = .360$

From Equation 6 $(x_A)_D = 1.0 - .567 - .360 = 0.73$

Check Using Equation 3 $(1,000)(.02) + 2,000(.10) = 3,000(.073)$

$20 + 200 = 219$

$220 \doteq 219$

Often in solving a material additional information may be given that provides
an independent relationship that can be used in the solution.

Example B-2:   A synthetic gasoline is composed of two components (A & B)
to be separated by distillation.   100 lbs of the raw gasoline (60% A) is
to be split into two equal portions.   Laboratory tests show that the ratio
of A/B in one stream is twice the ratio of A/B in the second stream.
a)   What are the compositions of A & B in the two streams?
b)   What is the yield of A defined as lbs of A in R stream/lbs of A in feed?

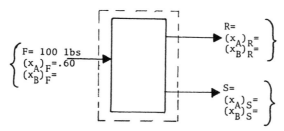

Unknowns: $(x_B)_F$, $(x_A)_R$, $(x_B)_R$, $(x_A)_S$, $(x_B)_S$, R, S - seven unknowns
Relations:

Mass Balance on A $(100)(.6) = R(x_A)_R + S(x_A)_S$ (1)

Mass Balance on B $(100)(x_B)_F = R(x_B)_R + S(x_B)_S$ (2)

Overall Mass Balance $100 = R + S$ (3)

Only two of these three are independent

$\Sigma(x)_F = 1$ $1 = .6 + (x_B)F$ (4)

$\Sigma(x)_R = 1$ $1 = (x_A)_R + (x_B)_R$ (5)

$\Sigma(x)_S = 1$ $1 = (x_A)_S + (x_B)_S$ (6)

The equations above provide five independent relations for the seven unknowns.
Two additional relationships are required.

Special Relations From Problem Statement:

   Stream Splits into equal portions: R = S   (7)

   Ratio A/B in streams is R/S: $(2/1) = [(x_A)_R/(x_B)_R]/[(x_A)_S/(x_B)_S]$ (8)

With these two additional relationships seven relationships have been provided
by the problem statement.

Solution:

From Equation 5 $(x_B)_F = 1 - .6 = .4$

From Equation 4 & 7 $100 = R + S; R = S$

                        $100 = 2R$

                        $R = 50$ lbs; $S = 50$ lbs

From Equation 5, 6, and 7

$$2 = \frac{(x_A)_R/(x_B)_R}{(x_A)_S/(x_B)_S} \; ; \; (x_B)_R = 1 - (x_A)_R; \; (x_B)_S = 1 - (x_A)_S$$

$$2 = \frac{(x_A)_R/[1 - (x_A)_R]}{(x_A)_S/[1 - (x_A)_S]}$$

$$(x_A)_S = \frac{(x_A)_R}{2 - (x_A)_R}$$

From Equation 9 and 1 $$60 = 50(x_A)_R + \frac{50(x_A)_R}{2 - (x_A)_R}$$

$$60 = 50(x_A)_R \left[1 + \frac{(x_A)_R}{2 - (x_A)_R}\right]$$

From Equation 5 $(x_A)_R = 0.75; \quad (x_B)_R = 0.25$

From Equation 9 $(x_A)_S = 0.75/(2 - 0.75) = 0.6$

From Equation 6 $(x_B)_S = 1 - 0.6 = 0.4$

Check using Equation 8 $2.0 = (0.75/0.25)/(0.60/0.40) = 3/1.5 = 2.0$

A consistent procedure for solving material balances without chemical reactions can be followed. The procedure used in the examples was:

   a) Reduce word problem to flow sheet showing all known information.

   b) Determine the unknown flow rates and compositions.

   c) Write down component material balances and sum of the mass fractions for each stream having one or more unknown compositions.

   d) Write down equations for any additional facts given.

   e) Solve the set of equations.

   f) Check solution.

Section B - Material Balances With Chemical Reactions

Most of the problems considered involve analysis of problems that involve chemical reactions. In these cases it is not correct to write a material balance for each chemical compound since compounds are formed or destroyed by the chemical reaction. In solving problems involving chemical reactions it is more convenient to work in moles as the number of moles involved in chemical reactions are related in simple ratios provided by the balanced chemical reaction. For example the reaction of methane with oxygen may be written:

$$CH_4 + 2O_2 \rightarrow CO_2 + 2H_2O$$

One mole of methane reacts with two moles of oxygen to yield one mole of CO_2 and 2 moles of water.

No mention was made as to what mole was involved, a kilogram-mole, a ton-mole, a lb-mole, etc. It makes no difference the ratios are the same. A mole is defined as the mass equal to the molecular weight of the material thus:

1 lb-mole of CH_4	12 + 4 or 16 lbs
1 kg-mole of O_2	2 x 16 or 32 kg
1 ton-mole of CO_2	12 + 2 x 16 or 44 tons.

A common problem is that of converting mass fractions to mole fractions or mole fractions to mass fractions. A standard procedure is given below.

Standard Procedure to Convert Mass Fraction to Mole Fraction

1 Mass Fraction	2 Molecular Weight	3 $\dfrac{\text{Column 1}}{\text{Column 2}}$	Mole Fraction $\dfrac{\text{Column 3}}{\Sigma\text{Column 3}}$
x_A—— x_B—— x_C—— x_D——			
		ΣColumn 3	

Mole Fraction to Mass Fraction

1 Mole Fraction	2 Molecular Weight	3 Column 1 x Column 2	Mass Fraction $\dfrac{\text{Column 3}}{\Sigma\text{Column 3}}$
x_A—— x_B—— x_C—— x_D——			
		ΣColumn 3	

Using these tables one can easily convert back and forth between mass fractions and mole fractions.

Example B-3: A gas contains 60% N_2, 30% O_2 and 10% CO_2 by weight. What is the mole fraction of the gases?

Component	Column 1 Mass Fraction	Column 2 Molecular Weight	Column 3 $\dfrac{\text{Column 1}}{\text{Column 2}}$	Column 4 Mole Fraction
N_2	0.60	28	0.0214	.65
O_2	0.30	32	0.0094	.28
CO_2	0.10	44	0.0023	.07
			Σ=0.0331	Σ=1.00

Example B-4: A gas contains 40% N_2, 18% H_2, 12% CO_2 and 30% CO by moles.
What is the weight fraction?

Component	Column 1 Mole Fraction	Column 2 Molecular Weight	Column 3 Column 1 x Column 2	Column 4 Mass Fraction
N_2	0.40	28	11.20	.444
H_2	0.18	2	0.36	.014
CO_2	0.12	44	5.28	.209
CO	0.30	28	8.40	.333
			Σ=25.24	Σ=1.00

In solving a problem involving chemicals it is essential to be able to write a balanced chemical reaction

$$aA + bB \rightleftharpoons cC + dD$$

It is convenient to divide by one of the coefficients. For example, if the equation above is divided by a

$$A + (b/a)B \rightleftharpoons (c/a)C + (d/a)D$$

By defining α as the moles of A that react the balanced chemical reaction states that $\alpha(b/a)$ moles of B react; $\alpha(c/a)$ moles of C are formed and $\alpha(d/a)$ moles of D are formed. The procedure for solving problems are similar to the one practiced before but the balance equation must include a term to represent the moles formed because of the reaction or lost because of the reaction. For each chemical compound in the chemical reaction equation the following relationships may be written.

Moles In + Moles Formed by Chemical Reaction - Moles Destroyed by
Chemical Reaction = Moles Out

Example B-5:   100[lb-moles/hr] of a gas composed of 35% CO, 20% CO_2, and 45% N_2 (mole %) is burned in 200 [lb-moles/hr] of air (21% O_2, 79% N_2) (mole %). What is the gas composition?

$$\begin{cases} G = Gas[100\ moles/hr] \\ x_{CO} = .35 \\ x_{CO_2} = .20 \\ x_{N_2} = .45 \\ A = Air[200\ moles/hr] \\ x_{O_2} = .21 \\ x_{N_2} = .79 \end{cases} \longrightarrow \boxed{} \longrightarrow \begin{cases} P = \\ x_{CO_2} = \\ x_{CO} = \\ x_{N_2} = \\ x_{O_2} = \end{cases}$$

The chemical reaction is $CO + 1/2O_2 \longrightarrow CO_2$
Let α = lb-moles of CO_2 formed by chemical reaction.

Component:	In	+ Gen	- Consumed	= Out	Equation
CO	0.35x100	0	α	0	1
CO_2	0.20x100	α	0	$P x_{CO_2}$	2
O_2	0.21x200	0	$1/2\alpha$	$P x_{O_2}$	3
N_2	0.79x200+.45x100	0	0	$P x_{N_2}$	4
Total	300	α	$3/2\alpha$	P	5

4 of the 5 equations from the table are independent

Number of unknowns = $5[x_{CO_2}, x_{O_2}, x_{N_2}, P, \alpha]$

For Stream P

$$\Sigma x = x_{CO_2} + x_{O_2} + x_{N_2} = 1 \tag{6}$$

Solution:

From Equation 1: α = 35

From Equation 5: $300 - 1/2\alpha = P$

 P = 282.5

From Equation 2: $20 + 35 = 300\ x_{CO_2}$

 $x_{CO_2} = 0.195$

From Equation 3: $42 \neq 35/2 = 300\ x_{O_2}$

 $x_{O_2} = 0.087$

From Equation 4: $158 + 45 = 300\ x_{N_2}$

 $x_{N_2} = 0.719$

Use Equation 6 to check solution: 1 = 0.195 + 0.087 + 0.719 = 1

In dealing with chemical reactions terms such as limiting reactant, excess reactant, and partial reactant are used. Each of these terms will be discussed.

For each chemical reaction, a balanced chemical reaction must be written. For example:

$$N_2 + 3H_2 \rightleftharpoons 2NH_3$$

This tells you that for every mole of N_2 that reacts, 3 moles of H_2 will also react and 2 moles of NH_3 will be formed. For a real process there is no reason to believe that the reactants will be added to the system in the ratio of 3 moles of H_2 to 1 mole of N_2 or that it will all react. Consider the process given below:

F_{N_2} = 1 lb-mole/hr

F_{H_2} = 4 lb-mole/hr

P_{N_2}

P_{H_2}

P_{NH_3} = 1 lb-mole/hr

It is seen that the actual ratio of H_2/N_2 for this process is 4/1 and is not in relation to the 3/1 ratio in the balanced chemical reaction. There is more H_2 present than would be required to react with all of the N_2. (For one mole of N_2, 3 moles of H_2 would be needed according to the chemical reaction). As there is more H_2 present than would be used to react with all of the N_2

H_2 is the excess reactant.
N_2 is the limiting reactant.

The values of excess reactant and limiting reactant depended upon the balanced chemical reaction and the input streams and was independent of the output. The values are not dependent upon whether or not the reaction takes place.

To evaluate which of the reactants are in excess and which is limiting, obtain the ratios of

Feed rate of reactant chemical component
Stoichiometric coefficient of chemical compound

for each reactant. The chemical component with the lowest value is limiting and the other components are in excess. Using this procedure for the ammonia problem above

For N_2: $F_{N_2}/1 = 1/1 = 1$

For H_2: $F_{H_2}/3 = 4/3 = 1.67$

N_2 is limiting and H_2 is excess. Once having determined which of the reactants are in excess the % excess is evaluated by,

% Excess = (Input of excess reactant - the theoretical amount of excess
reactant required to consume the limiting reactant)/Theoretical
amount required of excess reactant to consume the limiting
reactant (100)

For a reaction

$$aA + bB \rightleftharpoons cC + dD$$

Where

A - excess reactant

B - limiting reactant

The % excess is given by

$$\% \text{ excess A} = [F_A - (b/a)F_B]/[(b/a)F_B]$$

Applying this relation to the ammonia problem discussed above

$$\% \text{ excess H}_2 = [4 - (3/1)]/[3/1](100) = 33.3$$

The final term discussed is the conversion of a reactant that is consumed by the chemical reaction. For a reaction of the type

$$aA + bB \rightleftharpoons cC + dD$$

$$\text{Fractional Conversion} = \frac{\text{Moles of the reactant converted}}{\text{Moles of reactant fed}}$$

$$\text{Conversion of A} = \frac{\frac{a}{d} P_D}{F_A}(100) \text{ or } \frac{\frac{a}{c} P_C}{F_A}(100)$$

$$\text{Conversion of B} = \frac{\frac{b}{d} P_D}{F_B}(100) \text{ or } \frac{\frac{b}{c} P_C}{F_B}(100)$$

Again for the ammonia problem

$$\text{Conversion of N}_2 = \frac{1/2 P_{NH_3}}{F_{N_2}}(100) = \frac{(1/2)(1)}{1}(100) = 50\%$$

$$\text{Conversion of H}_2 = \frac{3/2 \, P_{NH_3}}{F_{H_2}}(100) = \frac{(3/2)(1)}{4}(100) = 37.5\%$$

If it is not stated in the problem when a conversion is given it is based upon the limiting reactant.

Example B-6: In heating your home with natural gas the gas is mixed with air for combustion. To obtain complete combustion it is desirable to add excess air to the system. Consider the following burner design.

1 lb-mole of methane, CH_4, is mixed with 3 lb-moles of air (79% N_2, 21% O_2).
Assume complete consumption of the limiting reactant. Obtain product gas
composition. Comment on the burner design.

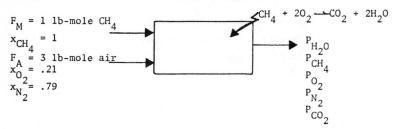

$$CH_4 + 2O_2 \longrightarrow CO_2 + 2H_2O$$

F_M = 1 lb-mole CH_4

x_{CH_4} = 1

F_A = 3 lb-mole air

x_{O_2} = .21

x_{N_2} = .79

P_{H_2O}

P_{CH_4}

P_{O_2}

P_{N_2}

P_{CO_2}

Determination of limiting reactant:

CH_4:   (1/1) = 1

O_2 :   [(3)(.21)]/2 = 0.315   The oxygen (or air) was the limiting reactant.

Determination of the excess CH_4

Excess CH_4 = [1 - 1/2(.21)(3)]/[1/2(.21)(3)] x 100 = 217%

Fraction CH_4 Converted = [1/2(.21)(3)]/[1] x 100 = 31.5%

Set α = moles of CH_4 consumed

Component:	In	+	Gen	-	Consumed	=	Out	Equation
CH_4	1		0		α		$P_{x_{CH_4}}$	(1)
O_2	0.21x(3)		0		-2α		$P_{x_{O_2}}$	(2)
N_2	0.79x(3)		0		0		$P_{x_{N_2}}$	(3)
H_2O	0		2α		0		$P_{x_{H_2O}}$	(4)
CO_2	0		α		0		$P_{x_{CO_2}}$	(5)
Total	4		3α		-3α		P	(6)

Since O_2 is limiting it is all consumed.

Therefore, x_{O_2} = 0; $P_{x_{O_2}}$ = 0

From Equation 2: 0.21 x (3) - 2α = 0

α = 0.315

From Equation 6: P = 4

From Equation 1: 1 - 0.315 = 4 x_{CH_4}

x_{CH_4} = 0.171

From Equation 3: 6.79 x 3 = 4 x_{N_2}

x_{N_2} = 0.593

From Equation 4: 2 x 0.315 = 4 x_{H_2O}

x_{H_2O} = 0.158

From Equation 5: 0.315 = 4 x_{CO_2}

x_{CO_2} = 0.079

Check: $\sum x = x_{CH_4} + x_{O_2} + x_{N_2} + x_{H_2O} + x_{CO_2} = 1$

$$0.171 + 0.593 + 0.158 + 0.079 = 1.001 = 1$$

The burner is poorly designed; only 31.5% of the methane is burned. The problem is one of insufficient air to the burner.

Often the values of excess reactant and conversion are given in the problem statement and the information is used in providing a numerical solution.

Example B-6: 1 lb-mole of propane (C_3H_8) is burned in 30% excess air. Only 95% of the propane reacts. Determine the air flow and determine the gas composition.

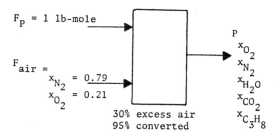

$$C_3H_8 + 5O_2 \longrightarrow 3CO_2 + 4H_2O$$

From excess reactant equation

$30\% = [F_{O_2} - (5/1)(1)]/[(5/1)(1)] \times 100$

$F_{O_2} = 6.5; \; F_{air} = 31.0$

Let α = moles of C_3H_8 consumed

From the statement that 95% was converted $\alpha = 0.95(1) = 0.95$

Component:	In	+	Gen	-	Consumed	=	Out	
C_3H_8	1		0		α		$P_{x_{C_3H_8}}$	(1)
O_2	6.5		0		-5α		$P_{x_{O_2}}$	(2)
N_2	0.79(31.0)		0		0		$P_{x_{N_2}}$	(3)
CO_2	0		3α		0		$P_{x_{CO_2}}$	(4)
H_2O	0		4α		0		$P_{x_{H_2O}}$	(5)
Total	32.0		7α		6α		P	(6)

$\alpha = 0.95$

From Equation 6: $32.0 + 7\alpha - 6\alpha = P$

$P = 32 + 1(.95) = 33.0$

From Equation 1: $1 - 0.95 = 33.0\ x_{C_3H_8}$

$x_{C_3H_8} = 0.002$

From Equation 2: $6.5 - 5(0.95) = 33.0\ x_{O_2}$

$x_{O_2} = 0.053$

From Equation 3: $0.79(31) = 33.0\ x_{N_2}$

$x_{N_2} = 0.742$

From Equation 4: $3(0.95) = 33.0\ x_{CO_2}$

$x_{CO_2} = 0.086$

From Equation 5: $4(0.95) = 33.0\ x_{H_2O}$

$x_{H_2O} = 0.115$

Check: $\sum x = x_{C_3H_8} + x_{O_2} + x_{N_2} + x_{CO_2} + x_{H_2O} = 1$

$= 0.002 + 0.053 + 0.742 + 0.086 + 0.115 = 0.998 = 1$

APPENDIX C-CONVERSION FACTORS

APPENDIX C

CONVERSION FACTORS & USEFUL CONSTANTS

Reproduced from "Selected Values of Properties of Chemical Compounds." Thermodynamics Research Center, Texas A&M University, College Station, Texas.

To convert the numerical value of a property expressed in one of the units in the left-hand column of a table to the numerical value of the same property in one of the units in the top row of the same table, multiply the former value by the factor in the block common to both units. The factors have been carried out to seven significant figures. However, this does not mean that the factors are always known to that accuracy. Factors written with fewer than seven significant digits should be taken as exact values. Numbers followed by an asterisk (*) are definitions of the relation between the two units.

521

TABLE 1 – THE MOL UNIT

Units		g-mol (Unified)	g-mol (Chemists')	g-mol (Physicists')	lb-mol (Unified)	lb-mol (Chemists')	lb-mol (Physicists')
1 g-mol (Unified)	=	1	0.9999570	0.9996821	2.204623×10^{-3}	2.204717×10^{-3}	2.2053237×10^{-3}
1 g-mol (Chemists')	=	1.000043	1	0.999725	2.204528×10^{-3}	2.204623×10^{-3}	2.205229×10^{-3}
1 g-mol (Physicists')	=	1.000318	1.000275	1	2.203922×10^{-3}	2.204017×10^{-3}	2.204623×10^{-4}
1 lb-mol (Unified)	=	453.5924	453.6119	453.7366	1	1.000043	1.000318
1 lb-mol (Chemists')	=	453.5729	453.5924	453.7171	0.9999570	1	1.000275
1 lb-mol (Physicists')	=	453.4482	453.4677	453.5924	0.9996821	0.999725	1

TABLE 2 – UNITS OF LENGTH

Units		cm	m	in.	ft	yd	mile
1 cm	=	1	0.01*	0.3937008	0.03280840	0.01093613	6.213712×10^{-6}
1 m	=	100.	1	39.37008	3.280840	1.093613	6.213712×10^{-4}
1 in.	=	2.54*	0.0254	1	0.08333333...	0.02777777...	1.578283×10^{-5}
1 ft	=	30.48	0.3048	12.*	1	0.3333333...	$1.893939 \times 10^{-4}...$
1 yd	=	91.44	0.9144	36.	3.*	1	$5.681818 \times 10^{-4}...$
1 mile	=	1.609344×10^{5}	1.609344×10^{3}	6.336×10^{4}	5280.*	1760.	1

TABLE 3 – UNITS OF AREA

Units		cm²	m²	in.²	ft²	yd²	mile²
1 cm²	=	1	10^{-4}*	0.155003	1.076391×10^{-3}	1.195990×10^{-4}	3.861022×10^{-11}
1 m²	=	10^{4}	1	1550.003	10.76391	1.195990	3.861022×10^{-7}
1 in.²	=	6.4516*	6.4516×10^{-4}	1	$6.944444 \times 10^{-3}...$	7.716049×10^{-4}	2.490977×10^{-10}
1 ft²	=	929.0304	0.09290304	144.*	1	0.1111111...	3.587007×10^{-8}
1 yd²	=	8361.273	0.8361273	1296.	9.*	1	3.228306×10^{-7}
1 mile²	=	2.589988×10^{10}	2.589988×10^{6}	4.014490×10^{9}	2.78784×10^{7}*	3.0976×10^{6}	1

*The electrical units in these tables are those in terms of which certification of standard cells, standard resistances, etc., is made by the National Bureau of Standards. Unless otherwise indicated, all electrical units are absolute.

TABLE 4 – UNITS OF VOLUME

Units		cm³	liter	in.³	ft³	qt	gal
1 cm³	=	1	10^{-3}	0.06102374	3.531467×10^{-5}	1.056688×10^{-3}	2.641721×10^{-4}
1 liter	=	1000.*	1	61.02374	0.03531467	1.056688	0.2641721
1 in.³	=	16.38706*	0.01638706	1	5.787037×10^{-4}	0.01731602	4.329004×10^{-3}
1 ft³	=	28316.85	28.31685	1728.*	1	2.992208	7.480520
1 qt	=	946.353	0.946353	57.75	0.0342014	1	0.25
1 gal (U.S.)	=	3785.412	3.785412	231.*	0.136806	4.*	1

TABLE 5 – UNITS OF MASS

Units		g	kg	oz	lb	metric ton	ton
1 g	=	1	10^{-3}	0.03527396	2.204623×10^{-3}	10^{-6}	1.102311×10^{-6}
1 kg	=	1000.	1	35.27396	2.204623	10^{-3}	1.102311×10^{-3}
1 oz (avdp)	=	28.34952	0.02834952	1	0.0625	2.834952×10^{-5}	$5. \times 10^{-4}$
1 lb (avdp)	=	453.5924	0.4535924	16.*	1	4.535924×10^{-4}	0.0005
1 metric ton	=	10^{6}	1000.*	35273.96	2204.623	1	1.102311
1 ton	=	907184.7	907.1847	32000.	2000.*	0.907847	1

TABLE 6 – UNITS OF DENSITY

Units		g cm⁻³	g l⁻¹	oz in.⁻³	lb in.⁻³	lb ft⁻³	lb gal⁻¹
1 g cm⁻³	=	1	1000.	0.5780365	0.03612728	62.42795	8.345403
1 g l⁻¹	=	10^{-3}	1	5.780365×10^{-4}	3.612728×10^{-5}	0.06242795	8.345403×10^{-3}
1 oz in.⁻³	=	1.729994	1729.994	1	0.0625	108.	14.4375
1 lb in.⁻³	=	27.67991	27679.91	16.	1	1728.	231.
1 lb ft⁻³	=	0.01601847	16.01847	9.259259×10^{-3}	5.787370×10^{-4}	1	0.1336806
1 lb gal⁻¹	=	0.1198264	119.8264	4.749536×10^{-3}	4.3290043×10^{-3}	7.480519	1

TABLE 7 – UNITS OF PRESSURE

Units	dyn cm⁻²	bar	atm	kg (wt) cm⁻²	mmHg (torr)	in. Hg	lb (wt) in.⁻²
1 dyn cm⁻² =	1	10^{-6}	9.869233×10^{-7}	1.019716×10^{-6}	7.500617×10^{-4}	2.952999×10^{-5}	1.450377×10^{-5}
1 bar =	10^{6*}	1	0.9869233	1.019716	750.0617	29.52999	14.50377
1 atm =	1013250.*	1.013250	1	1.033227	760.	29.92126	14.69595
1 kg (wt) cm⁻² =	980665.	0.980665	0.9678411	1	735.5592	28.95903	14.22334
1 mmHg (torr) =	1333.224	1.333224×10^{-3}	1.3157895×10^{-3}	1.3595099×10^{-3}	1	0.03937008	0.01933678
1 in. Hg =	33863.88	0.03386388	0.03342105	0.03453155	25.4	1	0.4911541
1 lb (wt) in.⁻² =	68947.57	0.06894757	0.06804596	0.07030696	51.71493	2.036021	1

TABLE 8 – UNITS OF ENERGY*

Units	g mass (energy equiv)	J	int J	cal	cal$_{IT}$	Btu$_{IT}$	kW hr	hp hr	ft-lb (wt)	cu ft-lb (wt) in.⁻²	l-atm
1 g mass (energy equiv) =	1	8.987554×10^{13}	8.986071×10^{13}	2.148077×10^{13}	2.146640×10^{13}	8.518558×10^{10}	2.496543×10^{7}	3.347919×10^{7}	6.628880×10^{13}	4.603399×10^{11}	8.870026×10^{11}
1 J =	1.112650×10^{-14}	1	0.999835	0.2390057	0.2388459	9.478172×10^{-4}	$2.777777... \times 10^{-7}$	3.725062×10^{-7}	0.7375622	5.121960×10^{-3}	9.869233×10^{-3}
1 int J =	1.112833×10^{-14}	1.0001165	1	0.2390452	0.2388853	9.479735×10^{-4}	2.778236×10^{-7}	3.725676×10^{-7}	0.7376639	5.122805×10^{-3}	9.870862×10^{-3}
1 cal =	4.655327×10^{-14}	4.184*	4.183310	1	0.9993312	3.965667×10^{-3}	1.1622222×10^{-6}	1.558562×10^{-6}	3.085960	2.143028×10^{-2}	0.04129287
1 cal$_{IT}$ =	4.658442×10^{-14}	4.1868*	4.186109	1.000669	1	3.968321×10^{-3}	1.163000×10^{-6}	1.559609×10^{-6}	3.088025	2.144462×10^{-2}	0.04132050
1 Btu$_{IT}$ =	1.173908×10^{-11}	1055.056	1054.882	252.1644	251.9958*	1	2.930711×10^{-4}	3.930148×10^{-4}	778.1693	5.403953	10.41259
1 kW hr =	4.005539×10^{-8}	3600000.*	3599406.	860420.7	859845.2	3412.142	1	1.341022	2655224	18439.06	35529.24
1 hp hr =	2.986930×10^{-8}	2684519.	2684077.	641615.6	641186.5	2544.33	0.7456998	1	1980000.*	13750.	26494.15
1 ft-lb (wt) =	1.508550×10^{-14}	1.355594	1.355818	0.3240483	0.3238315	1.285067×10^{-3}	3.766161×10^{-7}	$5.050505... \times 10^{-7}$	1	$6.944444... \times 10^{-3}$	0.01338088
1 cu ft-lb (wt) in.⁻² =	2.172313×10^{-12}	195.2378	195.2056	46.66295	46.63174	0.1850497	5.423272×10^{-5}	$7.272727... \times 10^{-5}$	144 *	1	1.926847
1 l-atm =	1.127392×10^{-12}	101.3250	101.3083	24.21726	24.20106	0.09603757	2.814583×10^{-5}	3.774419×10^{-5}	74.73349	0.5189825	1

TABLE 9 — UNITS OF MOLECULAR ENERGY*

Units		erg molecule⁻¹	J mol⁻¹	int J mol⁻¹	cal mol⁻¹	eV molecule⁻¹	int eV molecule⁻¹	wavenumber (cm⁻¹)
1 erg molecule⁻¹	=	1	6.022520×10^{16}	6.021526×10^{16}	1.439417×10^{16}	6.241808×10^{11}	6.239748×10^{11}	5.034474×10^{15}
1 J mol⁻¹	=	1.660435×10^{-17}	1	0.9998350	0.2390057	1.036411×10^{-5}	1.036069×10^{-5}	8.359414×10^{-2}
1 int J mol⁻¹	=	1.660709×10^{-17}	1.000165	1	0.2390452	1.036582×10^{-5}	1.036240×10^{-5}	8.360793×10^{-2}
1 cal mol⁻¹	=	6.947258×10^{-17}	4.1840	4.18331	1	4.336345×10^{-5}	4.334914×10^{-5}	0.3497579
1 eV molecule⁻¹	=	1.602100×10^{-12}	96486.79	96470.87	23060.90	1	0.9996701	8065.730
1 int eV molecule⁻¹	=	1.602393×10^{-12}	96518.63	96502.71	23068.51	1.000330	1	8068.392
1 wavenumber (cm⁻¹)	=	1.986305×10^{-16}	11.96256	11.96059	2.859121	1.239813×10^{-4}	1.239404×10^{-4}	1

TABLE 10 — UNITS OF SPECIFIC ENERGY*

Units		J g⁻¹	int J g⁻¹	cal g⁻¹	cal_IT g⁻¹	Btu_IT lb⁻¹	kW hr lb⁻¹
1 J g⁻¹	=	1	0.999835	0.2390057	0.2388459	0.4299226	1.259979×10^{-4}
1 int J g⁻¹	=	1.000165	1	0.2390452	0.2388853	0.4299936	1.260187×10^{-4}
1 cal g⁻¹	=	4.184*	4.18331	1	0.9993312	1.798796	5.271752×10^{-4}
1 cal_IT g⁻¹	=	4.1868*	4.186109	1.000669	1	1.8*	5.275279×10^{-4}
1 Btu_IT lb⁻¹	=	2.326000	2.325616	0.5559273	0.5555555....	1	2.390711×10^{-4}
1 kW hr lb⁻¹	=	7936.641	7935.332	1896.903	1895.643	3414.425	1

TABLE 11 – UNITS OF SPECIFIC ENERGY PER DEGREE*

Units		J g⁻¹ °C⁻¹	int J g⁻¹ °C⁻¹	cal g⁻¹ °C⁻¹	cal$_{IT}$ g⁻¹ °C⁻¹	Btu$_{IT}$ lb⁻¹ °F⁻¹	kW hr lb⁻¹ °F⁻¹
1 J g⁻¹ °C⁻¹	=	1	0.999835	0.2390057	0.2388459	0.2388459	6.999883 × 10⁻⁵
1 int J g⁻¹ °C⁻¹	=	1.000165	1	0.2390452	0.2388853	0.2388853	7.001037 × 10⁻⁵
1 cal g⁻¹ °C⁻¹	=	4.184*	4.183310	1	0.999312	0.9993312	2.928751 × 10⁻⁴
1 cal$_{IT}$ g⁻¹ °C⁻¹	=	4.1868*	4.186109	1.000669	;	1*	2.930711 × 10⁻⁴
1 Btu$_{IT}$ lb⁻¹ °F⁻¹	=	4.1868	4.186109	1.000669	1*	1	2.930711 × 10⁻⁴
1 kW hr lb⁻¹ °F⁻¹	=	14285.95	14283.60	3414.425	3412.42	3412.142	1

*The electrical units in these tables are those in terms of which certification of standard cells, standard resistances, etc., is made by the National Bureau of Standards. Unless otherwise indicated, all electrical units are absolute.

INDEX